METHODS IN MOLECULAR BIOLOGY™

Series Editor
John M. Walker
School of Life Sciences
University of Hertfordshire
Hatfield, Hertfordshire, AL10 9AB, UK

For other titles published in this series, go to
www.springer.com/series/7651

Androgen Action

Methods and Protocols

Edited by

Fahri Saatcioglu

Department of Molecular Biosciences, University of Oslo, Oslo, Blindern, Norway

Humana Press

Editor
Fahri Saatcioglu
Department of Molecular Biosciences
University of Oslo
Blindern, Oslo 0316, Norway
fahri.saatcioglu@imbv.uio.no

ISSN 1064-3745 e-ISSN 1940-6029
ISBN 978-1-61779-242-7 e-ISBN 978-1-61779-243-4
DOI 10.1007/978-1-61779-243-4
Springer New York Dordrecht Heidelberg London

Library of Congress Control Number: 2011932315

Printed on acid-free paper

Humana Press is part of Springer Science+Business Media (www.springer.com)

Preface

Androgens have a critical role in the development and maintenance of the male reproductive system and affect important physiological processes and pathological conditions, including the homeostasis of the normal prostate and prostate cancer. The effects of androgens are mediated by the androgen receptor (AR) which is a ligand-dependent transcription factor that belongs to the nuclear receptor superfamily. Like other steroid receptors, upon hormone binding, AR changes conformation, translocates to the nucleus, binds hormone response elements (HREs) in target promoters, and regulates gene transcription by the recruitment of chromatin modifying and remodelling complexes, coregulators, and other factors of the basal transcription apparatus. In addition, AR can associate with and directly affect other intracellular signalling pathways. This volume is designed with the aim to provide a tool box to study various phases of androgen action, from its entry to the cell to the phenotypic response that the cell mounts, with up-to-date techniques for biochemists, molecular biologists, cell biologists, geneticists, and pathologists.

This volume begins with two chapters that review the history of research on androgen action as well as perhaps the most widely studied area in this regard, androgen action and prostate carcinogenesis. This sets the stage for the ensuing chapters that present various methods to investigate androgen action, starting with **Chapters 3** and **4** which provide state-of-the-art methods to determine androgen levels in biological tissues and fluids; this is an area which has become all the more important with the recent demonstration that even under conditions of low circulating androgens, intra-tissue androgen levels can be quite high. **Chapters 5**, **6**, **7**, **8**, **9**, and **10** contain experimental procedures to study the activity of AR, including assessment of AR transactivation potential, identification of AR HREs, manipulation of AR levels in cells and xenografts, the role of interdomain interactions on AR function, and advanced microscopy methods to study AR interactions in living cells.

As most other proteins, AR has been shown to be modified posttranslationally in response to not only its natural ligands, but also various other stimuli. **Chapters 11**, **12**, **13**, and **14** focus on some of these modifications, including acetylation and SUMOylation, as well as interactions of AR with other proteins, which can also affect AR modifications, or vice versa. Next are **Chapters 15**, **16**, **17**, and **18** that present more recently developed methods and tools that have greatly facilitated androgen action research, including genome-wide identification of AR HREs, tissue-specific knockout of AR, and androgen-sensitive human prostate cancer xenograft work in mice, as well as novel approaches to analysis of AR function through automated microscopy methods.

As our knowledge of androgen signalling grew, for the phenotype to appear in any particular cell type, it became clear that androgen signalling interacts with various other signalling pathways. The last section of this volume is devoted to methodology to study some salient examples of these interactions. One chapter is devoted to the recently recognized androgen-regulated ETS fusion transcripts that may have a role in prostate cancer progression. The other chapters include methods to study androgen action on apoptotic pathways, crosstalk with Src signalling, as well as direct effects of androgens on lipid accumulation in prostate cancer cells. These methods provide endpoints that can be assessed when any particular aspect of androgen signalling is perturbed, experimentally or otherwise.

I would like to take this opportunity to thank all the contributors for graciously taking the time to provide their expertise in the excellent protocols; without them and their willingness to share so much of their time and hard-won expertise, this volume would not have materialized. I would also like to thank Dr. John Walker, the Editor-in-Chief of the *Methods in Molecular Biology* series, for his timely support during the editing process.

I believe that this volume provides the reader with a comprehensive overview of, and practical guidance on, the diverse methodologies that are propelling androgen action research forward, both in normal physiology and in disease states. I do hope that those consulting this volume will find it useful.

Fahri Saatcioglu

Contents

Contributors

LEGGY A. ARNOLD • *Department of Chemical Biology and Therapeutics, St. Jude Children's Research Hospital, Memphis, TN, USA*

YKE JILDOUW ARNOLDUSSEN • *Department of Molecular Biosciences, University of Oslo, Oslo, Norway*

FERDINANDO AURICCHIO • *Dipartimento di Patologia Generale, II Universita'di Napoli, Naples, Italy*

ARIA BANIAHMAD • *Institute of Human Genetics, Jena University Hospital, Jena, Germany*

RITA BOLLEN • *Department of Molecular Cell Biology, K.U. Leuven, Leuven, Belgium*

MICHAEL J. BOLT • *Department of Molecular and Cellular Biology, Baylor College of Medicine, Houston, TX, USA*

ALBERT O. BRINKMANN • *Department of Reproduction and Development, Erasmus MC, Rotterdam, The Netherlands*

GABRIELLA CASTORIA • *Dipartimento di Patologia Generale, II Universita'di Napoli, Napoli, Italy*

CHAWNSHANG CHANG • *George H. Whipple Lab, University of Rochester Medical Center, Rochester, NY, USA*

HELEN CHENG • *Vancouver Prostate Centre, University of British Columbia, Vancouver, BC, Canada*

FRANK CLAESSENS • *Molecular Endocrinology Laboratory, Department of Molecular Cell Biology, K.U. Leuven, Leuven, Belgium*

EVA COREY • *Department of Urology, University of Washington, Seattle, WA, USA*

REINHILDE DE BRUYN • *Department of Molecular Cell Biology and Therapeutics, St. Jude Children's Research Hospital, Memphis, TN, USA*

CLÉMENTINE FÉAU • *Department of Chemical Biology and Therapeutics, St. Jude Children's Research Hospital, Memphis, TN, USA*

DELILA GASI • *Department of Pathology, Josephine Nefkens Institute, Erasmus University Medical Centre, Rotterdam, The Netherlands*

CLAUDIA GERLACH • *Institute of Human Genetics, Jena University Hospital, Jena, Germany*

R. KIPLIN GUY • *Department of Chemical Biology and Therapeutics, St. Jude Children's Research Hospital, Memphis, TN, USA*

SEAN M. HARTIG • *Department of Molecular and Cellular Biology, Baylor College of Medicine, Houston, TX, USA*

TIINA JÄÄSKELÄINEN • *Institute of Biomedicine, University of Eastern Finland, Kuopio, Finland*

SANNA KAIKKONEN • *Institute of Biomedicine, University of Eastern Finland, Kuopio, Finland*

HATICE ZEYNEP KIRLI • *Department of Molecular Biosciences, University of Oslo, Oslo, Norway*

AARON KOSINSKI • *Department of Chemical Biology and Therapeutics, St. Jude Children's Research Hospital, Memphis, TN, USA*
ERIC LEBLANC • *Vancouver Prostate Centre, University of British Columbia, Vancouver, BC, Canada*
TZU-HUA LIN • *George H. Whipple Lab, University of Rochester Medical Center, Rochester, NY, USA*
TORSTEIN LINDSTAD • *Department of Molecular Biosciences, University of Oslo, Oslo, Norway*
THOMAS LINKE • *Institute of Human Genetics, Jena University Hospital, Jena, Germany*
HARRI MAKKONEN • *Institute of Biomedicine, University of Eastern Finland, Kuopio, Finland*
MICHAEL A. MANCINI • *Department of Molecular and Cellular Biology, Baylor College of Medicine, Houston, TX, USA*
MARCO MARCELLI • *Department of Molecular and Cellular Biology and Department of Medicine, Michael E. DeBakey VA Medical Center, Baylor College of Medicine, Houston, TX, USA*
CHARLES E. MASSIE • *Cancer Research UK, Cambridge Research Institute, Cambridge, UK*
ANTIMO MIGLIACCIO • *Dipartimento di Patologia Generale, II Universita'di Napoli, Napoli, Italy*
IAN G. MILLS • *Uro-Oncology Research Group, University of Oslo, Biotechnology Centre, Norway; Centre for Molecular Medicine (Norway), Nordic EMBL Partnership, University of Oslo, Oslo, Norway*
JUSTIN Y. NEWBERG • *Department of Molecular and Cellular Biology, Baylor College of Medicine, Houston, TX, USA*
HOLLY M. NGUYEN • *Department of Urology, University of Washington, Seattle, WA, USA*
JORMA J. PALVIMO • *Institute of Biomedicine, University of Eastern Finland, Kuopio, Finland*
RICHARD G. PESTELL • *Department of Cancer Biology, Kimmel Cancer Center, Thomas Jefferson University, Philadelphia, PA, USA*
MICHAEL POWELL • *Department of Cancer Biology, Kimmel Cancer Center, Thomas Jefferson University, Philadelphia, PA, USA*
PAUL S. RENNIE • *Vancouver Prostate Centre, University of British Columbia, Vancouver, BC, Canada; Department of Urologic Sciences, University of British Columbia, Vancouver, BC, Canada*
DANIELA ROELL • *Institute of Human Genetics, Jena University Hospital, Jena, Germany*
MIIA M. RYTINKI • *Institute of Biomedicine, University of Eastern Finland, Kuopio, Finland*
FAHRI SAATCIOGLU • *Department of Molecular Biosciences, University of Oslo, Oslo, Norway*
FRED SCHAUFELE • *Department of Medicine, University of California San Francisco, San Francisco, CA, USA*
MARTIN SCHOLTEN • *Department of Pediatrics, Jena University Hospital, Jena, Germany*

JØRGEN SIKKELAND • *Department of Molecular Biosciences, University of Oslo, Oslo, Norway*

PÄIVI SUTINEN • *Institute of Biomedicine, University of Eastern Finland, Kuopio, Finland*

ADAM T. SZAFRAN • *Department of Molecular and Cellular Biology, Baylor College of Medicine, Houston, TX, USA*

LIFENG TIAN • *Department of Cancer Biology, Kimmel Cancer Center, Thomas Jefferson University, Philadelphia, PA, USA*

DONALD J. TINDALL • *Departments of Urology, Biochemistry and Molecular Biology, Mayo Clinic College of Medicine, Rochester, MN, USA*

MARK TITUS • *Department of Urology, Roswell Park Cancer Institute, Buffalo, NY, USA*

KENNETH B. TOMER • *Laboratory of Structural Biology, NIEHS, Research Triangle Park, NC, USA*

JAN TRAPMAN • *Department of Pathology, Erasmus University Medical Centre, Rotterdam, The Netherlands*

CHENGUANG WANG • *Department of Stem Cell Biology and Regenerative Medicine, Kimmel Cancer Center, Thomas Jefferson University, Philadelphia, PA, USA*

DIPING WANG • *Departments of Urology, Biochemistry and Molecular Biology, Mayo Clinic College of Medicine, Rochester, MN, USA*

LING WANG • *Department of Molecular Biosciences, University of Oslo, Oslo, Norway*

ELIZABETH M. WILSON • *Laboratories for Reproductive Biology, Lineberger Comprehensive Cancer Center, Departments of Pediatrics, Biochemistry and Biophysics, University of North Carolina, Chapel Hill, NC, USA*

SHUYUAN YEH • *George H. Whipple Lab, University of Rochester Medical Center, Rochester, NY, USA*

Section I

Reviews

Chapter 1

Molecular Mechanisms of Androgen Action – A Historical Perspective

Albert O. Brinkmann

Abstract

Androgens and the androgen receptor (AR) are indispensable for expression of the male phenotype. The two most important androgens are testosterone and 5α-dihydrotestosterone. The elucidation of the mechanism of androgen action has a long history starting in the 19th century with the classical experiments by Brown-Séquard. In the 1960s the steroid hormone receptor concept was established and the AR was identified as a protein entity with a high affinity and specificity for testosterone and 5α-dihydrotestosterone. In addition, the enzyme 5α-reductase type 2 was discovered and found to catalyze the conversion of testosterone to the more active metabolite 5α-dihydrotestosterone. In the second half of the 1980s, the cDNA cloning of all steroid hormone receptors, including that of the AR, has been another milestone in the whole field of steroid hormone action. Despite two different ligands (testosterone and 5α-dihydrotestosterone), only one AR cDNA has been identified and cloned. The AR (NR3C4) is a ligand-dependent transcription factor and belongs to the family of nuclear hormone receptors which has 48 members in human. The current model for androgen action involves a multistep mechanism. Studies have provided insight into AR association with co-regulators involved in transcription initiation and on intramolecular interactions of the AR protein during activation. Knowledge about androgen action in the normal physiology and in disease states has increased tremendously after cloning of the AR cDNA. Several diseases, such as androgen insensitivity syndrome (AIS), prostate cancer and spinal bulbar muscular atrophy (SBMA), have been shown to be associated with alterations in AR function due to mutations in the AR gene or dysregulation of androgen signalling. A historical overview of androgen action and salient features of AR function in normal and disease states are provided herein.

Key words: Testosterone, transcription regulation, androgens, nuclear receptor, sexual differentiation, androgen insensitivity, functional domains, prostate cancer, mechanism of action.

F. Saatcioglu (ed.), *Androgen Action*, Methods in Molecular Biology 776,
DOI 10.1007/978-1-61779-243-4_1,

1. Introduction

Androgens are important hormones for expression of the male phenotype. They belong to the group of steroid hormones and have characteristic roles during male sexual differentiation, development and maintenance of secondary male characteristics, and initiation and maintenance of spermatogenesis (1). In this review, a short historical overview is presented with respect to the most important scientific achievements in uncovering the mechanism of androgen action starting from the experiment of Brown-Séquard in the late 19th century until the cloning of the androgen receptor (AR) cDNA in 1988. Subsequently the focus will be on the functional domain aspects of the AR and its crucial role in some pathological situations.

2. Historic Landmarks in Androgen Action

There is a long history on the elucidation of the nature, origin and mechanism of action of androgens. Already in 1889, the French physiologist and professor of medicine at the famous Collège de France in Paris, Charles Edward Brown-Séquard (1818–1894) (**Fig. 1.1**), communicated the first indirect evidence for androgen action via internal secretion after giving himself injections of a testicular extract. Brown-Séquard, at that time 72 years old, improved his health considerably from the injections as was established by himself and by a group of independent observers (2). Almost five decades later, in 1935 in the laboratory of professor Ernst Laqueur (**Fig. 1.2**) in the Netherlands, the active substance could be crystallized from an extract obtained from bull's testes. The crystalline compound appeared to fulfil all the criteria for a full androgenic hormone. It was chemically and physiologically characterized and was named by Laqueur and collaborators (2, 3) as "testosterone".

In the early 1950s the concept of receptors as mediators in the action of pharmacological compounds was postulated and established. It took another 15 years of research before this receptor concept could be successfully incorporated in the mechanism of action of steroid hormones. At the end of the 1950s, radioisotope labelling techniques for steroid hormones and technology for detection of tritiated compounds became available. These innovations facilitated the establishment of specific target tissue retention of hormones, and in particular steroid hormones.

Fig. 1.1. Charles Edward Brown-Séquard (1818–1894).

Fig. 1.2. Ernst Laqueur (1880–1947).

The important breakthrough was made at the University of Chicago in the research group of Elwood Jensen (4, 5) (**Fig. 1.3**). The radiolabelled steroid hormone was oestradiol. Jensen and collaborators showed that unaltered radioactive oestradiol was

Fig. 1.3. Elwood Jensen.

retained specifically and for relatively longer time periods in uterine and breast tissues than in non-oestrogen target tissues after injection in immature animals. These were the first indications of specific steroid hormone-binding entities in target tissues: steroid receptors.

The sucrose density gradient approach published in 1966 by David Toft and Jack Gorski established further in an elegant and convincing way the protein nature of the binding entity (6). This approach has successfully stimulated in the next decade the search for other steroid hormone receptors, including the AR. Important findings in this respect were the presence of receptor dimers and heat-shock proteins in the complexes and the steroid hormone-induced conformational changes of the receptor complexes. All these steroid receptor proteins had a common functional property: upon hormone binding, they became active transcription regulatory proteins, either positively or negatively.

The first papers on the protein nature and on the isolation of ARs from androgen target tissues were published at the end of the 1960s by several research groups. Evidence for a specific androgen-binding protein isolated from prostate tissue cytosol, was published by the research groups of Ian Mainwaring from the ICRF in London, Shutsung Liao (**Fig. 1.4**) at the University of Chicago and Étienne-Émile Baulieu (**Fig. 1.5**) at the INSERM Institute in Kremlin-Bicêtre in Paris (7–9). Furthermore, a dynamic nature could be attributed to the size of the complex in which ARs and more generally the sex steroid nuclear receptors sedimented in sucrose density gradients. In cytosolic fractions the receptor complexes sedimented as 8–10S complexes, while in nuclear extracts the sizes were smaller (4–5S).

Fig. 1.4. Shutsung Liao.

Fig. 1.5. Étienne-Émile Baulieu.

This difference in size indicated a hormone-induced conformational change of the complex. This conformational change was accompanied by a temperature-dependent translocation of the receptor to the nucleus. A "two-step" mechanism for steroid hormone action was proposed by Elwood Jensen. First step: binding of the steroid to an unoccupied cytoplasmic receptor protein complex, second step: dissociation upon hormone binding into a smaller complex and simultaneously translocation to the nucleus (10).

Another important finding in 1968 was the identification in the prostate of the role of 5α-dihydrotestosterone and the enzyme 5α-reductase by the groups of Jean Wilson in Dallas and Etienne-Emile Baulieu in Paris (11–13). The metabolism of testosterone to the more potent androgen 5α-dihydrotestosterone can be considered as another hallmark in the history of androgen action. Six years later in 1974 the specific role of 5α-dihydrotestosterone was established in the early differentiation of the urogenital sinus and tubercle (14).

During the 1970s and 1980s, much efforts were put in the purification and further characterization of ARs from different sources with the ultimate aim to generate antibodies as a tool for further studies (15–18). One of the main problems was the relatively low levels of AR protein in androgen target tissues as compared to other steroid receptors in their target tissues. A second complicating factor was the relatively high levels of proteolytic enzymes in some of the androgen target tissues (in particular prostatic tissue) (19). However, from these studies, a fairly true impression could still be obtained about the AR protein size despite the different sources. In addition, chemical- and photo-affinity labelling approaches have further contributed considerably to the characterization and the determination of the AR protein size (20–22).

In the second half of the1980s, the cloning of cDNAs of steroid receptors has been achieved by several research groups. The first paper was on the cDNA cloning of the human glucocorticoid receptor and was published by the group of Ron Evans in La Jolla, San Diego, in the "1985 Christmas issue of *Nature*" (23). The cloning of the glucocorticoid receptor cDNA was soon followed by the cloning of that of the human oestradiol receptor by the groups of Pierre Chambon in Strasbourg and of Geoffrey Greene of the University of Chicago in 1986 (24, 25). The information generated from these important steps in the characterization of steroid hormone receptors has stimulated the cDNA cloning of all steroid and nuclear receptors, including that of the human AR.

3. Testosterone Biosynthesis and Metabolism

The major circulating androgen is testosterone, which is synthesized from cholesterol in the Leydig cells of the testis. The biosynthetic conversion of cholesterol to testosterone involves several discrete steps, of which the first one includes the transfer of cholesterol from the outer to the inner mitochondrial membrane by the steroidogenic acute regulatory (StAR) protein and the subsequent side chain cleavage of cholesterol by the enzyme P450scc (26). This conversion, resulting in the synthesis of pregnenolone, is the rate-limiting step in testosterone biosynthesis. Subsequent steps require several enzymes including 3β-hydroxysteroid dehydrogenase, 17α-hydroxylase/C17-20-lyase and 17β-hydroxysteroid dehydrogenase type 3 (27).

Testosterone is metabolized in some target tissues to a more active metabolite: 5α-dihydrotestosterone. The irreversible conversion of testosterone to 5α-dihydrotestosterone is catalyzed by the microsomal enzyme 5α-reductase type 2 (SRD5A2) and is NADPH dependent (28). The cDNA of the gene for 5α-reductase type 2 codes for a protein of 254 amino acid residues with a predicted molecular mass of 28.4 kDa (29, 30). The NH_2-terminal part of the protein contains a subdomain suggested to be involved in testosterone binding, while the COOH-terminal region is involved in NADPH binding (31).

4. Physiological Effects of Androgens

As stated already before, the two most important androgens are testosterone and 5α-dihydrotestosterone. While acting through the same AR, each androgen has its own specific role during male sexual differentiation: testosterone is directly involved in development and differentiation of Wolffian duct-derived structures (epididymides, vasa deferentia, seminal vesicles and ejaculatory ducts), whereas 5α-dihydrotestosterone is the active ligand in a number of other androgen target tissues, such as the urogenital sinus and tubercle and their derived structures (prostate gland, scrotum, urethra and penis) (31, 32) (**Fig. 1.6**). The interaction of both androgens with the AR is different. The affinity of testosterone is twofold lower than that of 5α-dihydrotestosterone for the AR, while the dissociation rate of testosterone from the receptor is fivefold faster than that of 5α-dihydrotestosterone (33).

5. Androgen Action

5.1. The AR and the Nuclear Receptor Family

Actions of androgens are mediated by the AR (NR3C4; nuclear receptor subfamily 3, group C, gene 4). This ligand-dependent transcription factor belongs to the superfamily of 48 known nuclear receptors (34). The nuclear receptor family includes

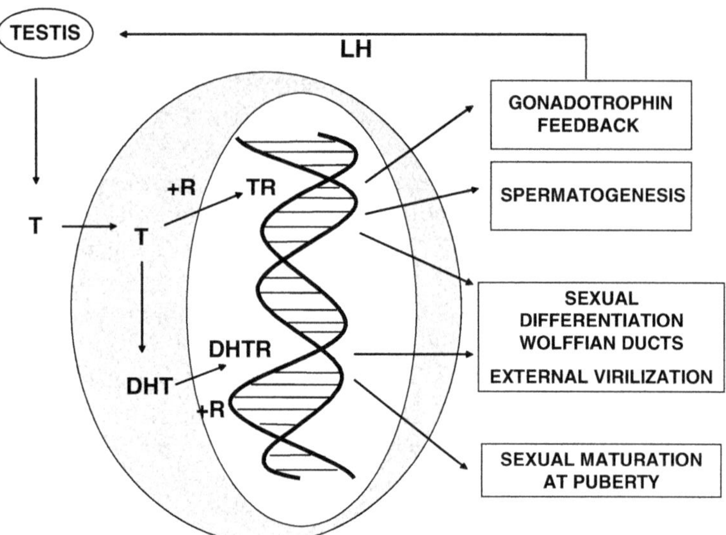

Fig. 1.6. Differential physiological actions of testosterone (T) and 5α-dihydrotestosterone (DHT) via the androgen receptor (R). LH, luteinizing hormone.

receptors for steroid hormones, thyroid hormones, all-*trans* and 9-*cis* retinoic acid, 1,25-dihydroxyvitamin D, ecdysone and peroxisome proliferator-activated receptors (35–37). Comparative structural and functional analysis of nuclear hormone receptors has revealed a common structural organization in four different functional domains.

The current model for androgen action involves a multistep mechanism (**Fig. 1.7**). Upon entry of testosterone into the androgen target cell, binding to the AR takes place either directly or after conversion to 5α-dihydrotestosterone. Binding to the receptor is followed by dissociation of heat-shock proteins in the cytoplasm, simultaneously accompanied by a conformational change of the receptor protein resulting in a transformation and a translocation to the nucleus. The receptor then dimerizes with a second molecule, binds to DNA and recruits further additional proteins (e.g. coactivators, general transcription factors and RNA polymerase II) resulting in specific activation or repression of transcription at discrete sites on the chromatin.

Interestingly, androgen signalling via the AR can also occur in a non-genomic, rapid and sex-nonspecific way by crosstalk with the Scr, Raf-1, Erk-2 pathway (38, 39).

5.2. Cloning and Structural Organization of AR Gene

Since the cloning of the human AR cDNA, our insights into the mechanism of androgen action have increased tremendously. The cloning of the human AR cDNA was published in 1988 and 1989 by several groups just a few years after cloning of the cDNAs of the human glucocorticoid receptor, the human

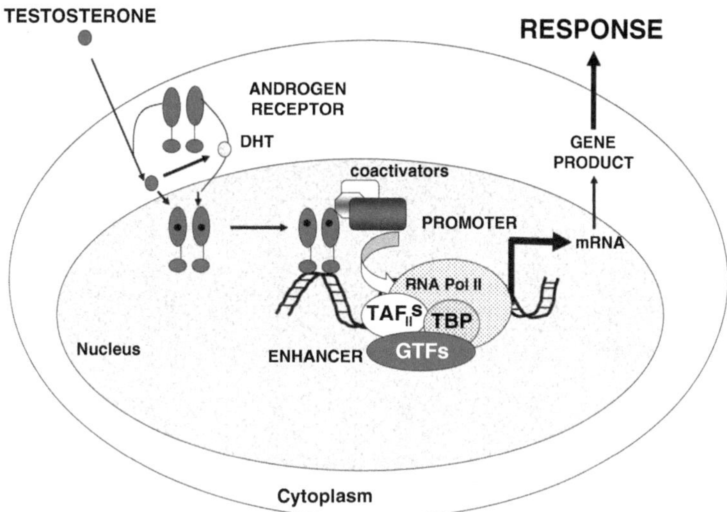

Fig. 1.7. Current model of androgen action. TAF_{II}s are TBP-associated factors; TBP is TATA-box-binding protein; GTFs are general transcription factors.

oestradiol receptor, the human mineralocorticoid receptor and the human progesterone receptor (40–43). Only one AR cDNA has been identified and cloned despite the two different ligands. The concept of two hormones and one receptor to explain the different actions of androgens has been generally accepted, and according to the information available from the human genome project, it is very unlikely that additional genes coding for a functional nuclear receptor with AR-like properties exist (37).

The *AR* gene is located on the human X chromosome at Xq11.2-q12 and consists of eight exons. The gene spans 186 kb in total (44, 45; www.genecards.org). The structural organization of the coding exons is essentially identical to that of the genes coding for the other steroid hormone receptors (i.e. exon/intron boundaries are highly conserved) (45, 46). As a result of differential splicing in the 3′-untranslated region, two AR mRNA species (8.5 and 11 kb, respectively) have been identified in several cell lines and both contain a 1.1-kb 5′-untranslated region (UTR) and a 2.7-kb open reading frame (ORF) (42, 47, 48) (**Fig. 1.8**).

The number of amino acid residues in the AR protein varies between individuals due to the polymorphic polyglutamine stretch and the less variable polyglycine stretch in the NH_2-terminal domain (NTD) (49, 50). Throughout this chapter, the numbering of the AR is based on the 919 amino acid residues according to the AR database (www.mcgill.ca/androgendb; 51). On SDS-PAGE, the AR appears as a 110–112-kDa doublet (52). However, in the presence of androgens, a 114-kDa band also

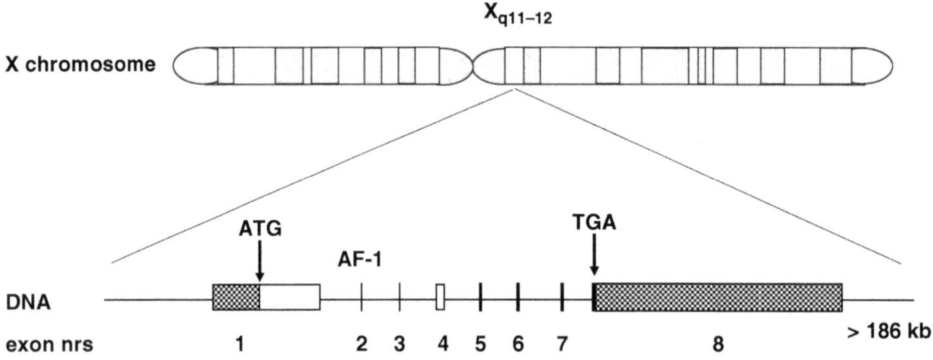

Fig. 1.8. Structural organization of the androgen receptor gene at the Xq11.2-q12 locus on the X chromosome and the derived protein size and structural organization.

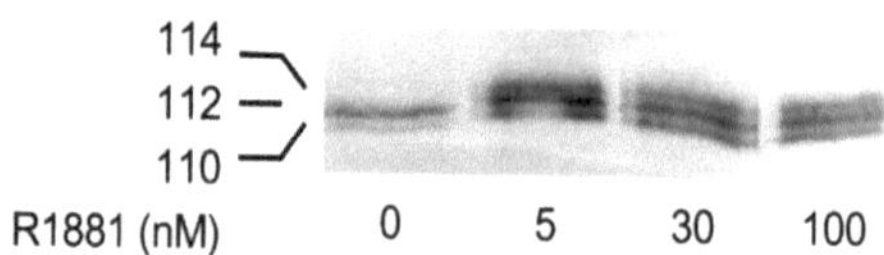

Fig. 1.9. Molecular size of androgen receptor isoforms on SDS-PAGE. The doublet in the absence of hormone (R1881, methyltrienolone) becomes a triplet upon hormone binding due to extra phosphorylation.

appears. These three bands represent different phosphorylated isoforms (53–56) (**Fig. 1.9**).

5.3. AR Polymorphisms

The AR DNA- and ligand-binding domains have a high homology with the corresponding domains of the other members in the steroid receptor subfamily. In contrast, there is a remarkably low homology between the AR NH_2-terminal domain and that of the other steroid receptors (23, 24, 50, 57–59). A polyglutamine stretch, encoded by a polymorphic $(CAG)_nCAA$ repeat, is present in the NH_2-terminal domain (60). The length of the repeat has been used for identification of X chromosomes for carrier detection in pedigree analyses (61, 62).

Variation in length (9–38 glutamine residues) is observed in the normal population and has been suggested to be associated with a very mild modulation of AR activity (63). This assumption is based on in vitro experiments after transient ectopic expression of AR cDNA-containing $(CAG)_nCAA$ repeats of different lengths (53, 64). Translating this finding to the in vivo situation, it can be envisaged that either shorter or longer repeat lengths can result in a relevant biological effect during the lifetime of the individual.

6. AR: Functional Domain Structure

6.1. The NH_2-Terminal Domain

The AR NH_2-terminal domain harbours the major transcription activation functions and several structural subdomains. Within its 538 amino acids, two independent activation domains have been identified: activation function 1 (AF-1) (located between residues 101 and 370) that is essential for transactivation potential of full-length AR and AF-5 (located between residues 360 and 485) that is required for transactivation potential of a constitutively active AR which lacks the LBD (65). Evidence is now available that the AF-5 region interacts with a glutamine-rich domain in p160 cofactors like SRC-1 and TIF2/GRIP1 and not with their LxxLL-like protein-interacting motifs (66). The NH_2-terminal domain is highly flexible: it has a structure between a fully unfolded state and a structured folded conformation, a molten-globule conformation (67).

Another function of the AR NH_2-terminal domain is its binding to the COOH-terminal LBD (N/C interaction) (68, 69). The NH_2-terminal regions required for the binding of the LBD have been mapped to two essential units – the first 36 amino acids and residues 370–494 (70). The hormone-dependent interaction of the NH_2-terminal domain with the ligand-binding domain can play a role in stabilization of the AR dimer complex and in stabilization of the ligand receptor complex by slowing down the rate of ligand dissociation and consequently decreasing receptor degradation (71, 72).

6.2. The DNA-Binding Domain

The DNA-binding domain is the best conserved among the members of the nuclear receptor superfamily. It is characterized by a high content of basic amino acids and by nine conserved cysteine residues. Detailed structural information has been published on the crystal structure of the DNA-binding domain of the glucocorticoid receptor complexed with DNA (73). Subsequently, 3D information became available for AR–DNA interaction at different types of response elements (74).

Briefly, the DNA-binding domain has a compact, globular structure in which three substructures can be distinguished: two zinc clusters and a more loosely structured carboxy-terminal extension (CTE) (75). Both zinc substructures centrally contain one zinc atom which interacts via coordination bonds with four cysteine residues. Both the zinc coordination centres are C-terminally flanked by an α-helix (73). The two zinc clusters are structurally and functionally different and are encoded by two different exons. The α-helix of the most N-terminally located zinc cluster interacts directly with nucleotides of the hormone response element in the major groove of the DNA. Three amino acid residues at the N terminus of this α-helix are responsible for the specific recognition of the DNA sequence of the responsive element. These three amino acid residues, the so-called P(roximal) box [Glycine-Serine-Valine], are identical in the androgen, progesterone, glucocorticoid and mineralocorticoid receptors, and differ from the residues at homologous positions in the oestradiol receptor. It is not surprising therefore that the androgen, progesterone, glucocorticoid and mineralocorticoid receptors can recognize the same response element.

6.3. The Hinge Region

Between the DNA-binding domain and the ligand-binding domain, a non-conserved hinge region is located, which is also variable in size in different steroid receptors. The hinge region can be considered as a flexible linker between the ligand-binding domain and the rest of the receptor molecule. The hinge region is important for nuclear localization and contains a bipartite nuclear localization signal. In some nuclear receptors, including the AR, acetylation can occur in the hinge region at a highly conserved acetylation consensus site [KLLKK] (76).

6.4. The Ligand-Binding Domain

Finally, the second-best conserved region in the AR is the hormone-binding domain. This domain is encoded by approximately 250 residues in the C-terminal end of the molecule (23, 24, 42, 57, 58, 77). The crystal structure of the human AR ligand binding in complex with the synthetic ligand methyltrienolone (R1881) and 5α-dihydrotestosterone has been determined (78, 79). The three-dimensional structure has the typical nuclear receptor ligand-binding domain fold. Interestingly, the ligand-binding pocket consists of 18 amino acid residues interacting more or less directly with the bound ligand (78). The ligand-binding pocket is somewhat flexible and can accommodate ligands with different structures. The structural data are being used in designing optimized selective AR modulators (SARMs) (80).

Crystallographic data on the ligand-binding domain complexed with agonist predict 11 helices (no helix 2) with two anti-parallel β-sheets arranged in the so-called helical sandwich pattern. In the agonist-bound conformation, the C-terminal helix 12 is positioned in an orientation allowing a closure of the ligand-binding pocket. The fold of the ligand-binding domain upon hormone binding results in a globular structure with an interaction surface for binding of interacting proteins such as coactivators. In this way the AR selectively recruits a number of proteins and can communicate with other partners of the transcription initiation complex. Crystallization studies of wild-type AR ligand-binding domain with antagonists are up till now unsuccessful (81).

The AF2 function in the ligand-binding domain is strongly dependent on the presence of nuclear receptor coactivators. In vivo experiments favour a ligand-dependent functional interaction between the AF-2 region in the ligand-binding domain with the NH_2-terminal domain (68, 70).

Deletions in the ligand-binding domain abolish hormone binding completely (82). Deletions in the N-terminal domain and DNA-binding domain do not affect hormone binding. Deletion of the ligand-binding domain leads to a constitutively active AR protein with transactivation capacity comparable to the full-length AR (82). Thus it appears that the hormone-binding domain acts as a repressor of the transactivation function in the absence of hormone. This regulatory function of the AR ligand-binding domain in the absence of hormone has also been reported for the glucocorticoid receptor (83).

6.5. Anti-androgens

AR antagonists are compounds that interfere in some way with the biological effects of androgens and are frequently used in the treatment of androgen-based pathologies. It was demonstrated with a limited proteolytic protection assay that binding of androgens by the AR results in two consecutive conformational changes of the receptor molecule. Initially, a fragment of 35 kDa, spanning the complete ligand-binding domain and part of the hinge region, is protected from digestion by the ligand. After prolonged incubation times with the ligand, a second conformational change occurs resulting in the protection of a smaller fragment of 29 kDa (84, 85). In the presence of several anti-androgens, only the 35-kDa fragment is protected from proteolytic digestion, and no smaller fragments are detectable upon longer incubations. This suggests that the 35-kDa fragment can be associated with an inactive conformation, whereas the second conformational change, only inducible by agonists and considered as the necessary step for transcription activation, is lacking upon binding of anti-androgens.

Based on the conformational changes of the AR ligand-binding domain, induced by androgens or anti-androgens, it can be concluded that the different transcriptional activities displayed by either full agonists, partial agonists or full antagonists are the result of recruitment of a different repertoire of co-regulators (coactivators or corepressors) as a consequence of these conformational changes. The differential recruitment of co-regulators can be considered as a special form of ligand-selective modulation of the AR ligand-binding domain and can be applied in a broader sense also to the tissue-selective modulation of androgen action, where levels of coactivators and corepressors may ultimately determine the final activity (86).

7. AR Disorders

7.1. Androgen Insensitivity Syndrome

It has been known for quite some time that defects in male sexual differentiation in 46,XY individuals have an X-linked pattern of inheritance. It was Reifenstein who reported in 1947 on families with severe hypospadias, infertility and gynecomastia (87). The end-organ resistance to androgens has been designated as androgen insensitivity syndrome (AIS) and is distinct from other XY disorders of sex development (XY, DSD; formerly named male pseudohermaphroditism) like 17β-hydroxysteroiddehydrogenase type 3 deficiency or 5α-reductase type 2 deficiency (31, 88–90). It is generally accepted that defects in the *AR* gene can prevent the normal development of both internal and external male structures in 46,XY individuals and information on the molecular structure of the human *AR* gene has facilitated the study of molecular defects associated with androgen insensitivity. Naturally occurring mutations in the *AR* gene are an interesting source for the investigation of receptor structure–function relationships. The variation in clinical phenotypes provides the opportunity to correlate a mutation in the AR structure with the impairment of a specific physiological function.

7.2. Genetics of Androgen Insensitivity Syndrome

Since the cloning of the AR cDNA in 1988 and the subsequent elucidation of the genomic organization of the *AR* gene, tools have become available for the molecular analysis of the *AR* gene in individuals with AIS (45, 46). In addition to endocrinological data, the most reliable approach is sequencing each individual AR exon and flanking intron sequences. In general, androgen insensitivity can be routinely analyzed and almost 400 different mutations in the *AR* gene have been reported. Differential diagnosis is now possible from entirely different syndromes presenting with similar phenotypes including testicular enzyme deficiencies, 5α-reductase type 2 deficiency and Leydig cell hypoplasia due to inactivating luteinizing hormone receptor mutations. Furthermore, in pedigree analysis, intragenic polymorphisms, such as the highly polymorphic $(CAG)_n$CAA repeat encoding a polyglutamine stretch, the polymorphic GGN repeat encoding a polyglycine stretch, the *Hin*dIII polymorphism (44) and the *Stu*I polymorphism (90), can be used as X-chromosomal markers

(61, 91–93). Extensive general information can be obtained at the Internet site www.genecards.org on the *AR* (NR3C4) gene and on the 375 identified single-nucleotide polymorphisms (SNPs).

7.3. Mutations in the AR Gene

In the *AR* gene, four different types of mutations have been detected in 46,XY individuals with AIS: single point mutations resulting in amino acid substitutions or premature stop codons, nucleotide insertions or deletions most often leading to a frame shift and premature termination, complete or partial gene deletions (>10 nucleotides), and intronic mutations in either splice/donor or splice/acceptor sites which affect the splicing of AR RNA (51). In general, in 70% of the cases, *AR* gene mutations are transmitted in an X-linked recessive manner, but in 30% of the cases, the mutations arise de novo. When de novo mutations occur after the zygotic stage, they result in somatic mosaicisms (94). The most recent update on *AR* gene mutations is available at http://www.mcgill.ca/androgendb/ (51).

7.4. Male Infertility

Several investigations into male infertile patients found an association between a longer $(CAG)_n$CAA repeat and the risk of defective spermatogenesis (95–97). This suggests that a less active AR, due to a moderate expanded repeat length, may be a factor in the aetiology of male infertility.

7.5. Spinal and Bulbar Muscular Atrophy

Kennedy's disease, also known as spinal and bulbar muscular atrophy (SBMA), is a slowly progressing degeneration of lower motor neurons, resulting in muscle weakness in adult males (98–100). La Spada and colleagues (101) were the first to demonstrate a direct correlation of SBMA with an extension of the $(CAG)_n$CAA repeat in exon 1, which encodes the polymorphic polyglutamine stretch in the AR NH_2-terminal domain. In the normal human population the repeat length is 34 or fewer, depending on the ethnic background (60, 102, 103). In SBMA, full penetrance alleles have a repeat length of 38 or more (**Fig. 1.10**). Reduced penetrance has been suggested for alleles with repeat lengths of 36 and 37. There is no consensus as to the clinical significance of alleles with a repeat length of 35 (114).

SBMA is associated with nuclear accumulation of the AR protein with the expanded polyglutamine stretch in motor neurons. Clinical symptoms usually manifest in the third to fifth decade and result from severe depletion of lower motor nuclei in the spinal cord and brainstem (63, 98, 104). In addition, SBMA patients frequently exhibit endocrinological abnormalities including testicular atrophy, infertility, gynecomastia and elevated LH, FSH and oestradiol levels. Sex differentiation proceeds normally and characteristics of mild androgen insensitivity appear later in life.

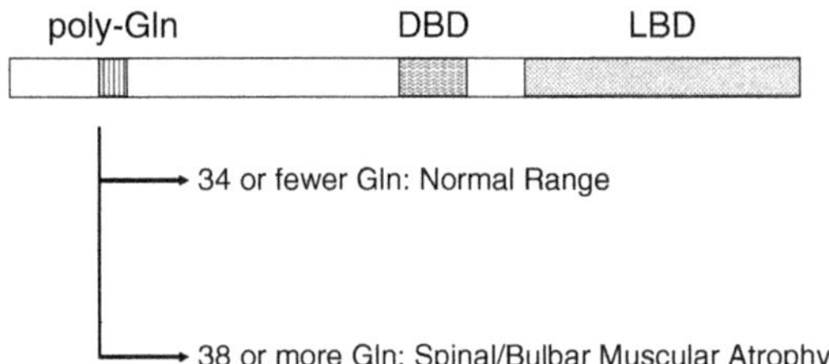

Fig. 1.10. Variation in the polyglutamine stretch encoded by the $(CAG)_nCAA$ repeat in exon 1. The normal range of this stretch is 34 or fewer glutamine residues, while in the motor neuron disease spinal/bulbar muscular atrophy (also called Kennedy's disease), the $(CAG)_nCAA$ repeat in exon 1 is expanded and full penetrance alleles have 38 or more glutamine residues (114).

7.6. Prostate Cancer

Prostate cancer is the second leading cause of cancer deaths in men in Western countries (105). Most prostate cancers express relatively high levels of AR protein (106). Initially, prostate cancer is androgen dependent, because removal of androgens or blocking the AR by anti-androgens results in growth arrest of the tumour. However, tumour growth arrest is only temporary and most tumours undergo a transition to an androgen-unresponsive state.

Despite many suggestions for a possible mechanism for the development of the androgen-unresponsive state of prostate tumours, the exact mechanism underlying the transition to androgen independency is still unclear (107, 108). Since the AR is expressed in androgen-independent prostate tumours, it is assumed that the AR is still involved in some way in tumour growth. One mechanism may be higher AR protein expression caused by amplification of the *AR* gene (109–111). Under extremely low androgen levels (by hormone deprivation), the AR can still be activated. Also somatic mutations in the *AR* gene can result in a more active receptor protein (www.mcgill.ca/androgendb) or may broaden the ligand specificity towards anti-androgens or other steroid hormones, such as those found for the AR mutant T877A (112, 113). This mutation is frequently found in androgen-independent prostate tumours. Another mechanism that has been proposed is the increased expression of AR-specific cofactors, resulting in an enhanced AR activity and consequently in enhanced tumour growth. Finally, a mechanism involving ligand-independent activation of the AR has been suggested. This might be achieved by crosstalk with other activated signal transduction pathways.

8. Conclusions

Androgens (testosterone and 5α-dihydrotestosterone) are important steroid hormones for expression of the male phenotype. Their actions are mediated by the AR. The development of the field of androgen action over the last century has culminated in the cloning of the AR cDNA and its cofactors and their implications in both physiological and pathological conditions. As evidenced by the various tools that have been developed to study androgen action, presented in the ensuing chapters in this volume, the future is bright to further unravel the molecular mechanisms of androgen action as they relate to different normal and disease states. The specific roles of the various AR co-regulators and the manner in which their actions are coordinated on androgen target genes will be a predictable subject of future investigations. Recently developed quantitative microscopic interaction assays will allow the investigation of the dynamic behaviour of interacting molecules with the androgen receptor in time and space in single living cells. In this respect, the genome-wide analysis of potential AR-binding sites and their characterization is also a prerequisite for a functional correlation of these sites with an androgen-dependent gene expression profile. Structure–function determinants of the androgen receptor underlying androgen action can stimulate development of patient-tailored therapeutics. Although we have gained great insight into androgen action until now, much more work is needed to learn how androgens work, e.g. in regulating metabolism, their effects on the nervous system and in female physiology, as well as in disease states. It is thus an exciting time when the collaborative efforts of basic laboratory and clinical scientists, equipped with the plethora of techniques described in this book, are required which will result in the uncovering of this important body of knowledge.

References

1. George, F.W., Wilson, J.D. 1994. Sex determination and differentiation. In: Knobil, E., Neill, J.D. (eds.) *The Physiology of Reproduction, Chapter 1.* pp. 3–28. Raven Press, New York, NY
2. Medvei, V.C. 1993. *A History of Clinical Endocrinology.* Parthenon Publishing Group, Carnforth, Lancashire, UK
3. Laqueur, E., de Jongh, S.E., Tausk, M. 1948. *Hormonologie.* Noord-Hollandsche Uitgevers Maatschappij, Amsterdam, The Netherlands
4. Jensen, E.V., Jacobson, H.I. 1960. Fate of steroidal estrogens in target tissues. In: Pincus, G., Volmer, E.P. (eds.) *Biological Activities of Steroids in Relation to Cancer,* pp. 161–174. Academic, New York, NY
5. Jensen, E.V., Jacobson, H.I., Walf, A.A., Frye, C.A. 2010. Estrogen action: a historic perspective on the implications of considering alternative approaches. *Physiol Behav* **99**: 151–162
6. Toft, D., Gorski, J. 1966. A receptor molecule for estrogens: isolation from the rat uterus and preliminary characterization. *Proc Natl Acad Sci USA* **55**: 1574–1581
7. Mainwaring, W.I. 1969. A soluble androgen receptor in the cytoplasm of rat prostate. *J Endocrinol* **45**: 531–541

8. Fang, S., Anderson, K.M., Liao, S. 1969. Receptor proteins for androgens. On the role of specific proteins in selective retention of 17-beta-hydroxy-5-alpha-androstan-3-one by rat ventral prostate in vivo and in vitro. *J Biol Chem* **244**: 6584–6595
9. Baulieu, E.E., Jung, I. 1970. A prostatic cytosol receptor. *Biochem Biophys Res Commun* **38**: 599–606
10. Jensen, E.V., Suzuki, T., Kawashima, T., Stumpf, W.E., Jungblut, P.W., DeSombre, E.R. 1968. A two-step mechanism for the interaction of estradiol with rat uterus. *Proc Natl Acad Sci USA* **59**: 632–638
11. Bruchovsky, N., Wilson, J.D. 1968. The conversion of testosterone to 5-alpha-androstan-17-beta-ol-3-one by rat prostate in vivo and in vitro. *J Biol Chem* **243**: 2012–2021
12. Bruchovsky, N., Wilson, J.D. 1968. The intranuclear binding of testosterone and 5-alpha-androstan-17-beta-ol-3-one by rat prostate. *J Biol Chem* **243**: 5953–5960
13. Baulieu, E.E., Lasnizki, I., Robel, P. 1968. Metabolism of testosterone and action of metabolites on prostate glands grown in organ culture. *Nature* **219**: 1155–1156
14. Imperato-McGinley, J., Guerrero, L., Gautier, T., Peterson, R.E. 1974. Steroid 5alpha-reductase deficiency in man: an inherited form of male pseudohermaphroditism. *Science* **186**: 1213–1215
15. Chang, C.H., Rowley, D.R., Lobl, T.J., Tindall, D.J. 1982. Purification and characterization of androgen receptor from steer seminal vesicle. *Biochemistry* **21**: 4102–4109
16. Wilson, E.M., Colvard, D.S. 1984. Factors that influence the interaction of androgen receptors with nuclei and nuclear matrix. *Ann N Y Acad Sci* **438**: 85–100
17. Foekens, J.A., Peerbolte, R., Mulder, E., van der Molen, H.J. 1981. Characterization and partial purification of androgen receptors from ram seminal vesicles. *Mol Cell Endocrinol* **23**: 173–186
18. van Loon, D., Voorhorst, M.M., Brinkmann, A.O., Mulder, E. 1988. Purification of the intact monomeric 110 kDa form of the androgen receptor from calf uterus to near homogeneity. *Biochim Biophys Acta* **970**: 278–286
19. Wilson, E.M., French, F.S. 1979. Effects of proteases and protease inhibitors on the 4.5 S and 8 S androgen receptor. *J Biol Chem* **254**: 6310–6319
20. Chang, C.H., Lobl, T.J., Rowley, D.R., Tindall, D.J. 1984. Affinity labeling of the androgen receptor in rat prostate cytosol with 17 beta-[(bromoacetyl)oxy]-5 alpha-androstan-3-one. *Biochemistry* **23**: 2527–2533
21. Brinkmann, A.O., Kuiper, G.G., de Boer, W., Mulder, E., van der Molen, H.J. 1985. Photoaffinity labelling of androgen receptors with 17 beta-hydroxy-17 alpha-[3H]methyl-4,9,11-estratrien-3-one. *Biochem Biophys Res Commun* **126**: 163–169
22. van Laar, J.H., Bolt-de Vries, J., Zegers, N.D., Trapman, J., Brinkmann, A.O. 1990. Androgen receptor heterogeneity and phosphorylation in human LNCaP cells. *Biochem Biophys Res Commun* **166**: 193–200
23. Hollenberg, S.M., Weinberger, C., Ong, E.S., Cerelli, G., Oro, A., Lebo, R., Thompson, E.B., Rosenfeld, M.G., Evans, R.M. 1985. Primary structure and expression of a functional human glucocorticoid receptor cDNA. *Nature* **318**: 635–641
24. Green, S., Walter, P., Kumar, V., Krust, A., Bornert, J.M., Argos, P., Chambon, P. 1986. Human oestrogen receptor cDNA: sequence, expression and homology to v-erb-A. *Nature* **320**: 134–139
25. Greene, G.L., Gilna, P., Waterfield, M., Baker, A., Hort, Y., Shine, J. 1986. Sequence and expression of human estrogen receptor complementary DNA. *Science* **231**: 1150–1154
26. Stocco, D.M., Clark, B.J. 1996. Regulation of the acute production of steroids in steroidogenic cells. *Endocr Rev* **17**: 221–244
27. Miller, W.L. 1988. Molecular biology of steroid hormone synthesis. *Endocr Rev* **9**: 295–318
28. Russell, D.W., Wilson, J.D. 1994. Steroid 5 alpha-reductase: two genes/two enzymes. *Annu Rev Biochem* **63**: 25–61
29. Andersson, S., Bishop, R.W., Russell, D.W. 1989. Expression cloning and regulation of steroid 5 alpha-reductase, an enzyme essential for male sexual differentiation. *J Biol Chem* **264**: 16249–16255
30. Andersson, S., Berman, D.M., Jenkins, E.P., Russell, D.W. 1991. Deletion of steroid 5 alpha-reductase 2 gene in male pseudohermaphroditism. *Nature* **354**: 159–161
31. Wilson, J.D., Griffin, J.E., Russell, D.W. 1993. Steroid 5 alpha-reductase 2 deficiency. *Endocr Rev* **14**: 577–593
32. Randall, V.A. 1994. Role of 5 alpha-reductase in health and disease. *Baillieres Clin Endocrinol Metab* **8**: 405–431
33. Grino, P.B., Griffin, J.E., Wilson, J.D. 1990. Testosterone at high concentrations interacts with the human androgen receptor similarly to dihydrotestosterone. *Endocrinology* **126**: 1165–1172

34. Committee, N.R.N. 1999. A unified nomenclature system for the nuclear receptor superfamily. *Cell* 97: 161–163
35. Evans, R.M. 1988. The steroid and thyroid hormone receptor superfamily. *Science* **240**: 889–895
36. Laudet, V., Hanni, C., Coll, J., Catzeflis, F., Stehelin, D. 1992. Evolution of the nuclear receptor gene superfamily. *EMBO J* **11**: 1003–1013
37. Robinson-Rechavi, M., Carpentier, A.S., Duffraisse, M., Laudet, V. 2001. How many nuclear hormone receptors are there in the human genome? *Trends Genet* **17**: 554–556
38. Migliaccio, A., Castoria, G., Di Domenico, M., de Falco, A., Bilancio, A., Lombardi, M., Barone, M.V., Ametrano, D., Zannini, M.S., Abbondanza, C., Auricchio, F. 2000. Steroid-induced androgen receptor–oestradiol receptor beta-Src complex triggers prostate cancer cell proliferation. *EMBO J* **19**: 5406–5417
39. Kousteni, S., Bellido. T., Plotkin, L.I., O'Brien, C.A., Bodenner, D.L., Han, L., Han, K., DiGregorio, G.B., Katzenellenbogen, J.A., Katzenellenbogen, B.S., Roberson, P.K., Weinstein, R.S., Jilka, R.L., Manolagas, S.C. 2001. Nongenotropic, sex-nonspecific signaling through the estrogen or androgen receptors: dissociation from transcriptional activity. *Cell* **104**: 719–730
40. Chang, C.S., Kokontis, J., Liao, S.T. 1988. Molecular cloning of human and rat complementary DNA encoding androgen receptors. *Science* **240**: 324–326
41. Lubahn, D.B., Joseph, D.R., Sullivan, P.M., Willard, H.F., French, F.S., Wilson, E.M. 1988. Cloning of human androgen receptor complementary DNA and localization to the X chromosome. *Science* **240**: 327–330
42. Trapman, J., Klaassen, P., Kuiper, G.G., van der Korput, J.A., Faber, P.W., van Rooij, H.C., Geurts van Kessel, A., Voorhorst, M.M., Mulder, E., Brinkmann, A.O. 1988. Cloning, structure and expression of a cDNA encoding the human androgen receptor. *Biochem Biophys Res Commun* **153**: 241–248
43. Tilley, W.D., Marcelli, M., Wilson, J.D., McPhaul, M.J. 1989. Characterization and expression of a cDNA encoding the human androgen receptor. *Proc Natl Acad Sci USA* **86**: 327–331
44. Brown, C.J., Goss, S.J., Lubahn, D.B., Joseph, D.R., Wilson, E.M., French, F.S., Willard, H.F. 1989. Androgen receptor locus on the human X chromosome: regional localization to Xq11-12 and description of a DNA polymorphism. *Am J Hum Genet* **44**: 264–269
45. Kuiper, G.G., Faber. P.W., van Rooij, H.C., van der Korput, J.A., Ris-Stalpers, C., Klaassen, P., Trapman, J., Brinkmann, A.O. 1989. Structural organization of the human androgen receptor gene. *J Mol Endocrinol* **2**: R1–R4
46. Lubahn, D.B., Brown, T.R., Simental, J.A., Higgs, H.N., Migeon, C.J., Wilson, E.M., French, F.S. 1989. Sequence of the intron/exon junctions of the coding region of the human androgen receptor gene and identification of a point mutation in a family with complete androgen insensitivity. *Proc Natl Acad Sci USA* **86**: 9534–9538
47. Faber, P.W., van Rooij, H.C., van der Korput, H.A., Baarends, W.M., Brinkmann, A.O., Grootegoed, J.A., Trapman, J. 1991. Characterization of the human androgen receptor transcription unit. *J Biol Chem* **266**: 10743–10749
48. Tilley, W.D., Marcelli, M., McPhaul, M.J. 1990. Expression of the human androgen receptor gene utilizes a common promoter in diverse human tissues and cell lines. *J Biol Chem* **265**: 13776–13781
49. Faber, P.W., Kuiper, G.G., van Rooij, H.C., van der Korput, J.A., Brinkmann, A.O., Trapman, J. 1989. The N-terminal domain of the human androgen receptor is encoded by one, large exon. *Mol Cell Endocrinol* **61**: 257–262
50. Sleddens, H.F., Oostra, B.A., Brinkmann, A.O., Trapman, J. 1993. Trinucleotide (GGN) repeat polymorphism in the human androgen receptor (AR) gene. *Hum Mol Genet* **2**: 493
51. Gottlieb, B., Beitel, L.K., Wu, J.H., Trifiro, M. 2004. The androgen receptor gene mutations database (ARDB): 2004 update. *Hum Mutat* **23**: 527–533
52. van Laar, J.H., Bolt-de Vries, J., Voorhorst-Ogink, M.M., Brinkmann, A.O. 1989. The human androgen receptor is a 110 kDa protein. *Mol Cell Endocrinol* **63**: 39–44
53. Jenster, G., de Ruiter, P.E., van der Korput, H.A., Kuiper, G.G., Trapman, J., Brinkmann, A.O. 1994. Changes in the abundance of androgen receptor isotypes: effects of ligand treatment, glutamine-stretch variation, and mutation of putative phosphorylation sites. *Biochemistry* **33**: 14064–14072
54. Kuiper, G.G., Brinkmann, A.O. 1995. Phosphotryptic peptide analysis of the human androgen receptor: detection of a hormone-induced phosphopeptide. *Biochemistry* **34**: 1851–1857
55. Kuiper, G.G., de Ruiter, P.E., Grootegoed, J.A., Brinkmann, A.O. 1991. Synthesis and post-translational modification of the

androgen receptor in LNCaP cells. *Mol Cell Endocrinol* **80**: 65–73

56. Wong, H.Y., Burghoorn, J.A., Van Leeuwen, M., De Ruiter, P.E., Schippers, E., Blok, L.J., Li, K.W., Dekker, H.L., De Jong, L., Trapman, J., Grootegoed, J.A., Brinkmann, A.O. 2004. Phosphorylation of androgen receptor isoforms. *Biochem J* **383**: 267–276
57. Misrahi, M., Atger, M., d'Auriol, L., Loosfelt, H., Meriel, C., Fridlansky, F., Guiochon-Mantel, A., Galibert, F., Milgrom, E. 1987. Complete amino acid sequence of the human progesterone receptor deduced from cloned cDNA. *Biochem Biophys Res Commun* **143**: 740–748
58. Arriza, J.L., Weinberger, C., Cerelli, G., Glaser, T.M., Handelin, B.L., Housman, D.E., Evans, R.M. 1987. Cloning of human mineralocorticoid receptor complementary DNA: structural and functional kinship with the glucocorticoid receptor. *Science* **237**: 268–275
59. Kuiper, G.G., Enmark, E., Pelto-Huikko, M., Nilsson, S., Gustafsson, J.A. 1996. Cloning of a novel receptor expressed in rat prostate and ovary. *Proc Natl Acad Sci USA* **93**: 5925–5930
60. Sleddens, H.F., Oostra, B.A., Brinkmann, A.O., Trapman, J. 1992. Trinucleotide repeat polymorphism in the androgen receptor gene (AR). *Nucleic Acids Res* **20**: 1427
61. Boehmer, A.L., Brinkmann, A.O., Niermeijer, M.F., Bakker, L., Halley, D.J., Drop, S.L. 1997. Germ-line and somatic mosaicism in the androgen insensitivity syndrome: implications for genetic counseling. *Am J Hum Genet* **60**: 1003–1006
62. Li, S.L., Ting, S.S., Lindeman, R., French, R., Ziegler, J.B. 1998. Carrier identification in X-linked immunodeficiency diseases. *J Paediatr Child Health* **34**: 273–279
63. Nance, M.A. 1997. Clinical aspects of CAG repeat diseases. *Brain Pathol* **7**: 881–900
64. Kazemi-Esfarjani, P., Trifiro, M.A., Pinsky, L. 1995. Evidence for a repressive function of the long polyglutamine tract in the human androgen receptor: possible pathogenetic relevance for the $(CAG)_n$-expanded neuronopathies. *Hum Mol Genet* **4**: 523–527
65. Jenster, G., van der Korput, H.A., Trapman, J., Brinkmann, A.O. 1995. Identification of two transcription activation units in the N-terminal domain of the human androgen receptor. *J Biol Chem* **270**: 7341–7346
66. Claessens, F., Denayer, S., Van Tilborgh, N., Kerkhofs, S., Helsen, C., Haelens, A. 2008. Diverse roles of androgen receptor (AR) domains in AR-mediated signaling. *Nucl Recept Signal* **6**: e008
67. Lavery, D.N., McEwan, I.J. 2008. Structural characterization of the native NH2-terminal transactivation domain of the human androgen receptor: a collapsed disordered conformation underlies structural plasticity and protein-induced folding. *Biochemistry* **47**: 3360–3369
68. Langley, E., Zhou, Z.X., Wilson, E.M. 1995. Evidence for an anti-parallel orientation of the ligand-activated human androgen receptor dimer. *J Biol Chem* **270**: 29983–29990
69. Doesburg, P., Kuil, C.W., Berrevoets, C.A., Steketee, K., Faber, P.W., Mulder, E., Brinkmann, A.O., Trapman, J. 1997. Functional in vivo interaction between the amino-terminal, transactivation domain and the ligand binding domain of the androgen receptor. *Biochemistry* **36**: 1052–1064
70. Berrevoets, C.A., Doesburg, P., Steketee, K., Trapman, J., Brinkmann, A.O. 1998. Functional interactions of the AF-2 activation domain core region of the human androgen receptor with the amino-terminal domain and with the transcriptional coactivator TIF2 (transcriptional intermediary factor2). *Mol Endocrinol* **12**: 1172–1183
71. Zhou, Z.X., Lane, M.V., Kemppainen, J.A., French, F.S., Wilson, E.M. 1995. Specificity of ligand-dependent androgen receptor stabilization: receptor domain interactions influence ligand dissociation and receptor stability. *Mol Endocrinol* **9**: 208–218
72. Centenera, M.M., Harris, J.M., Tilley, W.D., Butler, L.M. 2008. The contribution of different androgen receptor domains to receptor dimerization and signaling. *Mol Endocrinol* **22**: 2373–2382
73. Luisi, B.F., Xu, W.X., Otwinowski, Z., Freedman, L.P., Yamamoto, K.R., Sigler, P.B. 1991. Crystallographic analysis of the interaction of the glucocorticoid receptor with DNA. *Nature* **352**: 497–505
74. Shaffer, P.L., Jivan, A., Dollins, D.E., Claessens, F., Gewirth, D.T. 2004. Structural basis of androgen receptor binding to selective androgen response elements. *Proc Natl Acad Sci USA* **101**: 4758–4763
75. Jakob, M., Kolodziejczyk, R., Orlowski, M., Krzywda, S., Kowalska, A., Dutko-Gwozdz, J., Gwozdz, T., Kochman, M., Jaskolski, M., Ozyhar, A. 2007. Novel DNA-binding element within the C-terminal extension of the nuclear receptor DNA-binding domain. *Nucleic Acids Res* **35**: 2705–2718
76. Fu, M., Wang, C., Zhang, X., Pestell, R.G. 2004. Acetylation of nuclear receptors in cellular growth and apoptosis. *Biochem Pharmacol* **68**: 1199–1208

77. Brinkmann, A.O., Trapman, J. 2000. Genetic analysis of androgen receptors in development and disease. *Adv Pharmacol* **47**: 317–341
78. Matias, P.M., Donner, P., Coelho, R., Thomaz, M., Peixoto, C., Macedo, S., Otto, N., Joschko, S., Scholz, P., Wegg, A., Basler, S., Schafer, M., Ruff, M., Egner, U., Carrondo, M.A. 2000. Structural evidence for ligand specificity in the binding domain of the human androgen receptor: implications for pathogenic gene mutations. *J Biol Chem* **275**: 26164–26171
79. Sack, J.S., Kish, K.F., Wang, C., Attar, R.M., Kiefer, S.E., An, Y., Wu, G.Y., Scheffler, J.E., Salvati, M.E., Krystek, S.R., Weinmann, R., Einspahr, H.M. 2001. Crystallographic structures of the ligand-binding domains of the androgen receptor and its T877A mutant complexed with the natural agonist dihydrotestosterone. *Proc Natl Acad Sci USA* **98**: 4904–4909
80. Pereira de Jesus-Tran, K., Cote, P.L., Cantin, L., Blanchet, J., Labrie, F., Breton, R. 2006. Comparison of crystal structures of human androgen receptor ligand-binding domain complexed with various agonists reveals molecular determinants responsible for binding affinity. *Protein Sci* **15**: 987–999
81. Estebanez-Perpina, E., Arnold, L.A., Nguyen, P., Rodrigues, E.D., Mar, E., Bateman, R., Pallai, P., Shokat, K.M., Baxter, J.D., Guy, R.K., Webb, P., Fletterick, R.J. 2007. A surface on the androgen receptor that allosterically regulates coactivator binding. *Proc Natl Acad Sci USA* **104**: 16074–16079
82. Jenster, G., van der Korput, H.A., van Vroonhoven, C., van der Kwast, T.H., Trapman, J., Brinkmann, A.O. 1991. Domains of the human androgen receptor involved in steroid binding, transcriptional activation, and subcellular localization. *Mol Endocrinol* **5**: 1396–1404
83. Hollenberg, S.M., Evans, R.M. 1988. Multiple and cooperative trans-activation domains of the human glucocorticoid receptor. *Cell* **55**: 899–906
84. Kuil, C.W., Mulder, E. 1994. Mechanism of antiandrogen action: conformational changes of the receptor. *Mol Cell Endocrinol* **102:** R1–R5
85. Kuil, C.W., Berrevoets, C.A., Mulder, E. 1995. Ligand-induced conformational alterations of the androgen receptor analyzed by limited trypsinization. Studies on the mechanism of antiandrogen action. *J Biol Chem* **270**: 27569–27576
86. Berrevoets, C.A., Umar, A., Trapman, J., Brinkmann, A.O. 2004. Differential modulation of androgen receptor transcriptional activity by the nuclear receptor co-repressor (N-CoR). *Biochem J* **379**: 731–738
87. Reifenstein, E.C., Jr. 1947. Hereditary familial hypogonadism. *Proc Am Fed Clin Res* **3**: 86
88. Wilson, J.D., Harrod, M.J., Goldstein, J.L., Hemsell, D.L., MacDonald, P.C. 1974. Familial incomplete male pseudohermaphroditism, type 1. Evidence for androgen resistance and variable clinical manifestations in a family with the Reifenstein syndrome. *N Engl J Med* **290**: 1097–1103
89. Geissler, W.M., Davis, D.L., Wu, L., Bradshaw, K.D., Patel, S., Mendonca, B.B., Elliston, K.O., Wilson, J.D., Russell, D.W., Andersson, S. 1994. Male pseudohermaphroditism caused by mutations of testicular 17 beta-hydroxysteroid dehydrogenase 3. *Nat Genet* **7**: 34–39
90. Hughes, I.A. 2008. Disorders of sex development: a new definition and classification. *Best Pract Res Clin Endocrinol Metab* **22**: 119–134
91. Lu, J., Danielsen, M. 1996. A Stu I polymorphism in the human androgen receptor gene (AR). *Clin Genet* **49**: 323–324.
92. Ris-Stalpers, C., Verleun-Mooijman, M.C., de Blaeij, T.J., Degenhart, H.J., Trapman, J., Brinkmann, A.O. 1994. Differential splicing of human androgen receptor pre-mRNA in X-linked Reifenstein syndrome, because of a deletion involving a putative branch site. *Am J Hum Genet* **54**: 609–617
93. Davies, H.R., Hughes, I.A., Patterson, M.N. 1995. Genetic counselling in complete androgen insensitivity syndrome: trinucleotide repeat polymorphisms, single-strand conformation polymorphism and direct detection of two novel mutations in the androgen receptor gene. *Clin Endocrinol (Oxf)* **43**: 69–77
94. Kohler, B., Lumbroso, S., Leger, J., Audran, F., Grau, E.S., Kurtz, F., Pinto, G., Salerno, M., Semitcheva, T., Czernichow, P., Sultan, C. 2005. Androgen insensitivity syndrome: somatic mosaicism of the androgen receptor in seven families and consequences for sex assignment and genetic counseling. *J Clin Endocrinol Metab* **90**: 106–111
95. Dowsing, A.T., Yong, E.L., Clark, M., McLachlan, R.I., de Kretser, D.M., Trounson, A.O. 1999. Linkage between male infertility and trinucleotide repeat expansion in the androgen-receptor gene [In Process Citation]. *Lancet* **354**: 640–643

96. Mifsud, A., Sim, C.K., Boettger-Tong, H., Moreira, S., Lamb, D.J., Lipshultz, L.I., Yong, E.L. 2001. Trinucleotide (CAG) repeat polymorphisms in the androgen receptor gene: molecular markers of risk for male infertility. *Fertil Steril* **75**: 275–281
97. Wallerand, H., Remy-Martin, A., Chabannes, E., Bermont, L., Adessi, G., Bittard, H. 2001. Relationship between expansion of the CAG repeat in exon 1 of the androgen receptor gene and idiopathic male infertility. *Fertil Steril* **76**: 769–774
98. Kennedy, W.R., Alter, M., Sung, J.H. 1968. Progressive proximal spinal and bulbar muscular atrophy of late onset. A sex-linked recessive trait. *Neurology* **18**: 671–680
99. Arbizu, T., Santamaria, J., Gomez, J.M., Quilez, A., Serra, J.P. 1983. A family with adult spinal and bulbar muscular atrophy, X-linked inheritance and associated testicular failure. *J Neurol Sci* **59**: 371–382
100. Greenland, K.J., Zajac, J.D. 2004. Kennedy's disease: pathogenesis and clinical approaches. *Intern Med J* **34**: 279–286
101. La Spada, A.R, Wilson, E.M., Lubahn, D.B., Harding, A.E., Fischbeck, K.H. 1991. Androgen receptor gene mutations in X-linked spinal and bulbar muscular atrophy. *Nature* **352**: 77–79
102. Caskey, C.T., Pizzuti, A., Fu, Y.H., Fenwick, R.G., Jr., Nelson, D.L. 1992. Triplet repeat mutations in human disease. *Science* **256**: 784–789
103. Edwards, A., Hammond, H.A., Jin, L., Caskey, C.T., Chakraborty, R. 1992. Genetic variation at five trimeric and tetrameric tandem repeat loci in four human population groups. *Genomics* **12**: 241–253
104. Robitaille, Y., Lopes-Cendes, I., Becher, M., Rouleau, G., Clark, A.W. 1997. The neuropathology of CAG repeat diseases: review and update of genetic and molecular features. *Brain Pathol* **7**: 901–926
105. Gronberg, H. 2003. Prostate cancer epidemiology. *Lancet* **361**: 859–864
106. van der Kwast, T.H., Schalken, J., Ruizeveld de Winter, J.A., van Vroonhoven, C.C., Mulder, E., Boersma, W., Trapman, J. 1991. Androgen receptors in endocrine-therapy-resistant human prostate cancer. *Int J Cancer* **48**: 189–193
107. Feldman, B.J., Feldman, D. 2001. The development of androgen-independent prostate cancer. *Nat Rev Cancer* **1**: 34–45
108. Grossmann, M.E., Huang, H., Tindall, D.J. 2001. Androgen receptor signaling in androgen-refractory prostate cancer. *J Natl Cancer Inst* **93**: 1687–1697
109. Edwards, J., Krishna, N.S., Witton, C.J., Bartlett, J.M. 2003. Gene amplifications associated with the development of hormone-resistant prostate cancer. *Clin Cancer Res* **9**: 5271–5281
110. Edwards, J., Krishna, N.S., Grigor, K.M., Bartlett, J.M. 2003. Androgen receptor gene amplification and protein expression in hormone refractory prostate cancer. *Br J Cancer* **89**: 552–556
111. Visakorpi, T., Hyytinen, E., Koivisto, P., Tanner, M., Keinanen, R., Palmberg, C., Palotie, A., Tammela, T., Isola, J., Kallioniemi, O.P. 1995. In vivo amplification of the androgen receptor gene and progression of human prostate cancer. *Nat Genet* **9**: 401–406
112. Veldscholte, J., Ris-Stalpers, C., Kuiper, G.G., Jenster, G., Berrevoets, C., Claassen, E., van Rooij, H.C., Trapman, J., Brinkmann, A.O., Mulder, E. 1990. A mutation in the ligand binding domain of the androgen receptor of human LNCaP cells affects steroid binding characteristics and response to anti-androgens. *Biochem Biophys Res Commun* **173**: 534–540
113. Zhao, X.Y., Malloy, P.J., Krishnan, A.V., Swami, S., Navone, N.M., Peehl, D.M., Feldman, D. 2000. Glucocorticoids can promote androgen-independent growth of prostate cancer cells through a mutated androgen receptor. *Nat Med* **6**: 703–706
114. La Spada, AR. 2006. Spinal and bulbar muscular atrophy. In: Pagon, R.A., Bird, T.C., Dolan, C.R., Stephens, K. (eds.) *GeneReviews* [Internet]. University of Washington, Seattle, WA. (1993–1999 Feb 26 [updated 2006 Dec 28])

Chapter 2

Androgen Action During Prostate Carcinogenesis

Diping Wang and Donald J. Tindall

Abstract

Androgens are critical for normal prostate development and function, as well as prostate cancer initiation and progression. Androgens function mainly by regulating target gene expression through the androgen receptor (AR). Many studies have shown that androgen-AR signaling exerts actions on key events during prostate carcinogenesis. In this review, androgen action in distinct aspects of prostate carcinogenesis, including (i) cell proliferation, (ii) cell apoptosis, and (iii) prostate cancer metastasis will be discussed.

Key words: Androgen receptor, prostate cancer, androgen metabolism, androgen signaling, castration-resistant prostate cancer.

1. Androgen Signaling

Androgens are the male sex hormones, which control the differentiation and maturation of male reproductive organs, including the prostate gland. Testosterone is the principal androgen in circulation and is synthesized by Leydig cells in the testes, under the regulation of luteinizing hormone (LH), which is further regulated by gonadotropin-releasing hormone (GnRH). Adrenal glands also synthesize a small amount of androgens, such as dehydroepiandrosterone (DHEA) and androstenedione (4-dione) (1). Testosterone enters prostate cells by passive diffusion, where it is converted enzymatically by 5-α reductases to the more potent androgen dihydrotestosterone (DHT) (2). Binding of androgens to the androgen receptor (AR), a ligand-modulated transcription factor, induces a conformational change in the AR, causing release of heat shock proteins and translocation of the AR to the

F. Saatcioglu (ed.), *Androgen Action*, Methods in Molecular Biology 776,
DOI 10.1007/978-1-61779-243-4_2, © Springer Science+Business Media, LLC 2011

nucleus, where it transcriptionally regulates the expression of target genes (3).

In addition to the classic genomic effects of sex steroids, accumulating data have also shown the importance of nongenomic effects (4–7). For instance, androgen treatment results in the association of AR with Src kinase, and thereby activates Src/Raf-1/Erk pathway, leading to cell proliferation and survival (8). Androgens may also post-transcriptionally regulate gene expression by modulating the stability of mRNAs (6). Membrane androgen receptors may also account for some nongenomic effects of androgens (9).

2. Androgen and Prostate Carcinogenesis

Prostate cancer is the most frequently diagnosed cancer and the second leading cause of cancer-related death in US men. The American Cancer Society has estimated that in the USA the number of new cases diagnosed in 2009 was 192,280, and about 27,360 men died of this disease (10). The problem is even more substantial when viewed from a global perspective, with prostate cancer accounting for more than 220,000 deaths worldwide every year (11). Multiple signaling pathways have been demonstrated to be critical for prostate cancer initiation and progression (12, 13), with the androgen signaling pathway being one of the most prominent.

Since the landmark research of Huggins and Hodges in the 1940s, it has been postulated that androgens promote prostate carcinogenesis (14, 15). Although it is well accepted that androgens are critical for prostate cancer growth, it is still controversial whether androgens promote human prostate carcinogenesis in vivo. Indeed, there is no increased incidence of prostate cancer in men administered testosterone, there is no reduced risk of prostate cancer in men with low serum androgen levels, and there is no correlation between prostate cancer and serum androgen levels (16, 17). Taken together, these data suggest a 'saturation' model of androgen action on androgen-dependent growth (18). This model states that physiologic levels of androgen are important for both normal and malignant prostate cell proliferation, but excessive androgens alone do not lead to uncontrolled cell proliferation. On the other hand, ligand-independent activation of AR signaling plays a critical role in initiation and progression of prostate cancer, particularly following androgen ablation therapy (19, 20). The role of ligand-independent AR activation in prostate carcinogenesis and progression has been discussed in several excellent reviews (1, 21, 22). The present review focuses on the effects of androgen signaling during critical phases of prostate carcinogenesis.

3. Androgen Action on Prostate Cell Proliferation

Although epidemiologic data suggest that androgens alone are not sufficient to promote prostate carcinogenesis (23), abundant biological data have demonstrated that androgens promote prostate cancer cell proliferation. Androgens induce prostate epithelial cell proliferation via multiple ways, either directly or indirectly.

One of the most common genetic alterations in prostate cancer is the fusion between two genes, i.e., the TMPRSS2 gene and the ETS transcription factor genes, ERG or ETV1 (24). ETS transcription factors are involved in multiple processes, including cell proliferation and cancer cell invasion (25). Current data suggest that up to 72% of all prostate cancers harbor a TMPRSS2-ETS translocation (26–28). TMPRSS2 is a membrane-bound serine protease, which is regulated by androgens and overexpressed in prostate cancers (29, 30). In addition, TMPRSS2 expression is largely limited to prostate, more specifically, prostate luminal epithelial cells (31, 32). The ERG gene is the most commonly overexpressed proto-oncogene in prostate cancer but the underlying mechanism of ERG overexpression was not clear (33). The finding of TMPRSS2-ETS translocations suggests that androgens may promote the expression of ETV1 and ERG, contributing to prostate carcinogenesis. A recent study shows that androgens and irradiation synergistically induce the translocations of TMPRSS2-ERG and TMPRSS2-ETV1 (34). Liganded AR can induce juxtaposition of translocation loci by triggering intra- and interchromosomal interactions. Such interactions appear to promote stress-induced site-specific DNA double-stranded breaks at these translocation loci by recruiting activation-induced cytidine deaminase and LINE-1 repeat-encoded ORF2 endonuclease, which are critical for chromosomal translocations. These results suggest a potential mechanism by which androgens promote prostate carcinogenesis through inducing gene translocation.

Increased levels of growth factors, such as insulin-like growth factors (IGFs), fibroblast growth factors (FGFs), and epidermal growth factors (EGFs), are associated with prostate cancer (35–37). Increased expression of growth factors and their receptors promotes prostate cell proliferation, migration, and tumor angiogenesis, thereby facilitating prostate carcinogenesis and cancer progression. IGF-1 has been shown to promote prostate cancer cell proliferation in vitro and facilitate the progression of a prostate cancer xenograft to a castration-recurrent state (35). Androgens regulate the expression of both IGF-1 (38) and its receptor IGF-1R (39) (**Fig. 2.1**). Two androgen response elements (AREs) exist within the IGF-1 promoter, suggesting that androgens regulate IGF-1 via a direct transcriptional mechanism (38). On the other hand, the effects of androgen on the expression of IGF-1R appear to be through a nongenomic event. Quantitative RT-PCR demonstrated that IGF-1R induction is independent of AR DNA-binding activity. Rather, it appears to depend on the Src/MAPK pathway. Androgen treatment activates Erk1/2 within 5 min, and inhibition of Src/MAPK signaling pathway by various methods can block androgen-induced Erk1/2 activation and IGF-1R expression (39). Another way by which androgens modulate the IGF-1 signaling pathway is through regulation of IGF-binding protein (IGFBP) expression. For example, IGFBP-5 is transcriptionally regulated by androgens (40). Data on androgen regulation on IGFBP-3 are controversial. Some reports have shown that androgens suppress IGFBP-3 levels (41, 42), whereas other studies have reported that androgens increase IGFBP-3 expression (43).

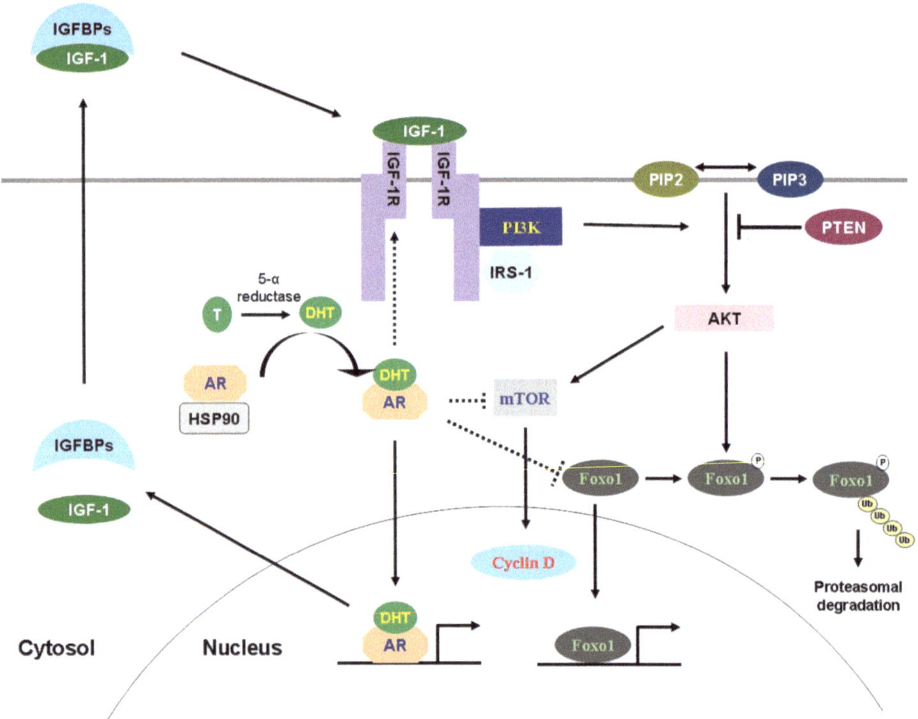

Fig. 2.1. Androgen action on IGF-1 signaling. IGF-1 interacts with IGF-1R to induce PI3K/AKT activation, which then regulates the activities of mTOR and FOXO1. The androgen–AR complex affects IGF-1 signaling via multiple ways, either in a transcription activity-dependent or transcription activity-independent manner. *Dashed line* indicates that the underlying mechanism is not fully elucidated.

FGF8 is another androgen-regulated growth factor (44). The tumorigenic effect of FGF8 has been demonstrated in both cell culture and transgenic mice (45, 46). FGF8 protein is overexpressed in human prostate cancers (47), and the level of expression correlates with tumor stage, pathological grade, and disease-specific survival (48). FGF8b protein expression is correlated with the AR expression in prostate cancer tissues and castration of mice significantly increases FGF8b expression in CWR22 prostate cancer xenografts, suggesting a role of androgen in the regulation of FGF8 expression. Indeed, androgens regulate FGF8 expression at the transcriptional level (49).

Prostate tumors also exhibit aberrant expression of EGF and EGF receptors (50–52). EGF signaling appears to be essential for the androgen-induced proliferation of LNCaP prostate cancer cells, since small molecule inhibitors against tyrosine kinase activity of the EGF receptor can completely suppress androgen-induced proliferation of these cells (52). Members of the EGF receptor family (ERBB1/EGFR, ERBB2, ERBB3, and ERBB4) may be differentially regulated by androgens. While androgens enhance the expression of EGFR, they reduce expression of ERBB2 in LNCaP cells (52, 53), suggesting that androgens regulate EGFR and ERBB2 via different mechanisms. The stimulation of EGFR gene expression by androgens appears to be at the transcriptional level, since androgen-induced EGFR upregulation does not require de novo protein synthesis. In contrast, androgen-induced reduction of ERBB2 does require de novo protein synthesis, suggesting that this repression is an indirect effect of androgens (53). Surprisingly, EGFR and ERBB2 expression in castration-recurrent 22Rv1 cells is not affected by androgen treatment. This discrepancy is most likely due to the presence of ARΔCTD, a constitutively active form of AR, in 22Rv1 cells (53). ARΔCTD is encoded by a splice variant mRNA of the AR gene, which has a novel exon 2b (54). Constitutively active AR variants promote expression of AR-regulated genes and cancer cell proliferation in the absence of androgens (54, 55). Taken together, these data suggest that AR regulates EGF/EGFR signaling during prostate cancer progression through either ligand-dependent or ligand-independent mechanisms.

Besides the regulation of growth factors and their receptors, androgens have also been shown to crosstalk with the downstream effectors of growth factor signaling, such as PI3K/AKT. The PI3K/AKT pathway is one of the most frequently altered signaling pathways in a variety of human cancers and plays a critical role in prostate carcinogenesis and its progression (56, 57). Constitutively activated AKT has been found frequently in prostate cancer cell lines and tissues, mostly due to loss of PTEN function (57). PTEN is a phosphatase which dephosphorylates PIP3, thereby inhibiting PI3K-induced AKT activation. Loss of function of PTEN may be due to gene mutations (58), loss of heterozygosity (59), or gene deletions (60) in prostate cancers. Loss of function of PTEN and activation of AKT are significantly correlated with the progression of prostate cancer (61). Androgen-independent prostate cancer cell proliferation is correlated with increased activity of PI3K/AKT, suggesting a role of AKT in progression to castration-recurrent prostate cancers (CRPC) (62). Moreover, heterozygous PTEN$^{+/-}$ mice develop low-grade PIN lesions in their prostates, while *PTEN*$^{hy/-}$ mutants (harboring a hypomorphic allele with decreased PTEN expression) develop high-grade PIN lesions and locally invasive carcinomas (63). Moreover, loss of function of both PTEN and Nkx3.1, a well-known androgen-regulated transcription factor (64, 65), synergistically promotes prostate carcinogenesis (66).

Androgen-induced proliferation and survival of androgen-sensitive LNCaP cells depend on the activation of PI3K/AKT. Inhibition of AKT with a dominant-negative AKT or a PI3K inhibitor significantly attenuates androgen-induced cell proliferation (4, 5). Androgen-induced AKT activation requires the AR, but deletion of the ligand-binding domain of AR does not abolish it, indicating that this regulation is independent of AR transcriptional activity. Indeed, androgen-bound AR physically binds to the p85α regulatory subunit of PI3K and thereby activates PI3K/AKT (4, 5). This is truly a two-way crosstalk, since AR expression and activity are also regulated by PI3K/AKT (67–69). Additional details about the crosstalk between androgen/AR and PI3K/AKT pathways have previously been discussed in several other reviews (61, 70).

The cell cycle, which is governed by the coordination of cyclins, cyclin-dependent kinases (CDKs) and CDK inhibitors, becomes deregulated in prostate cancer. Androgen/AR signaling is critical for normal prostate cell cycle progression, and dysregulation of this signaling may contribute to prostate cancer progression. Multiple mechanisms have been attributed to the effects of androgens on the cell cycle (71). For instance, androgens are important for cyclin D expression. Androgen-deprived prostate cancer cells arrest in early G1 phase, concomitant with loss of cyclin D, reduced CDK4 activity, and activated Rb. Addition of androgen rapidly increases cyclin D expression and promotes cell cycle progression (72, 73). Androgens increase cyclin D1 and D2 expression at the protein level but not at the mRNA level, suggesting that this regulation is at a post-transcriptional level. Androgen-induced cyclin D expression depends on activation of mammalian target of rapamycin (mTOR), which enhances the translation of cyclin D1 mRNA (73, 74). Although mTOR is a well-known target of AKT, androgen-induced mTOR activation is not mediated by AKT (73). The mechanism by which androgens activate mTOR remains unknown.

Androgens also regulate cell cycle progression through the CDK inhibitor $p21^{Waf1/Cip1}$. Androgens positively regulate $p21^{Waf1/Cip1}$ expression through an ARE within its promoter (75). Induction of $p21^{Waf1/Cip1}$ by androgens may promote the assembly of active cyclin D1/CDK4 or CDK6 complexes (71, 76). In contrast to the upregulation of $p21^{Waf1/Cip1}$, androgen treatment downregulates the expression of another CDK inhibitor $p27^{kip1}$, resulting in inhibition of cyclin E/CDK2 activity (77). This androgen-induced reduction of $p27^{Kip1}$ is mediated by decreased expression of the F-box protein SKP2 (78) which controls the ubiquitin-dependent degradation of $p27^{Kip1}$ (79, 80).

4. Androgen Action on Prostate Cell Apoptosis/Survival

Androgens are essential for the survival of both normal and malignant prostate epithelium (21). Androgen withdrawal in adult rodents and humans induces apoptosis in the secretory epithelium (81, 82). Because most human prostate cancers are initially androgen responsive, androgen deprivation therapy remains the standard treatment for advanced prostate cancer. Recent data have revealed important pathways through which androgens regulate prostate cell apoptosis (83, 84).

Two types of signaling pathways lead to cell apoptosis: intrinsic or extrinsic. The intrinsic pathway is initiated by a variety of cell stresses, which lead to changes in mitochondrial permeability and release of cytochrome C and Smac/DIABLO. Antiapoptotic members of the BH3 family such as Bcl-2 and Bcl-XL inhibit intrinsic apoptosis by preventing the leakage of cytochrome C and Smac/DIABLO from mitochondria. Elevated Bcl-2 expression is implicated in a variety of human malignancies, including prostate cancer (85, 86). Whereas Bcl-2 expression is positive in most CRPCs, only approximately 30% androgen-dependent prostate cancers express low levels of Bcl-2, suggesting that Bcl-2 may contribute to the development of CRPC (87–89). It has been shown that androgens suppress Bcl-2 expression but the underlying mechanism is still unclear. Huang et al. reported that the regulation of Bcl-2 by androgens might be an indirect effect of AR activation, probably mediated by the E2F1 protein through a putative E2F-binding site in the promoter of the Bcl-2 gene (90).

Bak1 is another member of the BH3 family, which is also known to be regulated by androgens (91). Unlike Bcl-2, Bak1 is a proapoptotic protein. Under normal circumstances, Bak1 forms a complex with Bcl-2 or Mcl-1, which restrains Bak1 activation. Elimination of Bcl-2 or Mcl-1 caused by an apoptotic stimulus, such as DNA damage, releases Bak1 from this inhibitory complex and promotes apoptosis (92). Bak-1 expression has been associated with prostate cancer progression. Bak1 is expressed in approximately 75% of primary and untreated localized prostate cancers, but in only approximately 33% of CRPCs (89). The suppression of Bak1 expression by androgens is mediated by a microRNA. MicroRNAs are a class of naturally occurring small RNAs, which do not encode proteins but regulate the expression of other genes. Androgens transcriptionally upregulate miR-125b, which then represses Bak1 expression (91). Thus, androgens suppress the expression of both Bcl-2 and Bak1 via different mechanisms. Further studies are essential to determine whether these androgen actions contribute to prostate carcinogenesis and cancer progression.

The extrinsic apoptotic pathway is mediated by death receptor signaling, which is triggered by ligands such as TNF-α, TRAIL, and FasL. Normal prostate cells are highly resistant to death receptor-induced cell apoptosis due to activation of NF-κB, which promotes cell proliferation and inhibits apoptosis (93). It has been reported that androgens inhibit TNF-α-induced NF-κB activation *via* multiple mechanisms (94–96) (**Fig. 2.2**). Keller et al. showed that androgens prevent the degradation of IκBα, which

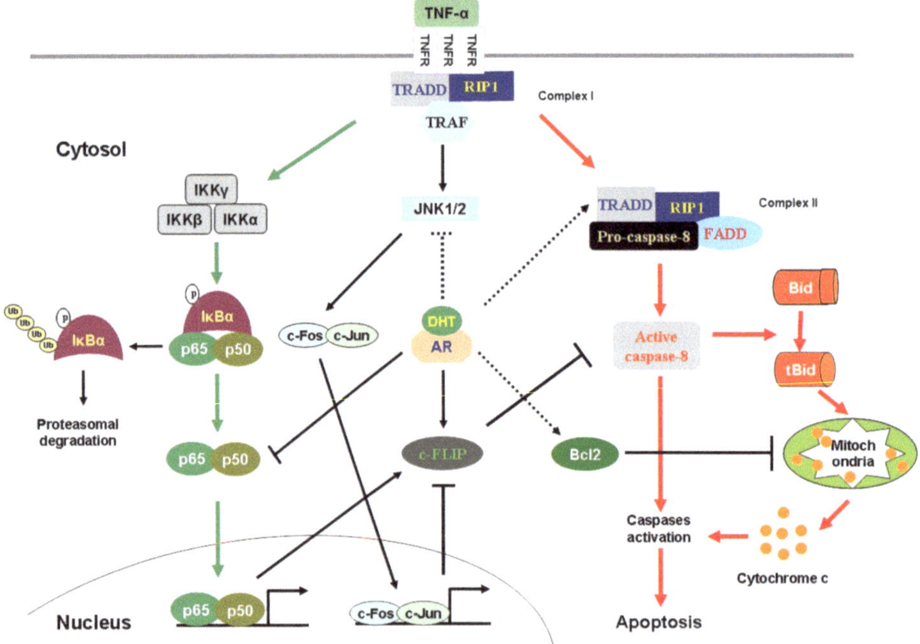

Fig. 2.2. Androgen action on TNF-α-induced signaling pathways. TNF-α binding to trimeric TNFR1 leads to assembly of complex I, which then results in activation of NF-κB, JNK, and/or the caspase cascade, depending on cellular context. All of these three TNF-α downstream signaling pathways can be regulated by androgen–AR via different mechanisms, as summarized in the text. *Dashed line* indicates that the underlying mechanism is not fully elucidated.

binds to NF-κB and inhibits its activation (95). AR activation also results in a decrease in RelA/p65, a subunit of NF-κB, thereby reducing its nuclear localization and transcriptional activity (97). Another mechanism whereby androgens affect NF-κB activity is through the formation of an AR-p65 complex via CREB-binding protein (CBP), which inhibits both p65 and AR transcriptional activities (94).

C-FLIP is a caspase-8 homologue, which functions as a dominant inhibitor of caspase-8 (98), thereby negatively regulating death receptor-induced apoptosis. In prostate cancer cells, increased c-FLIP expression is associated with increased resistance to death receptor-induced apoptosis (99, 100). LNCaP xenografts overexpressing c-FLIP are also resistant to castration-induced growth inhibition, suggesting a role of c-FLIP in the development of CRPC (99). It has also been observed that both c-FLIP mRNA and protein levels are reduced during progression to CRPC in animal models (101, 102). C-FLIP is directly regulated by androgens via a cluster of four AREs within a 156-bp region downstream from the transcription start site (99). Accordingly, both c-FLIP mRNA and protein levels are reduced following castration of rats in multiple tissues, including dorsolateral prostate and seminal vesicles (100). Unexpectedly, it has also been reported that androgen treatment downregulates c-FLIP in LNCaP cells, in which AKT is constitutively activated due to loss of PTEN. This discrepancy might be explained by the involvement of an AKT-regulated transcription factor, FOXO3a. Androgen induction of c-FLIP requires the presence of FOXO3a, which binds to the AR and potentially to the Forkhead-binding site within the c-FLIP promoter (102). Expression of FOXO3a TM, a constitutively active form of FOXO3a, rescues the androgen induction of c-FLIP in LNCaP cells, supporting a critical role of FOXO3a in androgen-induced c-FLIP upregulation (102).

FOXO3a belongs to the Forkhead transcription factor class-O family, members of which can act as tumor suppressors in a variety of malignancies (103, 104). Other members of the FOXO family include FOXO1, FOXO4, and FOXO6. AKT is a major regulator of the FOXO proteins. AKT phosphorylates the FOXO proteins, leading to their retention in the cytoplasm and proteasome-mediated degradation (105, 106) (**Fig. 2.1**). FOXO proteins inhibit cell proliferation and induce apoptosis in prostate cells. Thus, their activities are hypothesized to be reduced during prostate cancer progression (107, 108). Androgens negatively regulate the proapoptotic effects of FOXO1 through a physical interaction with it. Liganded AR blocks the DNA-binding activity of FOXO1 and impairs the ability of FOXO1 to induce Fas ligand expression and prostate cancer cell apoptosis (109). Moreover, androgen treatment results in a reduction of FOXO1 expression at the protein level via a proteolytic mechanism.

Treatment of LNCaP cells with androgens leads to FOXO1 protein cleavage and produces a truncated FOXO1, which lacks ~120 amino acid residues in the C-terminus. Ectopic expression of this truncated FOXO1 inhibits the transcriptional activity of the intact FOXO1, suggesting that androgen-induced FOXO1 protein cleavage results in reduction of FOXO1 transcriptional activity (107).

TRADD (TNF receptor-associated death domain), which is a transducer for death receptor-induced signaling, is also a target of androgens (110). TRADD mediates TNFR1-induced apoptosis as well as NF-κB activation (111). Overexpression of TRADD in a variety of cell lines leads to apoptosis; however, knockdown or knockout of TRADD expression in some cell lines does not inhibit apoptosis, suggesting that its function depends on the cellular context (112–114). TRADD protein is reduced in CRPC cells compared to androgen-responsive cells. Androgen deprivation reduces TRADD expression in prostate cancer cell lines, xenografts, and human tissues. Moreover, androgen treatment increases TRADD expression at both the mRNA and protein levels (110). Unpublished data (D. Wang, personal communication) suggest that this regulation is an indirect action of androgens because it requires AR and de novo protein synthesis. Reduced TRADD expression may account for the reduced sensitivity of CRPC cells to TNF-α.

5. Androgen Action on Prostate Cancer Metastasis

A pivotal problem of prostate cancer, as in other cancers, is its propensity to metastasize. The process of tumor metastasis includes the following: activation of epithelial–mesenchymal transition (EMT), remodeling of the extracellular matrix, neovascularization, and migration to specific secondary sites. Many of the androgen-regulated signaling pathways that were discussed above are also important for prostate cancer metastasis. For instance, NF-κB activity has been associated with many types of metastases. Inhibition of NF-κB activity in metastatic prostate cancer is associated with reduced expression of vascular endothelial growth factor, interleukin-8, and matrix metalloproteinase-9 and a concomitant decrease in angiogenesis, invasion, and metastasis in nude mice (115). Furthermore, nuclear NF-κB expression in primary prostate cancers is highly predictive for pelvic lymph node metastases (116). Thus, the regulation of prostate cancer metastasis by androgens may be achieved by crosstalk between androgen and NF-κB signaling pathways.

Cell adhesion molecules such as cadherins play a critical role in the activation of EMT (117). Cadherins are a class of type-1 transmembrane glycoproteins, of which E-cadherin and N-cadherin are the best characterized in prostate cancers. E-cadherin is an important tumor suppressor gene. Loss of E-cadherin expression disrupts cell–cell junctions and consequently promotes cell migration, leading to tumor metastasis. Multiple studies reported that loss of E-cadherin expression enhances the progression from nonmetastatic to metastatic carcinoma (118, 119). In primary prostate cancers, reduced E-cadherin expression has been correlated with increased tumor grade, bone metastasis, and poor prognosis (120, 121). In contrast, increased levels of N-cadherin and cadherin-11 are associated with poorly differentiated and metastatic prostate cancers (122, 123). In more aggressive prostate cancer specimens, N-cadherin expression is increased while E-cadherin is reduced. This phenomenon is called 'cadherin switching' (124). Interestingly, both E-cadherin and N-cadherin appear to be regulated by androgens (125, 126). Androgen deprivation therapy leads to elevated expression of E-cadherin in prostate cancer, suggesting that androgens repress E-cadherin expression (127). Moreover, androgen treatment reduces E-cadherin expression in the breast cancer cell lines MCF7 and T47D. Identification of AREs within the promoter of the E-cadherin gene further suggests that this repression is a direct transcriptional effect of AR (125). However, because these studies were not conducted in prostate cancer cell lines or models, it is still not clear whether similar mechanisms apply to prostate cancer cells. More studies are needed to elucidate the role of androgens on E-cadherin expression and function in prostate cancer.

In contrast to E-cadherin, N-cadherin is induced by androgen deprivation in experimental castration-recurrent prostate cancer models as well as in human prostate tumors (126). It has also been shown that testosterone increases N-cadherin expression in motor neurons (128). Taken together, these results suggest that androgens may positively regulate the expression of N-cadherin.

Cadherin-11 is a homophilic cell adhesion molecule that mediates osteoblast adhesion and thereby plays a critical role in the metastasis of prostate cancer to bone. Increased cadherin-11 expression correlates with development of CRPC (129). Cadherin-11 appears to be indirectly regulated by androgens, suggesting a role of androgen in the metastasis of prostate cancer to bone (130).

Maspin is a serine protease inhibitor with tumor suppressing properties. Loss of maspin is associated with a variety of tumors, such as breast and prostate cancers (131, 132). Consistently, re-expression of maspin in prostate cancer cells inhibits tumor growth and prostate cancer-induced bone remodeling (133).

Maspin exerts its anti-metastasis activities via multiple mechanisms. For instance, maspin blocks FGF and vascular endothelial growth factor-mediated endothelial cell migration in vitro. Maspin also inhibits prostate cancer growth and tumor angiogenesis in vivo, probably mediated by its ability to inhibit the degradation of extracellular matrix via tPA and pro-uPA (134–136). Maspin expression increases in prostate cancer cells and tissues following androgen deprivation while androgen treatment decreases maspin expression (132, 137). Identification of an ARE element in the promoter of the maspin gene has confirmed this regulation as a direct transcriptional effect of the androgen receptor (137). Therefore, androgens may affect prostate cancer progression and metastasis via regulation of maspin expression.

Table 2.1
Androgen-regulated factors associated with prostate carcinogenesis

Target	Function	Mechanism	References
TMPRSS2-ETS	Transcription factors	Direct transcriptional regulation	(24, 29)
IGF-1	Growth factor	Direct transcriptional regulation	(38)
IGF-I R	Growth factor receptor	Nongenomic effect	(39)
FGF8	Growth factor	Direct transcriptional regulation	(49)
EGFR	Growth factor reception	Direct transcriptional regulation	(53)
ERBB2	Growth factor receptor	Indirect transcriptional regulation	(53)
AKT	Protein kinase	Nongenomic effect	(4, 5)
Cyclin D	Cell cycle regulator	Indirect regulation mediated by mTOR	(73)
P21	CDK inhibitor	Direct transcriptional regulation	(75)
P27	CDK inhibitor	Indirect regulation mediated by SKP2	(78)
Bcl-2	Anti-apoptotic protein	Indirect transcriptional regulation	(90)
Bak1	Pro-apoptotic protein	Indirect transcriptional regulation mediated by miR-125b	(91)
C-FLIP	Anti-apoptotic protein	Direct transcriptional regulation	(99)
FOXO1	Transcription factor	Indirect effects on protein degradation	(107)
TRADD	Signaling transducer	Indirect transcriptional regulation[a]	(110)
E-cadherin	Cell adhesion molecule	Direct transcriptional regulation	(125)
Cadherin-11	Cell adhesion molecule	Indirect transcriptional regulation	(130)
Maspin	Serine protease inhibitor	Direct transcriptional regulation	(137)

[a] Personal communication with D. Wang

6. Conclusions

It has long been recognized that androgens play a critical role in prostate carcinogenesis. In this review we have highlighted several androgen-regulated signaling pathways and factors that may be involved in prostate carcinogenesis (**Table 2.1**). Although under certain circumstances androgens may inhibit cell proliferation or promote cell death, in general, it is well accepted that androgens are critical for prostate cancer cell proliferation and survival. However, the effects of androgens are diverse and complex, and focusing on one or two signaling pathways for delineating mechanisms is likely to be an oversimplification. More importantly, the downstream signaling pathways of androgen may also crosstalk with each other, making the contributions of androgen to prostate carcinogenesis more complicated. Nonetheless, recent development of more potent antiandrogens and inhibitors of androgen metabolism, which are in clinical trials (138–140), will aid in understanding the role of androgen signaling in prostate cancer cells. Taken together, better understanding of AR action is likely to lead to better clinical treatments for prostate cancer.

References

1. Attar, R. M., Takimoto, C. H., and Gottardis, M. M. (2009) Castration-resistant prostate cancer: locking up the molecular escape routes, *Clin Cancer Res* **15**, 3251–3255.
2. Evans, R. M. (1988) The steroid and thyroid hormone receptor superfamily, *Science* **240**, 889–895.
3. Dehm, S. M., and Tindall, D. J. (2006) Molecular regulation of androgen action in prostate cancer, *J Cell Biochem* **99**, 333–344.
4. Sun, M., Yang, L., Feldman, R. I., Sun, X. M., Bhalla, K. N., Jove, R., Nicosia, S. V., and Cheng, J. Q. (2003) Activation of phosphatidylinositol 3-kinase/Akt pathway by androgen through interaction of p85alpha, androgen receptor, and Src, *J Biol Chem* **278**, 42992–43000.
5. Baron, S., Manin, M., Beaudoin, C., Leotoing, L., Communal, Y., Veyssiere, G., and Morel, L. (2004) Androgen receptor mediates non-genomic activation of phosphatidylinositol 3-OH kinase in androgen-sensitive epithelial cells, *J Biol Chem* **279**, 14579–14586.
6. Sheflin, L. G., Zou, A. P., and Spaulding, S. W. (2004) Androgens regulate the binding of endogenous HuR to the AU-rich 3'UTRs of HIF-1alpha and EGF mRNA, *Biochem Biophys Res Commun* **322**, 644–651.
7. Foradori, C. D., Weiser, M. J., and Handa, R. J. (2008) Non-genomic actions of androgens, *Front Neuroendocrinol* **29**, 169–181.
8. Kousteni, S., Bellido, T., Plotkin, L. I., O'Brien, C. A., Bodenner, D. L., Han, L., Han, K., DiGregorio, G. B., Katzenellenbogen, J. A., Katzenellenbogen, B. S., Roberson, P. K., Weinstein, R. S., Jilka, R. L., and Manolagas, S. C. (2001) Nongenotropic, sex-nonspecific signaling through the estrogen or androgen receptors: dissociation from transcriptional activity, *Cell* **104**, 719–730.
9. Kampa, M., Papakonstanti, E. A., Hatzoglou, A., Stathopoulos, E. N., Stournaras, C., and Castanas, E. (2002) The human prostate cancer cell line LNCaP bears functional membrane testosterone receptors that increase PSA secretion and modify actin cytoskeleton, *FASEB J* **16**, 1429–1431.
10. Jemal, A., Siegel, R., Ward, E., Hao, Y., Xu, J., and Thun, M. J. (2009) Cancer statistics, 2009, *CA Cancer J Clin* **59**, 225–249.
11. Jemal, A., Thun, M. J., Ries, L. A., Howe, H. L., Weir, H. K., Center, M. M., Ward, E., Wu, X. C., Eheman, C., Anderson, R., Ajani, U. A., Kohler, B., and Edwards, B. K.

(2008) Annual report to the nation on the status of cancer, 1975–2005, featuring trends in lung cancer, tobacco use, and tobacco control, *J Natl Cancer Inst* **100**, 1672–1694.
12. De Marzo, A. M., DeWeese, T. L., Platz, E. A., Meeker, A. K., Nakayama, M., Epstein, J. I., Isaacs, W. B., and Nelson, W. G. (2004) Pathological and molecular mechanisms of prostate carcinogenesis: implications for diagnosis, detection, prevention, and treatment, *J Cell Biochem* **91**, 459–477.
13. Ramsay, A. K., and Leung, H. Y. (2009) Signalling pathways in prostate carcinogenesis: potentials for molecular-targeted therapy, *Clin Sci (Lond)* **117**, 209–228.
14. Huggins, C. (1967) Endocrine-induced regression of cancers, *Cancer Res* **27**, 1925–1930.
15. Huggins, C., and Hodges, C. V. (1972) Studies on prostatic cancer. I. The effect of castration, of estrogen and androgen injection on serum phosphatases in metastatic carcinoma of the prostate, *CA Cancer J Clin* **22**, 232–240.
16. Isbarn, H., Pinthus, J. H., Marks, L. S., Montorsi, F., Morales, A., Morgentaler, A., and Schulman, C. (2009) Testosterone and prostate cancer: revisiting old paradigms, *Eur Urol* **56**, 48–56.
17. Morgentaler, A. (2006) Testosterone and prostate cancer: an historical perspective on a modern myth, *Eur Urol* **50**, 935–939.
18. Morgentaler, A., and Traish, A. M. (2009) Shifting the paradigm of testosterone and prostate cancer: the saturation model and the limits of androgen-dependent growth, *Eur Urol* **55**, 310–320.
19. Debes, J. D., and Tindall, D. J. (2004) Mechanisms of androgen-refractory prostate cancer, *N Engl J Med* **351**, 1488–1490.
20. Dehm, S. M., and Tindall, D. J. (2006) Ligand-independent androgen receptor activity is activation function-2-independent and resistant to antiandrogens in androgen refractory prostate cancer cells, *J Biol Chem* **281**, 27882–27893.
21. Dehm, S. M., and Tindall, D. J. (2005) Regulation of androgen receptor signaling in prostate cancer, *Expert Rev Anticancer Ther* **5**, 63–74.
22. Dehm, S. M., and Tindall, D. J. (2007) Androgen receptor structural and functional elements: role and regulation in prostate cancer, *Mol Endocrinol* **21**, 2855–2863.
23. Hsing, A. W. (2001) Hormones and prostate cancer: what's next? *Epidemiol Rev* **23**, 42–58.
24. Tomlins, S. A., Rhodes, D. R., Perner, S., Dhanasekaran, S. M., Mehra, R., Sun, X. W., Varambally, S., Cao, X., Tchinda, J., Kuefer, R., Lee, C., Montie, J. E., Shah, R. B., Pienta, K. J., Rubin, M. A., and Chinnaiyan, A. M. (2005) Recurrent fusion of TMPRSS2 and ETS transcription factor genes in prostate cancer, *Science* **310**, 644–648.
25. Hsu, T., Trojanowska, M., and Watson, D. K. (2004) Ets proteins in biological control and cancer, *J Cell Biochem* **91**, 896–903.
26. Hermans, K. G., van Marion, R., van Dekken, H., Jenster, G., van Weerden, W. M., and Trapman, J. (2006) TMPRSS2:ERG fusion by translocation or interstitial deletion is highly relevant in androgen-dependent prostate cancer, but is bypassed in late-stage androgen receptor-negative prostate cancer, *Cancer Res* **66**, 10658–10663.
27. Soller, M. J., Isaksson, M., Elfving, P., Soller, W., Lundgren, R., and Panagopoulos, I. (2006) Confirmation of the high frequency of the TMPRSS2/ERG fusion gene in prostate cancer, *Genes Chromosomes Cancer* **45**, 717–719.
28. Rajput, A. B., Miller, M. A., De Luca, A., Boyd, N., Leung, S., Hurtado-Coll, A., Fazli, L., Jones, E. C., Palmer, J. B., Gleave, M. E., Cox, M. E., and Huntsman, D. G. (2007) Frequency of the TMPRSS2:ERG gene fusion is increased in moderate to poorly differentiated prostate cancers, *J Clin Pathol* **60**, 1238–1243.
29. Lin, B., Ferguson, C., White, J. T., Wang, S., Vessella, R., True, L. D., Hood, L., and Nelson, P. S. (1999) Prostate-localized and androgen-regulated expression of the membrane-bound serine protease TMPRSS2, *Cancer Res* **59**, 4180–4184.
30. Vaarala, M. H., Porvari, K., Kyllonen, A., Lukkarinen, O., and Vihko, P. (2001) The TMPRSS2 gene encoding transmembrane serine protease is overexpressed in a majority of prostate cancer patients: detection of mutated TMPRSS2 form in a case of aggressive disease, *Int J Cancer* **94**, 705–710.
31. Afar, D. E., Vivanco, I., Hubert, R. S., Kuo, J., Chen, E., Saffran, D. C., Raitano, A. B., and Jakobovits, A. (2001) Catalytic cleavage of the androgen-regulated TMPRSS2 protease results in its secretion by prostate and prostate cancer epithelia, *Cancer Res* **61**, 1686–1692.
32. Vaarala, M. H., Porvari, K. S., Kellokumpu, S., Kyllonen, A. P., and Vihko, P. T. (2001) Expression of transmembrane serine protease TMPRSS2 in mouse and human tissues, *J Pathol* **193**, 134–140.
33. Petrovics, G., Liu, A., Shaheduzzaman, S., Furusato, B., Sun, C., Chen, Y., Nau,

M., Ravindranath, L., Chen, Y., Dobi, A., Srikantan, V., Sesterhenn, I. A., McLeod, D. G., Vahey, M., Moul, J. W., and Srivastava, S. (2005) Frequent overexpression of ETS-related gene-1 (ERG1) in prostate cancer transcriptome, *Oncogene* **24**, 3847–3852.
34. Lin, C., Yang, L., Tanasa, B., Hutt, K., Ju, B. G., Ohgi, K., Zhang, J., Rose, D. W., Fu, X. D., Glass, C. K., and Rosenfeld, M. G. (2009) Nuclear receptor-induced chromosomal proximity and DNA breaks underlie specific translocations in cancer, *Cell* **139**, 1069–1083.
35. Kaaks, R., Lukanova, A., and Sommersberg, B. (2000) Plasma androgens, IGF-1, body size, and prostate cancer risk: a synthetic review, *Prostate Cancer Prostatic Dis* **3**, 157–172.
36. Kwabi-Addo, B., Ozen, M., and Ittmann, M. (2004) The role of fibroblast growth factors and their receptors in prostate cancer, *Endocr Relat Cancer* **11**, 709–724.
37. Byrne, R. L., Leung, H., and Neal, D. E. (1996) Peptide growth factors in the prostate as mediators of stromal epithelial interaction, *Br J Urol* 77, 627–633.
38. Wu, Y., Zhao, W., Zhao, J., Pan, J., Wu, Q., Zhang, Y., Bauman, W. A., and Cardozo, C. P. (2007) Identification of androgen response elements in the insulin-like growth factor I upstream promoter, *Endocrinology* **148**, 2984–2993.
39. Pandini, G., Mineo, R., Frasca, F., Roberts, C. T., Jr., Marcelli, M., Vigneri, R., and Belfiore, A. (2005) Androgens up-regulate the insulin-like growth factor-I receptor in prostate cancer cells, *Cancer Res* **65**, 1849–1857.
40. Yoshizawa, A., and Ogikubo, S. (2006) IGF binding protein-5 synthesis is regulated by testosterone through transcriptional mechanisms in androgen responsive cells, *Endocr J* **53**, 811–818.
41. Kojima, S., Mulholland, D. J., Ettinger, S., Fazli, L., Nelson, C. C., and Gleave, M. E. (2006) Differential regulation of IGFBP-3 by the androgen receptor in the lineage-related androgen-dependent LNCaP and androgen-independent C4-2 prostate cancer models, *Prostate* **66**, 971–986.
42. Le, H., Arnold, J. T., McFann, K. K., and Blackman, M. R. (2006) DHT and testosterone, but not DHEA or E2, differentially modulate IGF-I, IGFBP-2, and IGFBP-3 in human prostatic stromal cells, *Am J Physiol Endocrinol Metab* **290**, E952–960.
43. Peng, L., Wang, J., Malloy, P. J., and Feldman, D. (2008) The role of insulin-like growth factor binding protein-3 in the growth inhibitory actions of androgens in LNCaP human prostate cancer cells, *Int J Cancer* **122**, 558–566.
44. Tanaka, A., Miyamoto, K., Minamino, N., Takeda, M., Sato, B., Matsuo, H., and Matsumoto, K. (1992) Cloning and characterization of an androgen-induced growth factor essential for the androgen-dependent growth of mouse mammary carcinoma cells, *Proc Natl Acad Sci USA* **89**, 8928–8932.
45. Rudra-Ganguly, N., Zheng, J., Hoang, A. T., and Roy-Burman, P. (1998) Downregulation of human FGF8 activity by antisense constructs in murine fibroblastic and human prostatic carcinoma cell systems, *Oncogene* **16**, 1487–1492.
46. Daphna-Iken, D., Shankar, D. B., Lawshe, A., Ornitz, D. M., Shackleford, G. M., and MacArthur, C. A. (1998) MMTV-Fgf8 transgenic mice develop mammary and salivary gland neoplasia and ovarian stromal hyperplasia, *Oncogene* **17**, 2711–2717.
47. Tanaka, A., Furuya, A., Yamasaki, M., Hanai, N., Kuriki, K., Kamiakito, T., Kobayashi, Y., Yoshida, H., Koike, M., and Fukayama, M. (1998) High frequency of fibroblast growth factor (FGF) 8 expression in clinical prostate cancers and breast tissues, immunohistochemically demonstrated by a newly established neutralizing monoclonal antibody against FGF 8, *Cancer Res* **58**, 2053–2056.
48. Dorkin, T. J., Robinson, M. C., Marsh, C., Bjartell, A., Neal, D. E., and Leung, H. Y. (1999) FGF8 over-expression in prostate cancer is associated with decreased patient survival and persists in androgen independent disease, *Oncogene* **18**, 2755–2761.
49. Gnanapragasam, V. J., Robson, C. N., Neal, D. E., and Leung, H. Y. (2002) Regulation of FGF8 expression by the androgen receptor in human prostate cancer, *Oncogene* **21**, 5069–5080.
50. Sherwood, E. R., and Lee, C. (1995) Epidermal growth factor-related peptides and the epidermal growth factor receptor in normal and malignant prostate, *World J Urol* **13**, 290–296.
51. Li, Z., Szabolcs, M., Terwilliger, J. D., and Efstratiadis, A. (2006) Prostatic intraepithelial neoplasia and adenocarcinoma in mice expressing a probasin-Neu oncogenic transgene, *Carcinogenesis* **27**, 1054–1067.
52. Torring, N., Dagnaes-Hansen, F., Sorensen, B. S., Nexo, E., and Hynes, N. E. (2003) ErbB1 and prostate cancer: ErbB1 activity is essential for androgen-induced proliferation and protection from the apoptotic effects of LY294002, *Prostate* **56**, 142–149.

53. Pignon, J. C., Koopmansch, B., Nolens, G., Delacroix, L., Waltregny, D., and Winkler, R. (2009) Androgen receptor controls EGFR and ERBB2 gene expression at different levels in prostate cancer cell lines, *Cancer Res* **69**, 2941–2949.
54. Dehm, S. M., Schmidt, L. J., Heemers, H. V., Vessella, R. L., and Tindall, D. J. (2008) Splicing of a novel androgen receptor exon generates a constitutively active androgen receptor that mediates prostate cancer therapy resistance, *Cancer Res* **68**, 5469–5477.
55. Libertini, S. J., Tepper, C. G., Rodriguez, V., Asmuth, D. M., Kung, H. J., and Mudryj, M. (2007) Evidence for calpain-mediated androgen receptor cleavage as a mechanism for androgen independence, *Cancer Res* **67**, 9001–9005.
56. Yuan, T. L., and Cantley, L. C. (2008) PI3K pathway alterations in cancer: variations on a theme, *Oncogene* **27**, 5497–5510.
57. Vivanco, I., and Sawyers, C. L. (2002) The phosphatidylinositol 3-Kinase AKT pathway in human cancer, *Nat Rev Cancer* **2**, 489–501.
58. Yoshimoto, M., Cunha, I. W., Coudry, R. A., Fonseca, F. P., Torres, C. H., Soares, F. A., and Squire, J. A. (2007) FISH analysis of 107 prostate cancers shows that PTEN genomic deletion is associated with poor clinical outcome, *Br J Cancer* **97**, 678–685.
59. Suzuki, H., Freije, D., Nusskern, D. R., Okami, K., Cairns, P., Sidransky, D., Isaacs, W. B., and Bova, G. S. (1998) Interfocal heterogeneity of PTEN/MMAC1 gene alterations in multiple metastatic prostate cancer tissues, *Cancer Res* **58**, 204–209.
60. Verhagen, P. C., van Duijn, P. W., Hermans, K. G., Looijenga, L. H., van Gurp, R. J., Stoop, H., van der Kwast, T. H., and Trapman, J. (2006) The PTEN gene in locally progressive prostate cancer is preferentially inactivated by bi-allelic gene deletion, *J Pathol* **208**, 699–707.
61. Sarker, D., Reid, A. H., Yap, T. A., and de Bono, J. S. (2009) Targeting the PI3K/AKT pathway for the treatment of prostate cancer, *Clin Cancer Res* **15**, 4799–4805.
62. Murillo, H., Huang, H., Schmidt, L. J., Smith, D. I., and Tindall, D. J. (2001) Role of PI3K signaling in survival and progression of LNCaP prostate cancer cells to the androgen refractory state, *Endocrinology* **142**, 4795–4805.
63. Trotman, L. C., Niki, M., Dotan, Z. A., Koutcher, J. A., Di Cristofano, A., Xiao, A., Khoo, A. S., Roy-Burman, P., Greenberg, N. M., Van Dyke, T., Cordon-Cardo, C., and Pandolfi, P. P. (2003) Pten dose dictates cancer progression in the prostate, *PLoS Biol* **1**, E59.
64. He, W. W., Sciavolino, P. J., Wing, J., Augustus, M., Hudson, P., Meissner, P. S., Curtis, R. T., Shell, B. K., Bostwick, D. G., Tindall, D. J., Gelmann, E. P., Abate-Shen, C., and Carter, K. C. (1997) A novel human prostate-specific, androgen-regulated homeobox gene (NKX3.1) that maps to 8p21, a region frequently deleted in prostate cancer, *Genomics* **43**, 69–77.
65. Prescott, J. L., Blok, L., and Tindall, D. J. (1998) Isolation and androgen regulation of the human homeobox cDNA, NKX3.1, *Prostate* **35**, 71–80.
66. Kim, M. J., Cardiff, R. D., Desai, N., Banach-Petrosky, W. A., Parsons, R., Shen, M. M., and Abate-Shen, C. (2002) Cooperativity of Nkx3.1 and Pten loss of function in a mouse model of prostate carcinogenesis, *Proc Natl Acad Sci USA* **99**, 2884–2889.
67. Lin, H. K., Yeh, S., Kang, H. Y., and Chang, C. (2001) Akt suppresses androgen-induced apoptosis by phosphorylating and inhibiting androgen receptor, *Proc Natl Acad Sci USA* **98**, 7200–7205.
68. Lin, H. K., Hu, Y. C., Yang, L., Altuwaijri, S., Chen, Y. T., Kang, H. Y., and Chang, C. (2003) Suppression versus induction of androgen receptor functions by the phosphatidylinositol 3-kinase/Akt pathway in prostate cancer LNCaP cells with different passage numbers, *J Biol Chem* **278**, 50902–50907.
69. Taneja, S. S., Ha, S., Swenson, N. K., Huang, H. Y., Lee, P., Melamed, J., Shapiro, E., Garabedian, M. J., and Logan, S. K. (2005) Cell-specific regulation of androgen receptor phosphorylation in vivo, *J Biol Chem* **280**, 40916–40924.
70. Wang, Y., Kreisberg, J. I., and Ghosh, P. M. (2007) Cross-talk between the androgen receptor and the phosphatidylinositol 3-kinase/Akt pathway in prostate cancer, *Curr Cancer Drug Targets* **7**, 591–604.
71. Balk, S. P., and Knudsen, K. E. (2008) AR, the cell cycle, and prostate cancer, *Nucl Recept Signal* **6**, e001.
72. Knudsen, K. E., Arden, K. C., and Cavenee, W. K. (1998) Multiple G1 regulatory elements control the androgen-dependent proliferation of prostatic carcinoma cells, *J Biol Chem* **273**, 20213–20222.
73. Xu, Y., Chen, S. Y., Ross, K. N., and Balk, S. P. (2006) Androgens induce prostate cancer cell proliferation through mammalian target of rapamycin activation and

post-transcriptional increases in cyclin D proteins, *Cancer Res* **66**, 7783–7792.

74. Gera, J. F., Mellinghoff, I. K., Shi, Y., Rettig, M. B., Tran, C., Hsu, J. H., Sawyers, C. L., and Lichtenstein, A. K. (2004) AKT activity determines sensitivity to mammalian target of rapamycin (mTOR) inhibitors by regulating cyclin D1 and c-myc expression, *J Biol Chem* **279**, 2737–2746.
75. Lu, S., Liu, M., Epner, D. E., Tsai, S. Y., and Tsai, M. J. (1999) Androgen regulation of the cyclin-dependent kinase inhibitor p21 gene through an androgen response element in the proximal promoter, *Mol Endocrinol* **13**, 376–384.
76. Chen, Y., Robles, A. I., Martinez, L. A., Liu, F., Gimenez-Conti, I. B., and Conti, C. J. (1996) Expression of G1 cyclins, cyclin-dependent kinases, and cyclin-dependent kinase inhibitors in androgen-induced prostate proliferation in castrated rats, *Cell Growth Differ* **7**, 1571–1578.
77. Ye, D., Mendelsohn, J., and Fan, Z. (1999) Androgen and epidermal growth factor down-regulate cyclin-dependent kinase inhibitor p27Kip1 and costimulate proliferation of MDA PCa 2a and MDA PCa 2b prostate cancer cells, *Clin Cancer Res* **5**, 2171–2177.
78. Lu, L., Schulz, H., and Wolf, D. A. (2002) The F-box protein SKP2 mediates androgen control of p27 stability in LNCaP human prostate cancer cells, *BMC Cell Biol* **3**, 22.
79. Nakayama, K., Nagahama, H., Minamishima, Y. A., Matsumoto, M., Nakamichi, I., Kitagawa, K., Shirane, M., Tsunematsu, R., Tsukiyama, T., Ishida, N., Kitagawa, M., Nakayama, K., and Hatakeyama, S. (2000) Targeted disruption of Skp2 results in accumulation of cyclin E and p27(Kip1), polyploidy and centrosome overduplication, *EMBO J* **19**, 2069–2081.
80. Xu, K., Belunis, C., Chu, W., Weber, D., Podlaski, F., Huang, K. S., Reed, S. I., and Vassilev, L. T. (2003) Protein-protein interactions involved in the recognition of p27 by E3 ubiquitin ligase, *Biochem J* **371**, 957–964.
81. Kyprianou, N., and Isaacs, J. T. (1988) Activation of programmed cell death in the rat ventral prostate after castration, *Endocrinology* **122**, 552–562.
82. Buttyan, R., Shabsigh, A., Perlman, H., and Colombel, M. (1999) Regulation of Apoptosis in the Prostate Gland by Androgenic Steroids, *Trends Endocrinol Metab* **10**, 47–54.
83. Kimura, K., Markowski, M., Bowen, C., and Gelmann, E. P. (2001) Androgen blocks apoptosis of hormone-dependent prostate cancer cells, *Cancer Res* **61**, 5611–5618.
84. Guseva, N. V., Taghiyev, A. F., Rokhlin, O. W., and Cohen, M. B. (2004) Death receptor-induced cell death in prostate cancer, *J Cell Biochem* **91**, 70–99.
85. Raffo, A. J., Perlman, H., Chen, M. W., Day, M. L., Streitman, J. S., and Buttyan, R. (1995) Overexpression of bcl-2 protects prostate cancer cells from apoptosis in vitro and confers resistance to androgen depletion in vivo, *Cancer Res* **55**, 4438–4445.
86. Bruckheimer, E. M., Brisbay, S., Johnson, D. J., Gingrich, J. R., Greenberg, N., and McDonnell, T. J. (2000) Bcl-2 accelerates multistep prostate carcinogenesis in vivo, *Oncogene* **19**, 5251–5258.
87. McDonnell, T. J., Troncoso, P., Brisbay, S. M., Logothetis, C., Chung, L. W., Hsieh, J. T., Tu, S. M., and Campbell, M. L. (1992) Expression of the protooncogene bcl-2 in the prostate and its association with emergence of androgen-independent prostate cancer, *Cancer Res* **52**, 6940–6944.
88. Colombel, M., Symmans, F., Gil, S., O'Toole, K. M., Chopin, D., Benson, M., Olsson, C. A., Korsmeyer, S., and Buttyan, R. (1993) Detection of the apoptosis-suppressing oncoprotein bcl-2 in hormone-refractory human prostate cancers, *Am J Pathol* **143**, 390–400.
89. Yoshino, T., Shiina, H., Urakami, S., Kikuno, N., Yoneda, T., Shigeno, K., and Igawa, M. (2006) Bcl-2 expression as a predictive marker of hormone-refractory prostate cancer treated with taxane-based chemotherapy, *Clin Cancer Res* **12**, 6116–6124.
90. Huang, H., Zegarra-Moro, O. L., Benson, D., and Tindall, D. J. (2004) Androgens repress Bcl-2 expression via activation of the retinoblastoma (RB) protein in prostate cancer cells, *Oncogene* **23**, 2161–2176.
91. Shi, X. B., Xue, L., Yang, J., Ma, A. H., Zhao, J., Xu, M., Tepper, C. G., Evans, C. P., Kung, H. J., and deVere White, R. W. (2007) An androgen-regulated miRNA suppresses Bak1 expression and induces androgen-independent growth of prostate cancer cells, *Proc Natl Acad Sci USA* **104**, 19983–19988.
92. Cuconati, A., Mukherjee, C., Perez, D., and White, E. (2003) DNA damage response and MCL-1 destruction initiate apoptosis in adenovirus-infected cells, *Genes Dev* **17**, 2922–2932.
93. Van Antwerp, D. J., Martin, S. J., Kafri, T., Green, D. R., and Verma, I. M. (1996) Suppression of TNF-alpha-induced apoptosis by NF-kappaB, *Science* **274**, 787–789.

94. Palvimo, J. J., Reinikainen, P., Ikonen, T., Kallio, P. J., Moilanen, A., and Janne, O. A. (1996) Mutual transcriptional interference between RelA and androgen receptor, *J Biol Chem* **271**, 24151–24156.
95. Keller, E. T., Chang, C., and Ershler, W. B. (1996) Inhibition of NFkappaB activity through maintenance of IkappaBalpha levels contributes to dihydrotestosterone-mediated repression of the interleukin-6 promoter, *J Biol Chem* **271**, 26267–26275.
96. Norata, G. D., Tibolla, G., Seccomandi, P. M., Poletti, A., and Catapano, A. L. (2006) Dihydrotestosterone decreases tumor necrosis factor-alpha and lipopolysaccharide-induced inflammatory response in human endothelial cells, *J Clin Endocrinol Metab* **91**, 546–554.
97. Nelius, T., Filleur, S., Yemelyanov, A., Budunova, I., Shroff, E., Mirochnik, Y., Aurora, A., Veliceasa, D., Xiao, W., Wang, Z., and Volpert, O. V. (2007) Androgen receptor targets NFkappaB and TSP1 to suppress prostate tumor growth in vivo, *Int J Cancer* **121**, 999–1008.
98. Kreuz, S., Siegmund, D., Scheurich, P., and Wajant, H. (2001) NF-kappaB inducers upregulate cFLIP, a cycloheximide-sensitive inhibitor of death receptor signaling, *Mol Cell Biol* **21**, 3964–3973.
99. Gao, S., Lee, P., Wang, H., Gerald, W., Adler, M., Zhang, L., Wang, Y. F., and Wang, Z. (2005) The androgen receptor directly targets the cellular Fas/FasL-associated death domain protein-like inhibitory protein gene to promote the androgen-independent growth of prostate cancer cells, *Mol Endocrinol* **19**, 1792–1802.
100. Raclaw, K. A., Heemers, H. V., Kidd, E. M., Dehm, S. M., and Tindall, D. J. (2008) Induction of FLIP expression by androgens protects prostate cancer cells from TRAIL-mediated apoptosis, *Prostate* **68**, 1696–1706.
101. Gao, S., Wang, H., Lee, P., Melamed, J., Li, C. X., Zhang, F., Wu, H., Zhou, L., and Wang, Z. (2006) Androgen receptor and prostate apoptosis response factor-4 target the c-FLIP gene to determine survival and apoptosis in the prostate gland, *J Mol Endocrinol* **36**, 463–483.
102. Cornforth, A. N., Davis, J. S., Khanifar, E., Nastiuk, K. L., and Krolewski, J. J. (2008) FOXO3a mediates the androgen-dependent regulation of FLIP and contributes to TRAIL-induced apoptosis of LNCaP cells, *Oncogene* **27**, 4422–4433.
103. Huang, H., and Tindall, D. J. (2007) Dynamic FoxO transcription factors, *J Cell Sci* **120**, 2479–2487.
104. Fu, Z., and Tindall, D. J. (2008) FOXOs, cancer and regulation of apoptosis, *Oncogene* **27**, 2312–2319.
105. Brunet, A., Bonni, A., Zigmond, M. J., Lin, M. Z., Juo, P., Hu, L. S., Anderson, M. J., Arden, K. C., Blenis, J., and Greenberg, M. E. (1999) Akt promotes cell survival by phosphorylating and inhibiting a Forkhead transcription factor, *Cell* **96**, 857–868.
106. Huang, H., and Tindall, D. J. (2006) FOXO factors: a matter of life and death, *Future Oncol* **2**, 83–89.
107. Huang, H., Muddiman, D. C., and Tindall, D. J. (2004) Androgens negatively regulate forkhead transcription factor FKHR (FOXO1) through a proteolytic mechanism in prostate cancer cells, *J Biol Chem* **279**, 13866–13877.
108. Lynch, R. L., Konicek, B. W., McNulty, A. M., Hanna, K. R., Lewis, J. E., Neubauer, B. L., and Graff, J. R. (2005) The progression of LNCaP human prostate cancer cells to androgen independence involves decreased FOXO3a expression and reduced p27KIP1 promoter transactivation, *Mol Cancer Res* **3**, 163–169.
109. Li, P., Lee, H., Guo, S., Unterman, T. G., Jenster, G., and Bai, W. (2003) AKT-independent protection of prostate cancer cells from apoptosis mediated through complex formation between the androgen receptor and FKHR, *Mol Cell Biol* **23**, 104–118.
110. Wang, D., Montgomery, R. B., Schmidt, L. J., Mostaghel, E. A., Huang, H., Nelson, P. S., and Tindall, D. J. (2009) Reduced tumor necrosis factor receptor-associated death domain expression is associated with prostate cancer progression, *Cancer Res* **69**, 9448–9456.
111. Hsu, H., Xiong, J., and Goeddel, D. V. (1995) The TNF receptor 1-associated protein TRADD signals cell death and NF-kappa B activation, *Cell* **81**, 495–504.
112. Jin, Z., and El-Deiry, W. S. (2006) Distinct signaling pathways in TRAIL- versus tumor necrosis factor-induced apoptosis, *Mol Cell Biol* **26**, 8136–8148.
113. Ermolaeva, M. A., Michallet, M. C., Papadopoulou, N., Utermohlen, O., Kranidioti, K., Kollias, G., Tschopp, J., and Pasparakis, M. (2008) Function of TRADD in tumor necrosis factor receptor 1 signaling and in TRIF-dependent inflammatory responses, *Nat Immunol* **9**, 1037–1046.
114. Pobezinskaya, Y. L., Kim, Y. S., Choksi, S., Morgan, M. J., Li, T., Liu, C., and Liu, Z. (2008) The function of TRADD in signaling through tumor necrosis factor receptor 1

and TRIF-dependent Toll-like receptors, *Nat Immunol* **9**, 1047–1054.

115. Huang, S., Pettaway, C. A., Uehara, H., Bucana, C. D., and Fidler, I. J. (2001) Blockade of NF-kappaB activity in human prostate cancer cells is associated with suppression of angiogenesis, invasion, and metastasis, *Oncogene* **20**, 4188–4197.
116. Lessard, L., Karakiewicz, P. I., Bellon-Gagnon, P., Alam-Fahmy, M., Ismail, H. A., Mes-Masson, A. M., and Saad, F. (2006) Nuclear localization of nuclear factor-kappaB p65 in primary prostate tumors is highly predictive of pelvic lymph node metastases, *Clin Cancer Res* **12**, 5741–5745.
117. Xie, D., Gore, C., Liu, J., Pong, R. C., Mason, R., Hao, G., Long, M., Kabbani, W., Yu, L., Zhang, H., Chen, H., Sun, X., Boothman, D. A., Min, W., and Hsieh, J. T. (2010) Role of DAB2IP in modulating epithelial-to-mesenchymal transition and prostate cancer metastasis, *Proc Natl Acad Sci USA* **107**, 2485–2490.
118. Frixen, U. H., Behrens, J., Sachs, M., Eberle, G., Voss, B., Warda, A., Lochner, D., and Birchmeier, W. (1991) E-cadherin-mediated cell-cell adhesion prevents invasiveness of human carcinoma cells, *J Cell Biol* **113**, 173–185.
119. Handschuh, G., Candidus, S., Luber, B., Reich, U., Schott, C., Oswald, S., Becke, H., Hutzler, P., Birchmeier, W., Hofler, H., and Becker, K. F. (1999) Tumour-associated E-cadherin mutations alter cellular morphology, decrease cellular adhesion and increase cellular motility, *Oncogene* **18**, 4301–4312.
120. Cheng, L., Nagabhushan, M., Pretlow, T. P., Amini, S. B., and Pretlow, T. G. (1996) Expression of E-cadherin in primary and metastatic prostate cancer, *Am J Pathol* **148**, 1375–1380.
121. Umbas, R., Isaacs, W. B., Bringuier, P. P., Schaafsma, H. E., Karthaus, H. F., Oosterhof, G. O., Debruyne, F. M., and Schalken, J. A. (1994) Decreased E-cadherin expression is associated with poor prognosis in patients with prostate cancer, *Cancer Res* **54**, 3929–3933.
122. Bussemakers, M. J., Van Bokhoven, A., Tomita, K., Jansen, C. F., and Schalken, J. A. (2000) Complex cadherin expression in human prostate cancer cells, *Int J Cancer* **85**, 446–450.
123. Gravdal, K., Halvorsen, O. J., Haukaas, S. A., and Akslen, L. A. (2007) A switch from E-cadherin to N-cadherin expression indicates epithelial to mesenchymal transition and is of strong and independent importance for the progress of prostate cancer, *Clin Cancer Res* **13**, 7003–7011.
124. Jaggi, M., Nazemi, T., Abrahams, N. A., Baker, J. J., Galich, A., Smith, L. M., and Balaji, K. C. (2006) N-cadherin switching occurs in high Gleason grade prostate cancer, *Prostate* **66**, 193–199.
125. Liu, Y. N., Liu, Y., Lee, H. J., Hsu, Y. H., and Chen, J. H. (2008) Activated androgen receptor downregulates E-cadherin gene expression and promotes tumor metastasis, *Mol Cell Biol* **28**, 7096–7108.
126. Jennbacken, K., Tesan, T., Wang, W., Gustavsson, H., Damber, J. E., and Welen, K. (2010) N-cadherin increases after androgen deprivation and is associated with metastasis in prostate cancer, *Endocr Relat Cancer* **17**,469–479.
127. Patriarca, C., Petrella, D., Campo, B., Colombo, P., Giunta, P., Parente, M., Zucchini, N., Mazzucchelli, R., and Montironi, R. (2003) Elevated E-cadherin and alpha/beta-catenin expression after androgen deprivation therapy in prostate adenocarcinoma, *Pathol Res Pract* **199**, 659–665.
128. Monks, D. A., and Watson, N. V. (2001) N-cadherin expression in motoneurons is directly regulated by androgens: a genetic mosaic analysis in rats, *Brain Res* **895**, 73–79.
129. Chu, K., Cheng, C. J., Ye, X., Lee, Y. C., Zurita, A. J., Chen, D. T., Yu-Lee, L. Y., Zhang, S., Yeh, E. T., Hu, M. C., Logothetis, C. J., and Lin, S. H. (2008) Cadherin-11 promotes the metastasis of prostate cancer cells to bone, *Mol Cancer Res* **6**, 1259–1267.
130. Lee, Y. C., Cheng, C. J., Huang, M., Bilen, M. A., Ye, X., Navone, N. M., Chu, K., Kao, H. H., Yu-Lee, L. Y., Wang, Z., and Lin, S. H. (2010) Androgen depletion up-regulates cadherin-11 expression in prostate cancer, *J Pathol* **22**, 68–76
131. Zou, Z., Anisowicz, A., Hendrix, M. J., Thor, A., Neveu, M., Sheng, S., Rafidi, K., Seftor, E., and Sager, R. (1994) Maspin, a serpin with tumor-suppressing activity in human mammary epithelial cells, *Science* **263**, 526–529.
132. Zou, Z., Zhang, W., Young, D., Gleave, M. G., Rennie, P., Connell, T., Connelly, R., Moul, J., Srivastava, S., and Sesterhenn, I. (2002) Maspin expression profile in human prostate cancer (CaP) and in vitro induction of Maspin expression by androgen ablation, *Clin Cancer Res* **8**, 1172–1177.
133. Cher, M. L., Biliran, H. R., Jr., Bhagat, S., Meng, Y., Che, M., Lockett, J., Abrams, J.,

Fridman, R., Zachareas, M., and Sheng, S. (2003) Maspin expression inhibits osteolysis, tumor growth, and angiogenesis in a model of prostate cancer bone metastasis, *Proc Natl Acad Sci USA* **100**, 7847–7852.

134. Zhang, M., Volpert, O., Shi, Y. H., and Bouck, N. (2000) Maspin is an angiogenesis inhibitor, *Nat Med* **6**, 196–199.
135. Sheng, S., Truong, B., Fredrickson, D., Wu, R., Pardee, A. B., and Sager, R. (1998) Tissue-type plasminogen activator is a target of the tumor suppressor gene maspin, *Proc Natl Acad Sci USA* **95**, 499–504.
136. Yin, S., Lockett, J., Meng, Y., Biliran, H., Jr., Blouse, G. E., Li, X., Reddy, N., Zhao, Z., Lin, X., Anagli, J., Cher, M. L., and Sheng, S. (2006) Maspin retards cell detachment via a novel interaction with the urokinase-type plasminogen activator/urokinase-type plasminogen activator receptor system, *Cancer Res* **66**, 4173–4181.
137. He, M. L., Jiang, A. L., Zhang, P. J., Hu, X. Y., Liu, Z. F., Yuan, H. Q., and Zhang, J. Y. (2005) Identification of androgen-responsive element ARE and Sp1 element in the maspin promoter, *Chin J Physiol* **48**, 160–166.
138. Tran, C., Ouk, S., Clegg, N. J., Chen, Y., Watson, P. A., Arora, V., Wongvipat, J., Smith-Jones, P. M., Yoo, D., Kwon, A., Wasielewska, T., Welsbie, D., Chen, C. D., Higano, C. S., Beer, T. M., Hung, D. T., Scher, H. I., Jung, M. E., and Sawyers, C. L. (2009) Development of a second-generation antiandrogen for treatment of advanced prostate cancer, *Science* **324**, 787–790.
139. Scher, H. I., Beer, T. M., Higano, C. S., Anand, A., Taplin, M. E., Efstathiou, E., Rathkopf, D., Shelkey, J., Yu, E. Y., Alumkal, J., Hung, D., Hirmand, M., Seely, L., Morris, M. J., Danila, D. C., Humm, J., Larson, S., Fleisher, M., and Sawyers, C. L. (2010) Antitumour activity of MDV3100 in castration-resistant prostate cancer: a phase 1-2 study, *Lancet* **375**, 1437–1446.
140. Andriole, G. L., Bostwick, D. G., Brawley, O. W., Gomella, L. G., Marberger, M., Montorsi, F., Pettaway, C. A., Tammela, T. L., Teloken, C., Tindall, D. J., Somerville, M. C., Wilson, T. H., Fowler, I. L., and Rittmaster, R. S. (2010) Effect of dutasteride on the risk of prostate cancer, *N Engl J Med* **362**, 1192–1202.

Section II

Analysis of Androgens

Chapter 3

Androgen Quantitation in Prostate Cancer Tissue Using Liquid Chromatography Tandem Mass Spectrometry

Mark Titus and Kenneth B. Tomer

Abstract

Prostate cancer that recurs after androgen deprivation therapy is the second leading cause of cancer-related death in North American men. Clinical and experimental evidences indicate that the development of recurrent prostate cancer is dependent on re-activation of the androgen receptor signaling pathway. Androgen is required for androgen receptor translocation to the nucleus, interaction with androgen response elements, expression of target genes, and prostate cancer cell proliferation. The intra-tissue and serum testosterone and dihydrotestosterone levels are important biomarkers to monitor androgen deprivation therapy efficacy in prostate cancer and recurrent prostate cancer. We have measured testosterone and dihydrotestosterone in procured recurrent prostate cancer specimens using liquid chromatography tandem mass spectrometry. The measured androgen levels are sufficient to activate androgen receptor and suggest that the recurrent prostate cancer microenvironment is capable of intracrine androgen biosynthesis.

Key words: Androgen, testosterone, dihydrotestosterone, biomarker, castration-recurrent prostate cancer, metabonomics, prostate cancer, tandem mass spectrometry.

1. Introduction

Normal prostate development and function are dependent on the testicular androgen testosterone and its local conversion to dihydrotestosterone by steroid 5α-reductase (1). Dihydrotestosterone bound to androgen receptor, a more potent mediator of androgen action, promotes epithelial cell differentiation and expression of androgen-regulated prostate-specific androgen (PSA) in the prostate (2). Androgen receptor signaling is altered in prostate cancer to support cell proliferation via, among other

F. Saatcioglu (ed.), *Androgen Action*, Methods in Molecular Biology 776,
DOI 10.1007/978-1-61779-243-4_3, © Springer Science+Business Media, LLC 2011

pathways, transcription of fusion genes such as TMPRSS2-ERG and TMPRSS2-ETV-1 (3). Early clinical studies demonstrated that chemical or surgical castration reduced serum androgen levels, caused prostate tumors to regress, and improved patient survival (4, 5). Measured serum testosterone levels during androgen deprivation therapy are decreased ~15 fold to ≤50 ng/dL (6). However, intra-tissue androgen levels in recurrent prostate cancer remain at sufficient concentrations to reactivate androgen receptor signaling (7–9). The intracrine biosynthesis of androgens in castration-recurrent prostate cancer has led to renewed interest in inhibiting steroidogenic enzymes (10) and the accurate measurement of intra-tissue androgens.

Endogenous tissue androgens have been measured using radioimmunoassay (RIA) (11, 12), gas chromatography mass spectrometry (GC/MS) (13), and liquid chromatography tandem mass spectrometry (LC/MS/MS) (7–9). The methods used for steroid analysis have developed toward increased sensitivity, specificity, and reliability. An evaluation of RIA and MS assays from different laboratories that measure serum steroid levels consistently demonstrated higher steroid levels using RIA analysis (14). The increased mean steroid levels observed may be due to lack of RIA steroid hormone assay standardizations and/or antibody specificity for each steroid (15). Hence, analysis of semipolar androgens using mass spectrometry platforms offers accurate and reproducible high-throughput analysis. The selection of GC or LC to separate androgens merits consideration since each chromatographic system has unique sample preparation, injection, and ionization processes. However, improved LC/MS/MS testosterone and dihydrotestosterone detection limits using an atmospheric pressure photoionization (APPI) source have made derivatization of androgens unnecessary, thus increasing the accuracy and ability to quantitate androgens (16).

2. Materials

2.1. Chemicals and Reagents

1. Standards: Calibration standard polypropylene glycol (PPG) and analytical standards testosterone and dihydrotestosterone.
2. HPLC grade water (EMD Chemicals, Gibbstown, NJ) and methanol (Caledon Laboratories, Georgetown, Ontario, Canada).
3. Organic solvents: hexane, ethyl acetate, and acetic acid.

4. Deuterated (d_3-16,16,17α) testosterone and dihydrotestosterone internal standards (CDN Isotopes, Pionte-Claire, Quebec, Canada).

2.2. Instrumentation

1. HPLC: Agilent 1100 or 1200 series HPLC System, Binary Pumps (G1376A) and Vacuum Degassers (G1377A), Injection Systems and Thermostatted Auto-sampler, and Column Compartment (G1239A) (Santa Clara, CA).
2. Reverse phase column: An Agilent Zorbax SB-C18 (5 μm particles/80 Å pore size) with dimensions of 150 mm length × 0.5 mm internal diameter is used to separate testosterone and dihydrotestosterone analytes.
3. Solvents: mobile phase A: 0.1% acetic acid in water; mobile phase B: 0.1% acetic acid in methanol. Mobile phase solutions must be filtered and degassed using HPLC solvent filter/degasser assembly and 0.45 μm filter membranes to remove particulates (Agilent).
4. Ionization source: To ionize androgens eluting after HPLC chromatography an atmospheric pressure soft ionization technique, electrospray ionization (ESI, Turbo IonSpray, MDS Sciex), is used. The HPLC spray enters the ESI source through a charged stainless steel capillary tip. A highly charged sample aerosol is produced and the droplets are dispersed by the flow of co-axial nebulizing gas toward the mass spectrometer. Solvent evaporation transfers positive charge to testosterone and dihydrotestosterone molecules. The protonated $[M+H]^+$ positive ions pass through a sample cone and skimmer at intermediate vacuum before entering the mass spectrometer (**Fig. 3.1**).
5. Tandem mass spectrometer: The mass spectrometer used is a MDS Sciex API-3000 triple quadrupole instrument with Analyst™ Software to control instrument and analyze data (AB-Sciex, Foster City, CA) (*see* **Note 1**). The ions formed in the electrospray ionization source are resolved by the mass spectrometer according to their mass to charge ratio (*m/z*). In multiple reaction monitoring mode, the first (Q1) quadrupole performs as a mass filter to select specific parent $(M+H)^+$ androgen ions. The second (Q2) quadrupole acts as a collision cell where selected androgen ions are dissociated into product ions using nitrogen gas. The third (Q3) quadrupole selects for a unique, stable androgen product ion to detect and quantify. The mass transitions for testosterone and dihydrotestosterone are 289.4/97.2 and 291.3/255.2, respectively (**Fig. 3.2a, b**). Monitoring an androgen product ion provides increased assay specificity. However, prostate sample matrix is complex and more than one mass transition

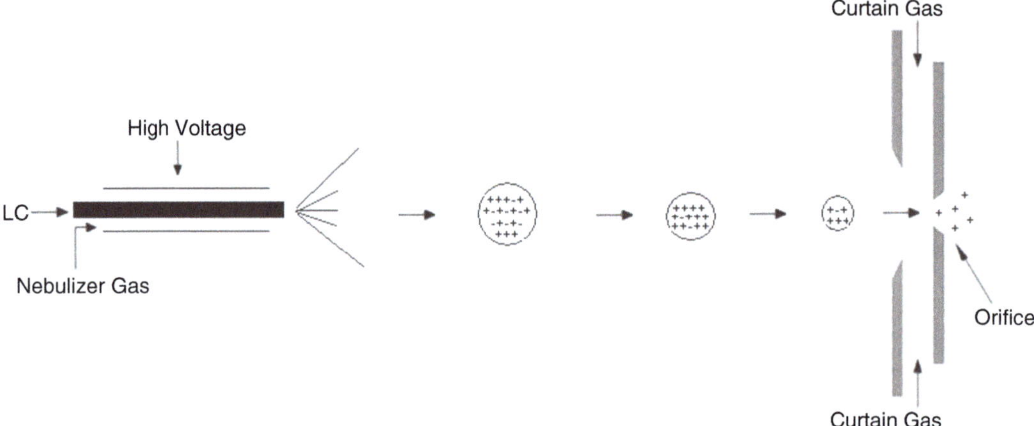

Fig. 3.1. Schematic diagram showing formation of positively charged ions during electrospray ionization. The sample in a volatile solvent is pumped through a capillary that has a high voltage (3–4 kV) applied at the capillary tip. The sample is dispersed into an aerosol of charged droplets by the nebulizing gas. As the droplets evaporate charge is transferred to sample ions that are introduced into vacuum by passing through a curtain gas.

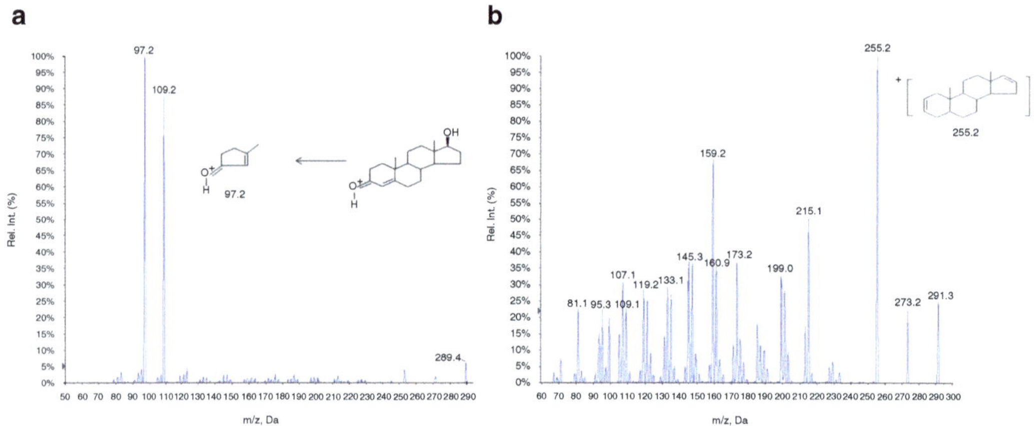

Fig. 3.2. Mass spectrometer fragmentation spectra (collision-induced spectra) of testosterone (**a**) and dihydrotestosterone (**b**).

may be used to confirm the identity of testosterone and dihydrotestosterone. The quadrupole mass analyzers were set to unit resolution.

6. Miscellaneous equipment: Ceramic mortar and pestle (Sigma Aldrich, St Louis, MO); Polytron homogenizer (Capital Scientific Inc., Austin, TX); CentriVap vacuum concentrator and cold trap (Labconco, Kansas City, MO) or Analytical nitrogen evaporator (Organomation, Thomson Instrument Co., Clear Brook, VA); Extraction manifold (Waters, Milford, MA); Kimble/Chase disposable 10 mL concentration tubes and test tubes (VWR); Amber glass vials

with Teflon-lined screw caps and deactivated glass polyspring inserts (Agilent); Solid-phase extraction cartridge C18AR (Varian, Palo Alto, CA).

3. Methods

The proper procurement of prostate specimens is critical for accurate androgen analysis using mass spectrometry. Prostate specimens that are immediately frozen after surgery using liquid nitrogen provide the most accurate androgen levels. Specimens can be stored at –80°C prior to androgen analysis. The procurement of prostate cancer specimens involves collaboration between clinical (Urology and Pathology) and administrative (IRB) departments for optimal specimen retrieval and to guarantee HIPAA compliance. A comprehensive protocol for procurement and storage of radical prostatectomy specimens has recently been published by James Mohler, M.D. (17).

Prostate tissue preparation, mass spectrometry analysis, and androgen quantitation involve tissue homogenation, liquid/ liquid extraction, semi-purification and concentration of extracted androgen, reconstitution of androgen sample, high-pressure liquid chromatography, mass analysis and detection, and manual integration of testosterone and dihydrotestosterone chromatogram peaks. The loss of endogenous testosterone and dihydrotestosterone is unavoidable during this process but can be accounted for by the addition of an internal standard in the tissue homogenization step and by use of a calibration curve (*see* **Note 2**). To ensure consistent analytical results, precautions must be taken to prevent contamination of solvents, glassware, internal standards, etc., and plastic pipette tips must be resistant to organic solvent. All steps involved in homogenation of tissue should be performed at 4°C. Sample preparations after homogenization are performed at room temperature.

3.1. Sample Preparation

1. Cryopreserved prostate specimens (50 mg) are pulverized in liquid nitrogen using mortar and pestle. Pulverized tissue is scraped into a glass test tube containing 1 mL HPLC grade water, vortexed, and placed on ice. Each sample is homogenized for 30 s (×3) and deuterated internal standard, testosterone-d3-16,16,17α (1.0 ng) and dihydrotestosterone-d3-16,16,17α (1.0 ng), added directly to homogenate (*see* **Note 3**).
2. Testosterone and dihydrotestosterone are extracted using 1.5 mL hexane/ethyl acetate (3:2 vol/vol). The sample is vortexed and centrifuged to separate water and organic

phases. A dry ice/acetone bath is used to freeze the water layer, and the organic layer is decanted into a concentration test tube. The extraction procedure is performed three times and the combined organic layers from each sample are evaporated using a vacuum concentrator cold trap or nitrogen evaporator.

3. Extracted testosterone and dihydrotestosterone are semi-purified using a solid-phase extraction cartridge. The solid-phase extraction cartridge is conditioned using 400 μL methanol followed by 400 μL of 0.1% acetic acid in water. The cartridge reservoir is filled with 400 μL of 0.1% acetic acid in water and sample residue from Step 2, solvated in 200 μL methanol, is added. The concentration tube is rinsed with 200 μL of 30% methanol and added to reservoir. The sample is loaded onto the C18 column, washed using 500 μL water and dried for 5 min at 15 mmHg vacuum. Testosterone and dihydrotestosterone are eluted using 1.0 mL methanol into a concentration tube and the column dried for 5 min at 15 mmHg vacuum. The sample is evaporated, capped, and stored at –80°C.

3.2. High-Pressure Liquid chromatography

1. The similar structure and molecular weight of testosterone and dihydrotestosterone require separation of the molecules using HPLC. Mobile phase A (0.1% acetic acid in water) and mobile phase B (0.1% acetic acid in methanol) solutions are filtered and degassed using 0.45 μm filter membranes. The HPLC is programmed using Analyst™ software to generate a gradient as follows: 60% B from 0.0 to 1.0 min, 70–100% B from 1.1 to 9.0 min, 100% B from 9.0 to 14.0 min and 100–60% B from 14.0 to 14.5 min at a flow rate of 20 μL/min. The reverse-phase column was equilibrated at 60% B for 12 min and maintained at 40°C. Testosterone and dihydrotestosterone peak shape and separation will determine the amount of time spent monitoring each *m/z*.
2. The semi-purified residue from the solid-phase extraction step is reconstituted in 100 μL 50% methanol and loaded into amber glass vials with Teflon screw caps and placed in auto-sampler tray. The HPLC is programmed to inject 15 μL of each sample into a 40 μL sample loop.

3.3. Electrospray Ionization Optimization

The electrospray ionization source parameters are optimized by direct infusion of a standard solution of testosterone and dihydrotestosterone at 5–10 μL/min using a syringe pump. The critical electrospray parameters are the ion source position and temperature, capillary tip alignment, nebulizing and curtain gas flow rates, and ion spray voltage (*see* **Note 4** and **Fig. 3.1**).

1. The nebulizing gas flow rate affects the ion signal for testosterone and dihydrotestosterone. The flow rate can be set at 6 L/min and adjusted to optimize sensitivity.
2. The curtain gas should be set to the highest flow rate without significant signal loss (12 L/min).
3. The ion spray potential should initially be set to 4.4 kV and the source temperature at 300°C. To achieve maximum ion production, the lowest voltage possible is used to prevent uneven spray, arcing, and in source androgen fragmentation.
4. Ion transmission into the mass spectrometer can be manipulated using the declustering potential (DP = 40 for testosterone and 50 for dihydrotestosterone) at the orifice plate and the focusing potential on the focusing ring (FP = 180 for testosterone and 240 for dihydrotestosterone).

3.4. Instrument Preparation

The performance of triple quadrupole mass spectrometry is dependent on optimization of the ion source, mass filter, and detector operational parameters. To adjust these components, a known standard, polypropylene glycol, is used to calibrate and tune the mass spectrometer. A series of ions of known exact mass from the tuning solution are monitored to ensure that relative ion abundances are detected accurately and to verify ion peak shape, resolution. and isotope ratios. Tuning can be performed automatically by Analyst software or by the instrument operator. After tuning, testosterone and dihydrotestosterone standards are used to optimize parent ion resolution/accuracy and peak shape.

1. The mass spectrometer quadrupole (Q) and lense voltages are adjusted to optimize testosterone and dihydrotestosterone ion detection and sensitivity (**Fig. 3.3**). The ion guide (Q0), quadrupole 1 (Q1), collision cell (Q2), and quadrupole 3 (Q3) radio frequency (Rf) voltages remain constant. The entrance potential (EP) is set at 10 V to draw testosterone and dihydrotestosterone ions into Ql where the direct current (DC) voltage is adjusted to select for resonant testosterone and dihydrotestosterone ions.
2. Collision-induced dissociation or fragmentation of testosterone and dihydrotestosterone parent ions occurs in Q2. The collision cell parameters are set to optimize production of unique testosterone and dihydrotestosterone product ions. The collision gas (N_2) flow rate is set at 4 L/min and collision energies are set at 25.8 for testosterone and 23.5 for dihydrotestosterone, respectively.
3. The collision exit potential is set at 17.7 to draw fragment ions into Q3 where a constant Rf/DC voltage selects for specific testosterone and dihydrotestosterone fragment ions. An electron multiplier detector amplifies electrons released

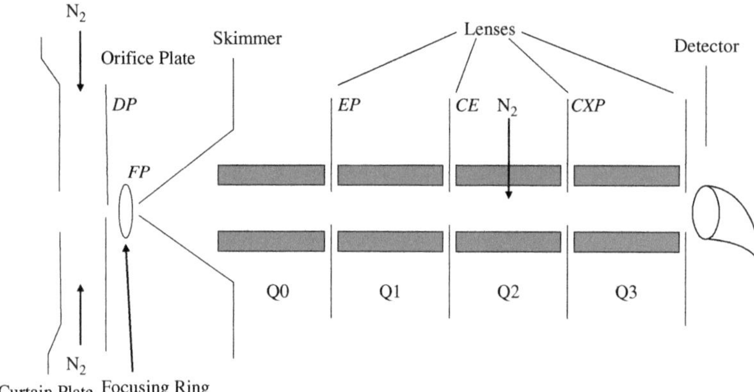

Fig. 3.3. A triple quadrupole mass spectrometer schematic diagram showing the entrance to the mass analyzer (orifice plate and focusing ring), the quadrupoles (Q0–Q3), lenses (EP, entrance potential; CE, collision energy; CXP, collision exit potential), and electron multiplier detector.

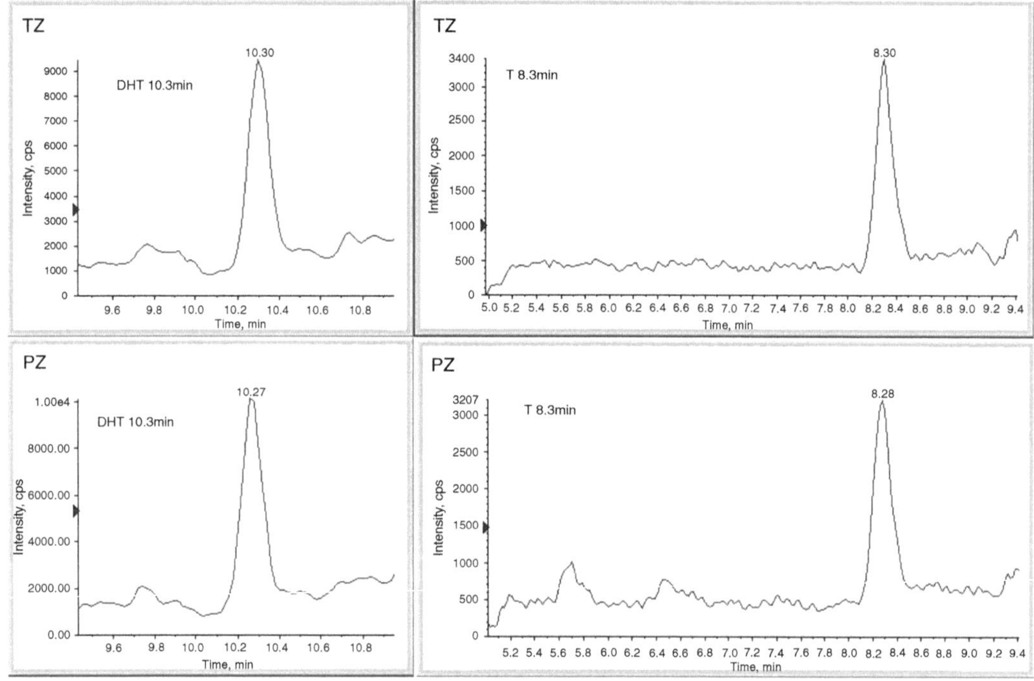

Fig. 3.4. Detection of dihydrotestosterone (DHT) and testosterone (T) in benign prostate transition (TZ) and peripheral (PZ) zones using selected reaction monitoring of *m/z* 289.4/97.2 for testosterone (T) and 291.4/255.2 for dihydrotestosterone (DHT).

after ion impact that are converted to testosterone and dihydrotestosterone fragmentation spectra as shown in **Fig. 3.4** or selected reaction monitoring chromatograms for peak integration and quantitation.

4. Notes

1. HPLC and mass spectrometry instrumentation is expensive. Hence, collaboration with University/College mass spectrometry core facilities or government institutes is useful. The androgen assay can be performed with modification using other HPLC and mass spectrometer systems, such as Shimadzu HPLC system (Shimadzu Scientific Instruments, Columbia, MD), the TSQ Quantum Ultra Triple Stage Quadrupole Mass Spectrometer (Thermo Scientific, Waltham, MA), or the Xevo TQ-S (Waters, Milford, MA), and others.
2. A calibration curve permits the determination of unknown prostate specimen's testosterone and dihydrotestosterone concentrations. Calibration curve samples should be a matrix that is similar to normal human prostate. Rat ventral prostate is an accessible substitute; however, endogenous testosterone and dihydrotestosterone must be removed. Rat ventral prostates (30) were combined, homogenized (50 mg/mL), and treated twice using dextran-770-activated charcoal to remove endogenous testosterone and dihydrotestosterone (**Fig. 3.5**). A series of 1.0 mL standard samples near the expected testosterone and dihydrotestosterone concentration (0.05 pg/μL to 12.0 pg/μL) were prepared. Deuterated testosterone (1 ng) and dihydrotestosterone (1 ng) were added and samples extracted, purified using solid-phase cartridge, and analyzed. The plot of signal (counts per second) versus concentration should be linear and establish the limit of detection (LOD), limit of quantitation (LOQ, signal to noise >3), the dynamic range of the calibration curve, and the limit of linearity (LOL). The testosterone and dihydrotestosterone concentrations calculated from the calibration curve incorporate extraction efficiency and solid-phase extraction recovery.
3. The crushed prostate tissue sample is transferred to a labeled test tube immediately after evaporation of liquid nitrogen. All glass test tubes and caps must be washed using detergent (Alconox), rinsed with water, rinsed with methanol, and dried to decrease background. The volume of the internal standard added to each prostate homogenate must be precise for accurate and reproducible quantitation of androgens.
4. The electrospray ion source settings described for testosterone and dihydrotestosterone should be considered starting settings that can be adjusted to optimize the source performance at individual facilities. The source settings may

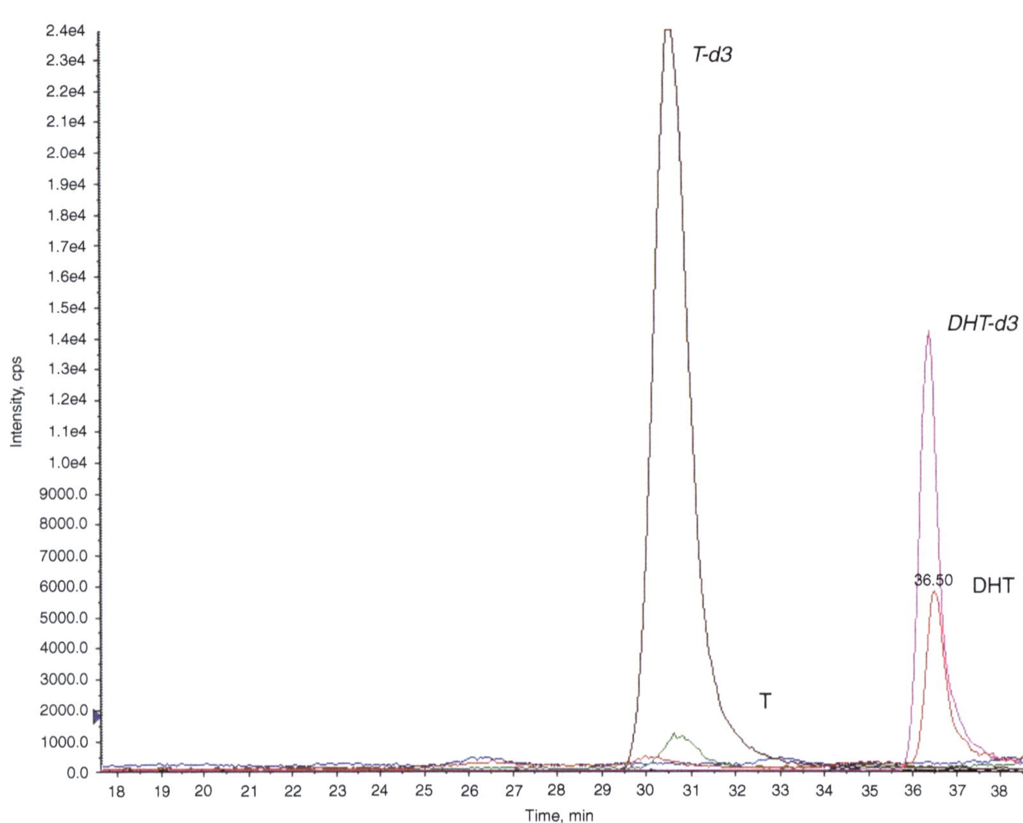

Fig. 3.5. Multiple reaction monitoring of endogenous dihydrotestosterone (DHT), testosterone (T), and deuterated dihydrotestosterone (DHT-d_3) and testosterone (T-d_3) from rat ventral prostate sample. The mass transitions monitored were *m/z* 289.2/97.2 for T, 292.2/97.2 for T-d_3, 291.2/255.2 for DHT and 294.2/258.2 for DHT-d_3.

require adjustment daily. If the electrospray source is used to analyze other biological fluids, such as urine, plasma, or cerebral spinal fluid, it should be cleaned prior to performing androgen assay.

Acknowledgments

The authors would like to thank Fred Lih for development of the androgen assay and analysis of clinical specimens kindly provided by James Mohler, M.D. This work was supported in part by the Intramural Research Program of the National Institute of Environmental Health Sciences/National institutes of Health (z050167) and NCI Grant PO1# CA-77739.

References

1. Imperato-McGinley, J., and Zhu, Y. S. (2002) Androgens and male physiology the syndrome of 5alpha-reductase-2 deficiency, *Mol Cell Endocrinol* **198**, 51–59.
2. Denmeade, S. R., and Isaacs, J. T. (2004) Development of prostate cancer treatment: the good news, *Prostate* **58**, 211–224.
3. Tomlins, S. A., Rhodes, D. R., Perner, S., Dhanasekaran, S. M., Mehra, R., Sun, X. W., Varambally, S., Cao, X., Tchinda, J., Kuefer, R., Lee, C., Montie, J. E., Shah, R. B., Pienta, K. J., Rubin, M. A., and Chinnaiyan, A. M. (2005) Recurrent fusion of TMPRSS2 and ETS transcription factor genes in prostate cancer, *Science* **310**, 644–648.
4. Waxman, J. (1985) Hormonal aspects of prostatic cancer: a review, *J R Soc Med* **78**, 129–135.
5. Hellerstedt, B. A., and Pienta, K. J. (2002) The current state of hormonal therapy for prostate cancer, *CA Cancer J Clin* **52**, 154–179.
6. Novara, G., Galfano, A., Secco, S., Ficarra, V., and Artibani, W. (2009) Impact of surgical and medical castration on serum testosterone level in prostate cancer patients, *Urol Int* **82**, 249–255.
7. Titus, M. A., Schell, M. J., Lih, F. B., Tomer, K. B., and Mohler, J. L. (2005) Testosterone and dihydrotestosterone tissue levels in recurrent prostate cancer, *Clin Cancer Res* **11**, 4653–4657.
8. Montgomery, R. B., Mostaghel, E. A., Vessella, R., Hess, D. L., Kalhorn, T. F., Higano, C. S., True, L. D., and Nelson, P. S. (2008) Maintenance of intratumoral androgens in metastatic prostate cancer: a mechanism for castration-resistant tumor growth, *Cancer Res* **68**, 4447–4454.
9. Nishiyama, T., Hashimoto, Y., and Takahashi, K. (2004) The influence of androgen deprivation therapy on dihydrotestosterone levels in the prostatic tissue of patients with prostate cancer, *Clin Cancer Res* **10**, 7121–7126.
10. Attard, G., Reid, A. H., Yap, T. A., Raynaud, F., Dowsett, M., Settatree, S., Barrett, M., Parker, C., Martins, V., Folkerd, E., Clark, J., Cooper, C. S., Kaye, S. B., Dearnaley, D., Lee, G., and de Bono, J. S. (2008) Phase I clinical trial of a selective inhibitor of CYP17, abiraterone acetate, confirms that castration-resistant prostate cancer commonly remains hormone driven, *J Clin Oncol* **26**, 4563–4571.
11. Geller, J., de la Vega, D. J., Albert, J. D., and Nachtsheim, D. A. (1984) Tissue dihydrotestosterone levels and clinical response to hormonal therapy in patients with advanced prostate cancer, *J Clin Endocrinol Metab* **58**, 36–40.
12. Mohler, J. L., Gregory, C. W., Ford, O. H., 3rd, Kim, D., Weaver, C. M., Petrusz, P., Wilson, E. M., and French, F. S. (2004) The androgen axis in recurrent prostate cancer, *Clin Cancer Res* **10**, 440–448.
13. Boucher, E., Provost, P. R., Devillers, A., and Tremblay, Y. (2010) Levels of dihydrotestosterone, testosterone, androstenedione, and estradiol in canalicular, saccular, and alveolar mouse lungs, *Lung* **188**, 229–233.
14. Hsing, A. W., Stanczyk, F. Z., Belanger, A., Schroeder, P., Chang, L., Falk, R. T., and Fears, T. R. (2007) Reproducibility of serum sex steroid assays in men by RIA and mass spectrometry, *Cancer Epidemiol Biomarkers Prev* **16**, 1004–1008.
15. Wang, C., Shiraishi, S., Leung, A., Baravarian, S., Hull, L., Goh, V., Lee, P. W., and Swerdloff, R. S. (2008) Validation of a testosterone and dihydrotestosterone liquid chromatography tandem mass spectrometry assay: Interference and comparison with established methods, *Steroids* **73**, 1345–1352.
16. Lih, F. B., Titus, M. A., Mohler, J. L., and Tomer, K. B. (2010) Atmospheric pressure photoionization tandem mass spectrometry of androgens in prostate cancer, *Anal Chem* **82**, 6000–6007.
17. Morrison, C., Cheney, R., Johnson, C. S., Smith, G., and Mohler, J. L. (2009) Central quadrant procurement of radical prostatectomy specimens, *Prostate* **69**, 770–773.

Chapter 4

Ligand Competition Binding Assay for the Androgen Receptor

Clémentine Féau, Leggy A. Arnold, Aaron Kosinski, and R. Kiplin Guy

Abstract

Evaluating endocrine activities of environmental chemicals or screening for new small molecule modulators of the androgen receptor (AR) transcription activity requires standardized and reliable assay procedures. Scintillation proximity assays (SPA) are sensitive and reliable techniques that are suitable for ligand competition binding assays. We have utilized a radiolabeled ligand competition binding assay for the androgen receptor (AR) that can be carried out in a 384-well format. This standardized, highly reproducible and low-cost assay has been automated for high-throughput screening (HTS) purposes.

Key words: Androgen receptor, nuclear receptor, hormone competition assay, endocrine disrupting chemicals, scintillation proximity assay, high-throughput screening.

1. Introduction

Androgenic characteristics, including lean body mass regulation, maintenance of sexual male characteristics, and development of the prostate gland, are governed by the androgen receptor (AR). Like other nuclear hormone receptors (NRs), AR is a transcription factor that becomes active upon binding to its endogenous ligand, dihydrotestosterone (DHT) (1), and regulates its target genes. Accordingly, small molecules with the ability to inhibit DHT binding may have the potential to modulate gene transcription regulated by AR. Exposure to these compounds during development can lead to abnormalities, especially in males (2). Several natural and synthetic pesticides have been identified as

F. Saatcioglu (ed.), *Androgen Action*, Methods in Molecular Biology 776,
DOI 10.1007/978-1-61779-243-4_4, © Springer Science+Business Media, LLC 2011

endocrine disrupting chemicals (EDC) which possess antiandrogenic properties including vinclozolin (3), DDT, and its metabolites (4). On the other hand, antiandrogens like flutamide (5) and bicalutamide (6) are currently used in prostate cancer treatments. Despite the great progress in treating prostate cancer, there is still a compelling need to discover new compounds to overcome the drawbacks posed by the current antiandrogens. Therefore, formally standardized AR binding assays to find out new exogenous AR ligands are highly useful.

Traditionally, potential AR ligands have been detected by either cell-based transcription assays measuring the inhibition of AR transcriptional activity or biochemical competition assays measuring blockade of ligand-binding AR by small molecules. With the purpose of seeking for direct AR binders, biochemical assays are preferred, as opposed to cell-based assays, since they focus on direct interaction with the ligand-binding pocket.

Historically, biochemical assays have had low throughput due to the lack of necessary amounts of pure and functional AR protein. Purification of AR is complicated due to low solubility, instability in the absence of androgen, and problems of aggregation (7). The earliest assays used cytosolic preparations of animal tissue containing AR in combination with radiolabeled ligands (4, 8) (e.g., Hershberger assay) or AR from cultured cells, e.g., LnCaP (9) and MCF-7 (10). In cytosolic preparations, other steroid receptors besides AR are present causing cross-reactivity (11). Recombinant full length AR (recAR) or AR-LBD (ligand-binding domain) from transfected bacterial (12, 13), insect (11, 14), or mammalian cells (15) have been used as animal-free alternatives. Recently, full-length biotin-tagged AR has been successfully expressed in Sf9 insect cells (14). Most of these AR protein expression methods have disadvantages including low yield, unsatisfactory purity, low solubility, and insufficient stability. We have described a high-yield expression of His_6-tagged AR-LBD in *Escherichia coli* in the presence of DHT that takes advantage of improved stability of AR-LBD once it is bound to its natural ligand. In addition, purification of AR-LBD as a histidine (His)-tagged protein is advantageous, because, unlike protein tags such as glutathione-*S*-transferase (GST), the short His sequence has minimal effects on protein structure and binding affinity. Furthermore, it can efficiently bind metal-chelating columns which simplifies the development of solid-phase supported assays (7).

For ligand competition assays, radioligands are preferred as they most closely mimic the natural ligand in contrast to fluorescently labeled ligands. Among ligand-binding assays developed for AR (16–19), few radiometric binding assays have been described in the 96-well format for the AR (16, 19). Among ligand-binding assays developed for NRs (20–23), only scintillation proximity assays (SPAs) are safe, reproducible and truly amenable to high-throughput screening (HTS) applications. SPA technology is a radioisotopic homogeneous assay system that does not require any separation step. In our case, we used commercially available 384-well FlashPlates® precoated with a fluoroscintillant coating and with Ni-chelate able to trap our recHis$_6$-AR-LBD. When the radioligand [^{3}H]-DHT binds to the plate-bound AR-LBD, it is in turn close enough to the fluoroscintillant coating to transfer energy from its β-particles, resulting in photon emission. A luminescent signal is measured that reflects binding events. In this SPA competition assay, the radiolabeled ligand is added at the very last step to reduce handling of radioactive material and to enable semi- or full automation of this method.

Herein we report an AR competition ligand-binding assay using DHT-liganded AR-LBD protein expressed in *E. coli* and SPA technique in a 384-well format. This standardized and highly reproducible assay is safe in terms of radioactive contamination and can be easily automatable for high-throughput screening (HTS) campaigns (24).

2. Materials

2.1. Protein Expression

1. cAR-LBD (His$_6$; residues 663–919) expression plasmid cloned in pKBU553 vector (Addgene, http://genome-www.stanford.edu/vectordb//vector_descrip/PKBU5.html).
2. OneShot BL21 Star (DE3) chemically competent *E. coli.*
3. Dihydrotestosterone (DHT, CAS 521-18-6), stored in stock 10 mM solution in dimethylsulfoxide (DMSO) at 4°C.
4. Shaking incubator.
5. Ready made isopropyl-β-D-thiogalactoside (IPTG, CAS 367-93-1, Sigma) solution is a 0.2 μm filter sterilized solution of 200 mM IPTG. This reagent remains liquid at its storage temperature (–20°C), avoid freeze–thaw cycles. IPTG is used at 60 μM (0.6 mL of ready made solution for 2 L of culture medium).

2.2. Protein Purification

1. Buffer 1 (*cell suspension buffer*): 50 mM Tris pH 7.5, 150 mM NaCl, 10 μM DHT, 0.1 mM phenylmethanesulfonyl-fluoride (PMSF, CAS 329-98-6), 10 mg/L Lysozyme (Roche), and complete ethylenediaminetetraacetic acid (EDTA)-free protease inhibitor cocktail tablet (Roche) (**Notes 1** and **2**).
2. Digital Sonifier (Branson).

3. Ultra-centrifuge (Beckman Coulter).
4. 50 mL conical tube (BD Falcon).
5. Talon resin (Clontech).
6. Buffer 2 (*resin washing buffer*): 46.75 mM dibasic sodium phosphate pH 8 (Na_2HPO_4, CAS 7558-79-4), 3.25 mM monobasic sodium phosphate pH 8 (NaH_2PO_4, CAS 10049-21-5) [50 mM phosphate buffer, pH 8], 300 mM sodium chloride (NaCl, CAS 7647-14-5), 10% glycerol (CAS 56-81-5), 0.2 mM Tris(2-carboxyethyl)phosphine hydrochloride (TCEP, CAS 51805-45-9), 0.1 mM PMSF (CAS 329-98-6), 0.002 mM DHT (CAS 521-18-6).
7. Rotisserie shaker (ThermoFisher).
8. Buffer 3 (*protein washing buffer*): 46.75 mM Na_2HPO_4 pH 8 (CAS 7558-79-4, Sigma), 3.25 mM NaH_2PO_4 pH 8 [50 mM phosphate buffer, pH 8], 300 mM NaCl (CAS 7647-14-5), 10% glycerol (CAS 56-81-5), 0.2 mM TCEP (CAS 51805-45-9), 0.1 mM PMSF (CAS 329-98-6), 0.002 mM DHT (CAS 521-18-6), 10 mM imidazole (CAS 288-32-4).
9. Buffer 4 (*wash out buffer*): 46.75 mM Na_2HPO_4 pH 8 (CAS 7558-79-4, Sigma), 3.25 mM NaH_2PO_4 pH 8 [50 mM phosphate buffer, pH 8], 300 mM NaCl (CAS 7647-14-5), 10% glycerol (CAS 56-81-5), 0.2 mM TCEP (CAS 51805-45-9), 0.1 mM PMSF (CAS 329-98-6), 0.002 mM DHT (CAS 521-18-6), 10 mM imidazole (CAS 288-32-4), 2 mM adenosine 5′-triphosphate disodium salt (ATP disodium salt, CAS 51963-61-2), 10 mM magnesium chloride ($MgCl_2$, CAS 7786-30-3).
10. Buffer 5 (*elution buffer*): 46.75 mM Na_2HPO_4, pH 8 (CAS 7558-79-4), 3.25 mM NaH_2PO_4 pH 8 [50 mM phosphate buffer, pH 8], 300 mM NaCl (CAS 7647-14-5), 10% glycerol (CAS 56-81-5), 0.2 mM TCEP (CAS 51805-45-9), 0.1 mM PMSF (CAS 329-98-6), 0.002 mM DHT (CAS 521-18-6), 250 mM imidazole (CAS 288-32-4), 100 mM potassium chloride (KCl, CAS 7447-40-7).
11. Buffer 6 (*dialysis buffer*): 50 mM 4-(2-hydroxyethyl)-1-piperazineethanesulfonic acid (HEPES, CAS 7365-45-9) pH 7.2, 150 mM lithium sulfate (Li_2SO_4, CAS 10377-48-7), 10% glycerol (CAS 56-81-5), 0.2 mM TCEP (CAS 51805-45-9), 20 μM DHT (CAS 521-18-6).
12. SDS-polyacrylamide gel electrophoresis (SDS-PAGE): NuPAGE® Tris-Acetate SDS Running Buffer is used with NuPAGE® Novex Tris-Acetate Gels (Invitrogen).
13. Akta Fast Protein Liquid Chromatography (FPLC) (Amersham).

14. Bradford protein assay kit (Pierce/ThermoFisher).
15. BCA protein assay reagent kit (Pierce/ThermoFisher).
16. Dialysis cassette (Slide-a-Lyzer®, Pierce/ThermoFisher).

2.3. Ligand Competition Assay

1. 384-well Ni-chelate Flashplates® (PerkinElmer).
2. [1,2,4,5,6,7-3*H*(*N*)]-5α-Androstan-17β-ol-3-one ([^{3}H]-DHT), 110 Ci/mmol (PerkinElmer).
3. Buffer 7 (*Assay buffer*): 50 mM HEPES, 150 mM lithium sulfate (Li_2SO_4, CAS 10377-48-7), 0.2 mM TCEP (CAS 51805-45-9), 10% glycerol (CAS 56-81-5), and 0.01% Triton X-100 (CAS 9002-93-1) (pH 7.2).
4. Multiscreen Sealing clear tape (Millipore).
5. TopCount Microplate Scintillation and Luminescence Counter (Packard Instrument Company).
6. GraphPad Prism 4.03 (GraphPad Software, San Diego, CA). Although we use GraphPad Prism software, any graphing and data analysis software package will be sufficient.
7. Tested panel of AR binders: dihydrotestosterone (DHT, CAS 521-18-6, Sigma), 17β-estradiol (CAS 50-28-2, Sigma), flutamide (CAS 13311-84-7, Sigma), progesterone (CAS 57-83-0, Sigma), hydroxyflutamide (CAS 52806-53-8, LKT Laboratories), cyproterone acetate (CAS 427-51-0, LKT Laboratories), dexamethasone (CAS, LKT Laboratories), methyltrienolone (R1881, CAS 965-93-5, PerkinElmer), and bicalutamide (CAS 90357-06-5, Toronto Research Chemicals).

3. Methods

cAR-LBD (His_6; residues 663–919) was expressed in *E. coli* and purified to homogeneity in the presence of DHT using a modified version of published protocols (**Note 3**) (12). DHT-bound AR-LBD was directly used in a SPA radioligand competition assay. This assay has proven to be flexible (96- or 384-well format, assay plates can be incubated at room temperature or at 4°C), highly reproducible, can be semi- or full automated and adapted to other NRs (**Note 4**) (24).

3.1. Protein Expression

1. Add 1 μL of cAR-LBD expression plasmid (1 ng/μL) to 50 μL of OneShot BL21 Star (DE3) chemically competent E. coli in ice-cold Eppendorf tubes. Gently mix by flicking with your finger (do not mix up and down with your pipet) and incubate the transformation mixture on ice for 30 min.

Put the tube at 42°C for exactly 90 s. Return the cells to ice for 2 min. Pipette the transformation mixtures onto a LB agar + 1X carbenicillin (100 μg/ml) plate and spread using sterile glass beads. Place upside down in a 37°C incubator dedicated to bacteria growth.

2. After incubating for 16–20 h, pick a single colony from this plate and begin a liquid culture in a 500 mL flask containing 1X carbenicillin LB agar medium. The seed culture is then grown up overnight at 37°C in a shaking incubator.
3. The next day seed 2 L of 2X LB agar medium +1X carbenicillin and 10 μM DHT at 0.1 optical density (OD) and grown at 25°C with shaking until OD reaches 0.6–0.8.
4. Induce expression the next day by adding to the culture medium isopropyl-β-D-thiogalactoside at 60 μM, and grow for 14–16 h at 17°C. Pellet the cells by spinning for 20 min at 5000×*g*. Transfer the pelleted cells into a 50 mL conical tube (**Note 5**).

3.2. Protein Purification

1. Add 30 mL of freshly prepared Buffer 1 to pelleted cells and resuspend using a spatula. Next, sonicate the sample using a Branson Digital Sonifier on ice for 2 min intervals at 30% amplitude followed by a 2 min break. After six or more repetitions the suspension should no longer be "gooey". Pellet insoluble cellular debris by ultracentrifugation (2 × 30 min at 100,000×*g* at 4°C).
2. Meanwhile, add Talon resin beads (1 mL per liter cell culture) to a 50 mL conical tube and wash two times with 15 mL of freshly prepared Buffer 2. Apply the supernatant from the ultracentrifugation to the 50 mL conical tube containing the washed beads. Cap the tube and seal with parafilm. Rotate the sample overnight at 4°C.
3. The next day pellet the beads by spinning the 50 mL conical tube for 20 min at 3000×*g*. Discard supernatant. Wash the beads five times with 10 ml Buffer 3. Additionally, wash the beads five times with 10 ml Buffer 4. Elution is carried out in fractions equal to or less then bed volume using Buffer 5.
4. Assess protein purity by SDS-PAGE (**Note 6, Fig. 4.1**) and analytical size exclusion FPLC (**Note 7**). Measure protein concentrations using both Bradford (**Note 8**) and BCA protein assays (**Notes 9** and **10**).
5. Dialyze the protein overnight against Buffer 6 and store at –80°C in presence of two equivalents of DHT, to ensure fully liganded AR-LBD.

3.3. Ligand Competition Assay

1. To each well of a 384-well Ni-chelate-coated Flashplate add 50 μL of 5 μM purified DHT-bound AR-LBD in Buffer 7

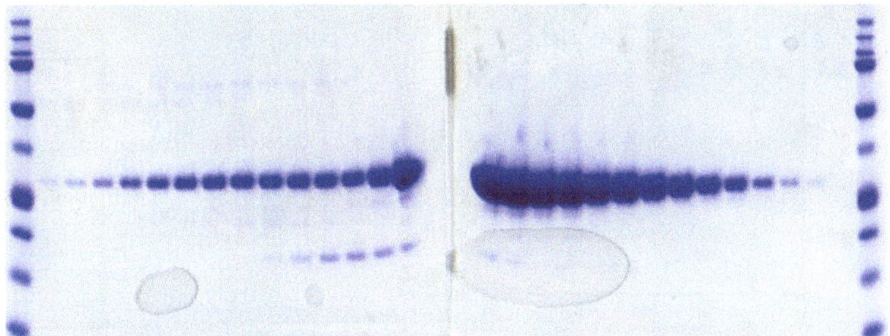

Fig. 4.1. SDS-PAGE gel and Coomassie stain of fractions containing AR-LBD after purification on analytical size exclusion FPLC (*see* **Notes 6** and **7**). Molecular weight markers are included in the *first* and the *last lanes*.

(**Note 11**). Incubate the protein solution for 1 h (**Note 12**). Discard the protein solution and wash at least twice with 50 μL of buffer 7 (**Note 13**).

2. Add 25 μL of serially diluted small molecules in buffer 7 containing 10% DMSO into each well, followed by addition of 25 μL of a [^{3}H]-DHT solution in buffer 7 (20 nM final concentration, **Fig. 4.2a**). The final assay solution contains 5% DMSO.
3. Seal the plates with clear tape and equilibrate for 1–24 h at room temperature or 4°C (**Fig. 4.2b**, **Note 14**).
4. Measure radiocounts using a TopCount Microplate Counter.
5. Analyze data by fitting data to the following equation (sigmoidal dose–response (variable slope)): y = bottom + (top – bottom)/(1 + 10*((log IC_{50} – x) × Hillslope)), where x is the logarithm of concentration, and y is the response. To

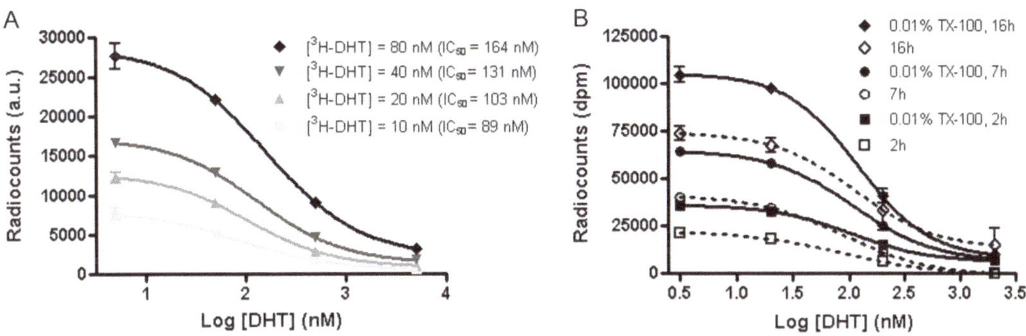

Fig. 4.2. Optimization of assay conditions. (**a**) Influence of the [^{3}H]-DHT concentration on the signal window and binding affinity measurements. Specific binding was measured for experiments carried out with 5 μM AR for the incubation step using serially diluted unlabeled DHT in the presence of variable concentration of [^{3}H]-DHT after 1 h. (**b**) Influence of the detergent Triton X-100 (TX-100) over time. Specific binding was measured for experiments carried out with 5 μM AR for incubation step, serially diluted DHT in the presence of 20 nM [^{3}H]-DHT after various time points.

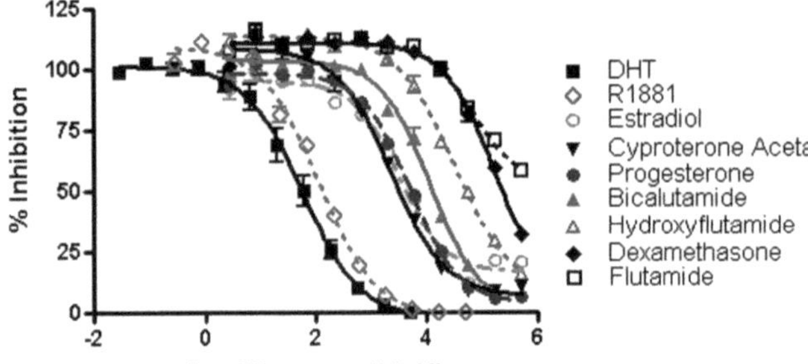

	EC50
Hydroxyflutamide	25023 to 41151
Estradiol	2716 to 3852
DHT	43.07 to 79.05
Bicalutamide	9447 to 14667
Flutamide	44087 to 102722
Cyproterone Acetate	1908 to 2789
Dexamethasone	87210 to 290178
Progesterone	4423 to 5892
R1881	98.30 to 133.5

Fig. 4.3. Dose–response experiments with known AR binders. 5 μM AR solution was incubated within a Ni-chelate-coated FlashPlate®; after two consecutive washes, serially diluted drugs were added immediately followed by the addition of a [^{3}H]-DHT solution (final concentration was 20 nM). The assay was carried out with a maximum of 5% DMSO. Radiocounts measured after 5 h were normalized to maximum binding of [^{3}H]-DHT (treatment with DMSO) and full inhibition (treatment with 10 μM DHT). IC50 and RBA were calculated (*see* **Section 3.3**, Steps 5 and 6).

obtain reliable potency measurements, carry out at least two independent experiments, in triplicate, for each compound (**Fig. 4.3**)

6. For potential comparison with other binding assays, relative binding affinities (RBA) can be calculated relative to DHT-binding affinity. RBA = (IC_{50} DHT/IC_{50} Drug) × 100

4. Notes

1. Unless stated otherwise, all buffer solutions should be prepared in water that has been purified through a three-step process provided by Millipore purification system. (1) Plug-in Q-Gard® purification packs are tailored to the feedwater source. (2) Quantum application-specific cartridges remove ionic and organic contaminants down to trace levels. (3) Final purification of ultrapure water is carried out at the point of use by a pharmaceutical grade, absolute 0.22 μm Millipak® membrane filter that is recommended for most analytical applications.
2. Unless stated otherwise, all buffer solutions were sterilized by filtration over "Steritop" Sterile Vacuum Bottle, Millipore, Billerica, MA, and stored at 4°C.
3. DHT access is regulated. R1881 has proven to be also a good ligand for AR expression and purification (12).
4. This assay was also optimized for other nuclear receptors using conditions as followed. For the hPPARγ assay, [^{3}H]-rosiglitazone was at 40 nM, and the assay buffer contained 50 mM Tris (pH 8.0), 25 mM KCl, 2 mM DTT, 10% glycerol, and 0.01% Triton X-100 (pH 7.2). For the hTR assays, [^{125}I]-T3 was at 1 nM, and the assay buffer

contained 50 mM HEPES, 100 mM NaCl, 1 mM DTT, 0.1% bovine serum albumin (BSA), 10% glycerol, and 0.01% Triton X-100 (pH 7.2).

5. At that stage pelleted cells can be flash frozen in liquid nitrogen and stored at −80°C. Before starting the purification process, the frozen cells must be thawed on ice.
6. General protocols for using the NuPAGE® electrophoresis system can be found at http://tools.invitrogen.com/content/sfs/manuals/nupage_tech_man.pdf.
7. General information for using the Akta Fast Protein Liquid Chromatography can be found at http://www.biocompare.com/Articles/ProductReview/465/GE-Healthcares-AKTA-FPLC-System.html.
8. General information and protocols for Bradford protein assay kit can be found at http://www.piercenet.com/Files/1601669_PAssayFINAL_Intl.pdf (p24).
9. General information and protocols for BCA protein assay kit can be found at http://www.piercenet.com/Files/1601669_PAssayFINAL_Intl.pdf (p19).
10. Usually 6–8 mg protein per liter cell culture is obtained (>90% pure).
11. AR concentration has been optimized with 5 μM being the saturated concentration (24). This competition assay has been optimized in 96 format and can be used with the following conditions: 100 μL of 5 μM purified DHT-bound AR-LBD in buffer 7 and 100 μL of final assay mixture with same concentrations described for 384-format assay.
12. We observed that this incubation step could be performed at 4°C for 1 h or at room temperature for 30 min without any change on the further measurements (24).
13. We observed that two washing steps were sufficient to remove significantly the protein present in excess (24).
14. We monitored the performance of the assay at both temperatures, no particular effect on the measured IC_{50} was observed, and temperature can be chosen at will of the user (24).

Acknowledgments

This work was supported by the American Lebanese Syrian Associated Charities (ALSAC), St. Jude Children's Research Hospital (SJCRH), the NIH (DK58080) and the Department of Defense Prostate Cancer Research Program (PC060344-W81XWH-07-1-0073).

References

1. Chang C (2002) Androgens and androgen receptor: mechanisms, functions, and clinical applications. Kluwer: Norwell.
2. Gray LE, Ostby J, Furr J et al (2001) Effects of environmental antiandrogens on reproductive development in experimental animals. *Hum Reprod Update* 7(3):248–64.
3. Kelce WR, Monosson E, Gamcsik MP et al (1994) Environmental hormone disruptors: evidence that vinclozolin developmental toxicity is mediated by antiandrogenic metabolites. *Toxicol Appl Pharmacol* **126**(2): 276–85.
4. Kelce WR, Stone CR, Laws SC et al (1995) Persistent DDT metabolite p,p′-DDE is a potent androgen receptor antagonist. *Nature* **375**(6532):581–5.
5. Seftel A (2005) Comparison of the pharmacological effects of a novel selective androgen receptor modulator (SARM), the 5alpha-reductase inhibitor finasteride, and the antiandrogen hydroxyflutamide in intact rats: new approach for benign prostate hyperplasia (BPH). *J Urol* **173**(4):1279.
6. Iversen P, Wirth MP, See WA et al (2004) Is the efficacy of hormonal therapy affected by lymph node status? Data from the bicalutamide (Casodex) Early Prostate Cancer program. *Urology* **63**(5):928–33.
7. Liao M, Wilson EM (2001) Production and purification of histidine-tagged dihydrotestosterone-bound full-length human androgen receptor. *Methods Mol Biol* **176**:67–79.
8. Breiner M, Romalo G, Schweikert HU (1986) Inhibition of androgen receptor binding by natural and synthetic steroids in cultured human genital skin fibroblasts. *Klin Wochenschr* **64**(16):732–7.
9. Sonnenschein C, Olea N, Pasanen ME et al (1989) Negative controls of cell proliferation: human prostate cancer cells and androgens. *Cancer Res* **49**(13):3474–81.
10. Schoonen WG, Deckers G, de Gooijer ME et al (2000) Contraceptive progestins. Various 11-substituents combined with four 17-substituents: 17alpha-ethynyl, five- and six-membered spiromethylene ethers or six-membered spiromethylene lactones. *J Steroid Biochem Mol Biol* **74**(3):109–23.
11. Bauer ER, Daxenberger A, Petri T et al (2000) Characterisation of the affinity of different anabolics and synthetic hormones to the human androgen receptor, human sex hormone binding globulin and to the bovine progestin receptor. *Apmis* **108**(12):838–46.
12. Matias PM, Donner P, Coelho R et al (2000) Structural evidence for ligand specificity in the binding domain of the human androgen receptor. Implications for pathogenic gene mutations. *J Biol Chem* **75**(4):26164–71.
13. Roehrborn CG, Zoppi S, Gruber J A et al (1992) Expression and characterization of full-length and partial human androgen receptor fusion proteins. Implications for the production and applications of soluble steroid receptors in Escherichia coli. *Mol Cell Endocrinol* **84** (1–2):1–14.
14. Juzumiene D, Chang CY, Fan D et al (2005) Single-step purification of full-length human androgen receptor. *Nucl Recept Signal* **3**:e001.
15. Quarmby VE, Kemppainen JA, Sar M et al (1990) Expression of recombinant androgen receptor in cultured mammalian cells. *Mol Endocrino* **4**(9):1399–407.
16. Bauer ER, Bitsch N, Brunn H et al (2002) Development of an immuno-immobilized androgen receptor assay (IRA) and its application for the characterization of the receptor binding affinity of different pesticides. *Chemosphere* **46**(7):1107–15.
17. Fang H, Tong W, Branham WS et al (2003) Study of 202 natural, synthetic, and environmental chemicals for binding to the androgen receptor. *Chem Res Toxicol* **16**(10): 1338–58.
18. Yamasaki K, Sawaki M, Noda S et al (2004) Comparison of the Hershberger assay and androgen receptor binding assay of twelve chemicals. *Toxicology* **195**(2–3):177–86.
19. Freyberger A, Ahr HJ (2004) Development and standardization of a simple binding assay for the detection of compounds with affinity for the androgen receptor. *Toxicology* **195** (2–3):113–26.
20. Nichols JS, Parks DJ, Consler TG et al (1998) Development of a scintillation proximity assay for peroxisome proliferator-activated receptor gamma ligand binding domain. *Anal Biochem* **257**(2):112–9.
21. Janowski BA, Grogan MJ, Jones SA et al (1999) Structural requirements of ligands for the oxysterol liver X receptors LXRα and LXRβ. *Proc Natl Acad Sci* **96**: 266–71.
22. Allan GF, Hutchins A, Clancy J (1999) An ultrahigh-throughput screening assay for estrogen receptor ligands. *Anal Biochem* **275**(2):243–7.
23. Wu B, Gao J, Wang M (2005) Development of a complex scintillation proximity assay for high-throughput screening of PPARγ modulators. *Acta Pharma Sinica* **26**(3):339–44.
24. Feau C, Arnold L A, Kosinski A, Guy RK (2009) A high-throughput ligand competition binding assay for the androgen receptor and other nuclear receptors. *JBS* **14**(1): 43–48.

Section III

Androgen Receptor Biology

Chapter 5

Analysis of Androgen Receptor Activity by Reporter Gene Assays

Harri Makkonen, Tiina Jääskeläinen, Miia M. Rytinki, and Jorma J. Palvimo

Abstract

Androgen receptor (AR) acts as a ligand-regulated transcription factor that conveys the message of both natural and synthetic androgens directly to the level of gene programs. Reporter gene assays provide a convenient and quantitative way to measure the transcriptional activity of AR and the functionality of its binding sites (AREs) in DNA. Many reporter genes and different transfection methods can be used for this purpose. In this chapter, we describe the use of firefly luciferase gene-based reporters and transfection protocols for the measurement of AR activity in heterologous COS-1 cells cotransfected with an AR expression vector and in VCaP prostate cancer cells expressing endogenous AR. We also discuss the suitability of different reporter constructs and transfection methods for different cell types and how reporter gene assays can be employed to complement chromatin immunoprecipitation assays.

Key words: Androgen receptor, transcription, reporter gene, luciferase, androgen response element (ARE), enhancer, transfection, cell culture, prostate cancer cell.

1. Introduction

Androgen receptor (AR) mediates the effects of the male sex steroids, testosterone (T) and 5α-dihydrotestosterone (DHT), that play a crucial role in the maintenance and development of the male sexual characteristics (1). Defective AR signaling leads to a wide array of androgen insensitivity disorders, and deregulated AR function is involved in the growth and progression of prostate cancer (2). AR is a member of the steroid receptor family, a subgroup of the nuclear receptor superfamily. The modular structure of AR, like all nuclear receptors, is comprised of a less conserved

F. Saatcioglu (ed.), *Androgen Action*, Methods in Molecular Biology 776,
DOI 10.1007/978-1-61779-243-4_5,

N-terminal domain, a C-terminal hormone-binding domain and, in between of them, a highly conserved DNA-binding domain (DBD) (1).

In keeping with the similarities between the DBDs of AR, glucocorticoid receptor, progesterone receptor, and that of mineralocorticoid receptor, they all can recognize with a similar affinity so-called non-selective response elements that are composed of 3-nt-spaced inverted repeats of a 5′-TGTTCT-3′-type half-site (3). However, the predicted androgen response elements (AREs) that bind the AR in vitro do not necessarily function in vivo, since epigenetic regulation of chromatin structure and packaging dictates their accessibility (4). The AREs are mandatory for AR-mediated transcriptional activation, but not for transcriptional repression by the receptor (5). In addition (to basal transcription machinery and RNA polymerase II), activation of genes and transcriptional programs by the AR requires coregulator proteins, coactivators and corepressors, that augment or attenuate the AR-dependent transcription (6). *Prostate-specific antigen* (*PSA*), also known as *kallikrein 3 (KLK3),* is established as a canonical AR target gene. The *PSA* gene contains two AREs in its proximal promoter and a more distal compound ARE containing an enhancer at ~4.2 kb 5′ from the transcription start site (TSS) (7). However, PSA mRNA may not be a very sensitive marker of androgen levels and, for example, FKBP51 (FKBP5) mRNA better reflects the androgen concentration and activity in human prostate (8). Activation of the *FKBP51* locus by androgens in vivo interestingly relies on very distal, compound ARE-containing enhancers at ~34 kb 5′ and ≥90 kb 3′ of the TSS (9).

Here, we describe the use of the *PSA* promoter- and *FKBP51* enhancer-driven firefly luciferase (LUC) gene reporters and different transfection protocols for measurement of AR activity in COS-1 cells in the presence of an AR expression construct and in VCaP prostate cancer cells with endogenous AR. The VCaP cells are derived from a bone metastasis and express amplified levels of wild-type AR, resembling the situation in hormone-refractory prostate cancer (10).

2. Materials

2.1. Reporter Genes and AR Expression Construct

1. Plasmid constructs: Reporters pPSA5.8-LUC (11), pTATA-LUC, pFKBP51(-3)-LUC (12), pFKBP51(11)-LUC, pFKBP51(12)-LUC (9) (in pGL3-Basic backbone, Promega, Madison, WI, USA), pCMVβ (Clontech Laboratories Inc., Mountain View, CA, USA) encoding β-galactosidase, and pSG5-hAR for AR expression.

2. The plasmids are purified from *E. coli* JM109 strain cultures using QIAGEN Plasmid Maxi Kit according to manufacturer's instructions (Qiagen, Hilden, Germany). The purified plasmids are dissolved in TE buffer (10 mM Tris–HCl pH 8, 1 mM EDTA pH 8) at 1 μg/μl and stored at 4°C (for frequent use) or –20°C (long-term storage).

2.2. Cell Culture and Transfection

1. Cell lines used: African Green monkey kidney (COS-1) cells and Vertebral Cancer of the Prostate (VCaP) cells from ATCC (Manassas, VA, USA).
2. For maintenance of the COS-1 cells: Dulbecco's Modified Eagle's Medium (DMEM) (Gibco®, Invitrogen, part of Life Technologies, Carlsbad, CA, USA) supplemented with 10% (v/v) fetal bovine serum (FBS), 25 U/ml penicillin and 25 μg/ml streptomycin. For maintenance of the VCaP cells: DMEM containing 10% (v/v) US defined FBS (HyClone, Logan, Utah, USA), 25 U/ml penicillin, and 25 μg/ml streptomycin in a 5% CO_2 atmosphere at 37°C.
3. JetPEI transfection dilution solution: Sterile-filtered 150 mM NaCl.
4. For hormone treatment of the cell lines: DMEM supplemented with 2.5% (v/v) charcoal-stripped FBS.
5. Charcoal-stripping of FBS: Add 10 g of charcoal (charcoal-activated GR for analysis; Merck, Darmstadt, Germany) and 1 g of dextran to 100 ml of DMEM in 1000-ml flask. Mix the DMEM–charcoal–dextran slurry well before adding 500 ml of FBS to the flask. Magnetic stir at room temperature for 1 h. Centrifuge in two 500-ml bottles at ~17,000×*g* for 20 min. Pour supernatants to clean centrifuge bottles and repeat the centrifugation step. Sterile filter, aliquot, and store at –20°C.
6. Trypsin (0.25% w/v) with 1 mM ethylenediamine tetraacetic acid (EDTA).
7. Transfection reagents: TransIT®-LT1 (Mirus Bio LLC, Madison, WI, USA) and jetPEI™ (Polyplus-transfection SA, Illkirch, France).
8. AR ligands: R1881 (Perkin-Elmer, Waltham, MA, USA) is stored in glass tube as 10 mM stock in absolute ethanol (EtOH) at –20°C (*see* **Note 1**).
9. Cell washing buffer: Phosphate-buffered saline (PBS, 137 mM NaCl, 2.7 mM KCl, 100 mM Na_2HPO_4, 2 mM KH_2PO_4, pH 7.4).

2.3. Cell Lysis and Enzyme Assays

1. Passive reporter lysis buffer (Promega, Madison, WI, USA).
2. Luciferase Assay System (Promega).

3. β-Galactosidase reaction buffer: 1 mM $MgCl_2$, 50 mM β-mercaptoethanol, and 1 mg/ml *o*-nitrophenyl-β-D-galactopyranoside (ONPG) in 100 mM sodium phosphate, pH 7.
4. Reaction stop buffer: 1 M Na_2CO_3.
5. Protein Assay Reagent (Bio-Rad, Hercules, CA, USA).
6. Luminoskan Ascent (Thermo Fisher Scientific, Helsinki, Finland) luminometer and Multiskan EX Plate reader (Thermo Fisher Scientific) (*see* **Note 2**).
7. White, flat bottom 96-well polystyrene microplates (medium binding) for LUC and clear, flat bottom 96-well polystyrene microplates for β-galactosidase and protein assays.

3. Methods

Many reporter genes and constructs can be utilized for quantification of AR activity in cultured cells. Firefly luciferase (LUC) constructs driven by AR-regulated promoters, such as rat probasin promoter fragment (from –285 to +32 relative to TSS, encompassing three AREs) or 5.8-kb long 5′ flanking region of human PSA (containing both the upstream enhancer AREs and the proximal AREs), offer sensitive means to measure the activity of AR in cells of both non-prostate and prostate origin. Synthetic minimal constructs, e.g., pARE$_2$TATA-LUC driven by two AREs in front of a TATA sequence generally yield low LUC activity, but often (due to their low basal activities) give high androgen inductions in heterologous cells, such as COS-1 cells cotransfected with an AR expression plasmid. Delivery of plasmid DNA into mammalian cells for gene expression can be accomplished by various reagents or means. Mixing plasmid DNA with cationic lipids (lipofection) is often suitable for efficient transfer of DNA into a broad spectrum of cell lines. However, the efficiencies of transfection protocols and reagents can vary a lot between cell types. Therefore, different reagents and protocols may need to be compared for choosing the optimal protocol for the cell line of interest. We have found the TransIT®-LT1 Transfection Reagent, containing a proprietary blend of a polyamine, a histone, and a lipid, suitable for efficient transfection of COS-1, HeLa, and HEK293 cells. However, in our experience, this reagent is not optimal for transfection of VCaP or LNCaP cells. For these prostate cancer cells, jetPEI™ transfection reagent, containing a linear polyethylenimine derivative, provides a more effective and reproducible gene delivery. Stable cell lines harboring integrated reporter genes should in principle offer a practical and more reproducible means to measure AR activity. However, in our hands, cell lines with integrated reporter genes have yielded poor androgen inductions which decline during the passage of the cells. Also transient transfection-based reporter assays can be easily scaled to 96-well plates and (with the aid of robotics to 384-well plates) used for high-throughput applications, such as for screening novel AR agonists and antagonists. Reporter gene assays can also be utilized to complement the AR-DNA binding data from chromatin immunoprecipitation (ChIP) assays; for example, to prove that the identified, often very distally located, AR-binding regions can indeed function as androgen-induced enhancers, i.e., that these DNA regions confer androgen induction to a reporter gene. The protocol below describes the usage of ~300-bp enhancer regions of *FKBP51* containing compound AREs (identified through ChIP assays) as drivers of LUC and compares their activity to that of a LUC construct driven by the 5.8-kb 5′-flanking region of the *PSA*.

3.1. Cell Culture and Transfection

1. COS-1 cells are passaged (1:8) by trypsinization at near confluency (twice a week) for new maintenance cultures on 10-cm dishes. Experimental cultures are seeded onto 12-well plates (140,000 cells with 2 ml medium/well). The seeded experimental cultures are left to settle in the maintenance culture media with 10% FBS for 24 h. The culture media is replaced with 1 ml of the transfection culture media containing 2.5% charcoal-stripped FBS. The cells are transfected 4 h after the medium replacement. VCaP cells are passaged (1:2) by trypsinization at near confluency (once a week) for new maintenance cultures on 10-cm dishes. Remove trypsin by centrifugation at 69 × *g* for 4 min and resuspend in fresh medium before seeding. Fresh maintenance medium is changed once a week to the maintenance cultures. Experimental cultures are seeded onto 12-well plates (320,000 cells/well) in 1 ml of medium. The seeded experimental cultures are left to settle in maintenance culture media with 10% FBS for 48 h. The culture media is replaced with the 1 ml transfection culture media containing 2.5% charcoal-stripped FBS 4 h prior to transfection.
2. Transfection of COS-1 cells is performed using TransIT®-LT1 transfection reagent. Dilute the plasmid stocks (1 μg/μl) 1:10 with H_2O, if needed, so that the pipetting volumes are practical. The following volumes and DNA amounts are required for one sample (well): Mix well by pipetting gently 1.5 μl of TransIT®-LT1 with 97.6 μl of DMEM in a 1.5-ml microcentrifuge tube and incubate for 5 min. Add 0.2 μl of 1:10 diluted pSG5-hAR, 0.2 μl

of 1:10 diluted pCMVβ (to yield 20 ng of both plasmids per well), and 0.5 μl (500 ng) of LUC construct (*see* **Note 3**). Mix by gentle vortexing and incubate for 20 min at room temperature. Distribute the transfection mixture evenly by dropwise pipetting to the cells and incubate the cells for 24 h (*see* **Note 4**). Transfection of VCaP cells is carried out using jetPEI™ transfection reagent according to manufacturer's instructions. The following volumes and DNA amounts are required for one sample (well): Mix well by pipetting gently 4 μl of jetPEI™ with 46 μl of 150 mM NaCl in a 1.5-ml microcentrifuge tube. Pipette 1.8 μl (1.8 μg) of LUC construct and 0.2 μl (0.2 μg) pCMVβ to another 1.5-ml microcentrifuge tube, and add 48 μl of 150 mM NaCl. Mix by vortexing briefly. Pipet the jetPEI™–NaCl mixture to plasmid-containing tube, mix by gentle vortexing, and incubate for 15 min at room temperature (*see* **Note 5**). Distribute the transfection mixture evenly by dropwise pipetting to the cells and incubate the cells for 24 h.

3. Androgen treatment: Dilute 10 μM R1881 stock in DMEM (1.2 μl 10 μM R1881 and 98.8 μl DMEM) and add 100 μl of the dilution (or the same volume of DMEM containing the same volume of vehicle, EtOH) to the well. After 24-h incubation, the cells are ready for the reporter assays.

3.2. Measurement of Reporter Activity

1. Harvesting of the cells: Wash once with 1 ml of PBS (*see* **Note 6**). Add 100 μl of Passive Reporter Lysis Buffer to each well and lyse the cells by shaking (300 rpm, Heidolph Unimax 1010, Shire Hill, UK) for 15 min at room temperature. Transfer the lysates to 1.5-ml microcentrifuge tubes and centrifuge for 5 min at 16,100×*g* at room temperature (*see* **Note 7**).
2. Measurement of LUC activity: Pipette 10 μl of the above supernatants and three reagent blanks (10 μl of lysis buffer) to white 96-well plates. Add with 12-channel pipette 30 μl of luciferase assay system to each well and mix by gently tapping the plate, so that the reagent covers the whole bottom of the well. Transfer the plate to luminometer and measure the LUC activity with reading time 1 s and PMT voltage 924 (default) (*see* **Note 8**).
3. Measurement of β-galactosidase activity: Pipette 10 μl of each sample and three reagents blanks (10 μl of lysis buffer) to clear 96-well plates. Add with 12-channel pipette 65 μl of assay reagent to each well and mix by gentle shaking. Incubate the plate at 37°C until the samples become clearly

yellow (*see* **Note 9**). Add 125 μl of the reaction stop buffer by tapping the plate and measure the absorbance in a plate reader at 420 nm.

4. Measurement of protein concentration: Pipette 10 μl of each sample and three reagent blanks (10 μl of lysis buffer) to clear 96-well plates. Add with 12-channel pipette 200 μl of 1:5 sterile H_2O-diluted assay reagent to each well and mix by gentle shaking. Incubate the plate for 5 min at room temperature and measure the absorbance in microplate reader at 595 nm (*see* **Note 10**).
5. Calculation of relative (β-galactosidase- or protein-normalized) LUC activity: First subtract the mean of the blank measurements from sample values for the reduction of background. Divide the background-corrected LUC values by the corresponding β-galactosidase or protein values to normalize the differences in the transfection efficiency (β-galactosidase activity) or in the cell numbers (protein) between the samples (*see* **Note 11**). **Figure 5.1** shows results with four different AR-regulated LUC reporters in COS-1 and VCaP cells. The relative LUC activities of the three *FKBP51* enhancers studied differ between the two cell lines.

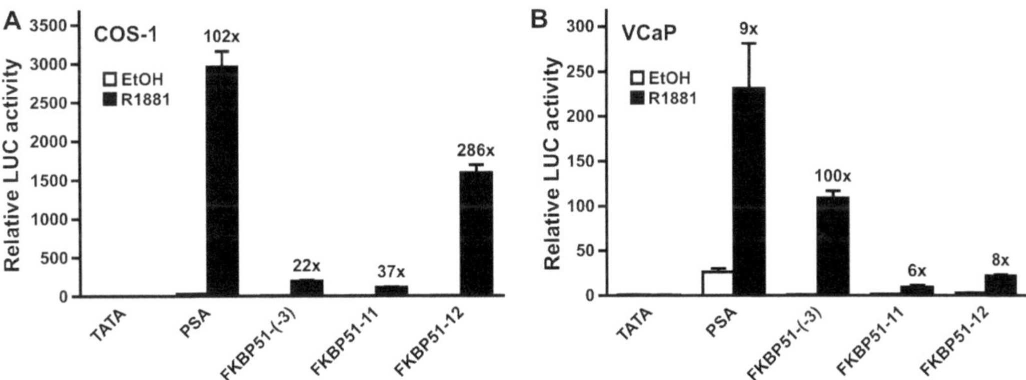

Fig. 5.1. Transcriptional activity of AR as assessed by luciferase reporter gene assays. (**a**) COS-1 or (**b**) VCaP cells were transfected with pPSA5.8-LUC, pTATA-LUC (without added AREs) or LUC constructs driven by different 0.3-kb long *FKBP51* enhancer fragments (–3, 11, and 12) harboring compound AREs. pCMVβ was cotransfected as an internal control. For COS-1 (**a**) analyses, pSG5-hAR was cotransfected with the reporter constructs. The cells were treated with vehicle (EtOH) or 10 nM R1881 for 24 h before harvesting the cells for reporter analyses. β-galactosidase activity was used for normalization of transfection efficiency. Results are shown as relative LUC activity, with the activity of pTATA-LUC in the absence of R1881 set as 1, and fold inductions of androgen-treated samples in the relation to the activity of ethanol-treated samples are shown above the columns (*see* **Note 12**). Columns represent the mean ± SD of three independent experiments.

4. Notes

1. Natural androgens, testosterone, and DHT, for AR induction can be purchased from, e.g., Steraloids Inc. (Newport, RI, USA).
2. There is a wide selection of suitable luminometers and plate readers on the market.
3. Renilla luciferase-encoding plasmids (e.g., pRLUC-C1, BioSignal Packard, Montreal, Canada) can be used instead of β-galactosidase-encoding plasmids as internal control reporters. Renilla LUC activity can be assayed in the presence of firefly LUC by using, e.g., Promega's Dual-Luciferase Reporter Assay System. The assay is very sensitive, and 1 ng of pRLUC-C1 is a sufficient transfection dose for a 12-well plate well.
4. Prepare one TransIT®-LT1-DMEM mixture that is sufficient for all the wells to be transfected. Depending on the experimental setup, e.g., if different LUC constructs are compared in the same experiment, it may be practical to produce a master mix of pSG5-hAR and pCMVβ, and divide the produced master mix (99.5 μl/well) into 1.5-ml microcentrifuge tubes for each LUC construct.
5. Prepare one jetPEI™–NaCl mixture that is sufficient for all the wells to be transfected. Depending on the experimental setup, e.g., if LUC constructs are compared in the same experiment, it may be practical to prepare a pCMVβ-containing 150 mM NaCl solution, and divide the produced mix (48.2 μl/well) into 1.5-ml microcentrifuge tubes for each LUC construct.
6. One may skip this washing step without significantly compromising the reporter gene activity. If the wells are not washed with PBS, carefully drain the medium.
7. A centrifuge capable of handling 96-well plates is practical, if available. Instead of passive lysis buffer, reporter lysis buffer (Promega) can be used, but make sure to follow the protocol provided by the manufacturer for the latter reagent.
8. The LUC assay gives linear results over several orders of magnitude of enzyme concentration. The luciferase assay reagent generates light that is nearly constant for several minutes and a luminometer equipped with an automatic reagent injector is not necessary. However, after the addition of the reagent, it is recommended to measure the plate as soon as possible.

9. The 420-nm filter is optimal for the measurement of the reaction absorbance, but the measurement can also be carried out using a 405-nm filter. The color development in the assay is linear up to absorbance 1.4. Stopping of the reactions is not mandatory, but it increases the sensitivity and accuracy of the assay. If the color development occurs very rapidly, the plate can be incubated at a lower temperature using pre-cooled reaction buffer.
10. The linear range of the protein assay is from 0.05 to ~0.5 mg/ml. At least, if the absorbance is above 1.5 absorbance units, the assay should be rerun with less amount of sample (~2.5–5 μl).
11. Our Multiskan plate readers are connected to a PC and the measurement data are transferred to an Excel spreadsheet application that directly subtracts the reagent background and calculates both β-galactosidase and protein-normalized luciferase values. If coregulator activity is studied in reporter gene assays by cotransfecting expression plasmids, the coexpressed coregulator sometimes influences the internal reporter activity. If this is the case, the protein concentration offers a more reliable means to normalize the LUC activity.
12. Since the transfection efficiency can vary between experiments, especially if they are carried out during a longer period of time, direct enzyme activity values cannot necessarily be used to compile data from different experiments. The usage of relative reporter (LUC) activity values can be used to circumvent this obstacle. It is of note that the reporter inductions by androgen are usually stable between experiments. In LNCaP cells, the pPSA5.8-LUC yields ~100 times stronger androgen induction than in VCaP cells. This is largely due to lower "basal" (in the absence of androgen) activity of the reporter in LNCaP cells.

Acknowledgments

This work was supported by grants from the Academy of Finland, Finnish Cancer Organisations, and Sigrid Jusélius Foundation. We thank Merja Räsänen for skilful technical assistance.

References

1. Gao, W., Bohl, C.E., Dalton, J.T. (2005) Chemistry and structural biology of androgen receptor. *Chem Rev* **105**:3352–3370
2. Knudsen, K.E., Scher, H.I. (2009) Starving the addiction: new opportunities for durable suppression of AR signaling in prostate cancer. *Clin Cancer Res* **15**:4792–4798
3. Claessens, F., Verrijdt, G., Schoenmakers, E., Haelens, A., Peeters, B., Verhoeven, G., Rombauts, W. (2001) Selective DNA binding by the androgen receptor as a mechanism for hormone-specific gene regulation. *J Steroid Biochem Mol Biol* **76**: 23–30
4. Kouzarides, T. (2007) Chromatin modifications and their function. *Cell* **128:**693–705
5. Kallio, P.J., Poukka, H., Moilanen, A., Jänne, O.A., Palvimo, J.J. (1995) Androgen receptor-mediated transcriptional regulation in the absence of direct interaction with a specific DNA element. *Mol Endocrinol* **9**: 1017–1028
6. Heemers, H.V., Tindall, D.J. (2007) Androgen receptor (AR) coregulators: a diversity of functions converging on and regulating the AR transcriptional complex. *Endocr Rev* **28**:778–808
7. Cleutjens, K.B., van der Korput, H.A., van Eekelen, C.C., van Rooij, H.C., Faber, P.W., Trapman, J. (1997). An androgen response element in a far upstream enhancer region is essential for high, androgen-regulated activity of the prostate-specific antigen promoter. *Mol Endocrinol* **11**:148–161
8. Mostaghel, E.A., Page, S.T., Lin, D.W., Fazli, L., Coleman, I.M., True, L.D., Knudsen, B., Hess, D.L., Nelson, C.C., Matsumoto, A.M. et al. (2007) Intraprostatic androgens and androgen-regulated gene expression persist after testosterone suppression: Therapeutic implications for castration-resistant prostate cancer. *Cancer Res* **67**:5033–5041
9. Makkonen, H., Kauhanen, M., Paakinaho, V., Jääskeläinen, T., Palvimo, J.J. (2009) Long-range activation of FKBP51 transcription by the androgen receptor via distal intronic enhancers. *Nucleic Acids Res* **37**:4135–4148
10. Korenchuk, S., Lehr, J.E., Mclean, L., Lee, Y.G., Whitney, S., Vessella, R., Lin, D.L., Pienta, K.J. (2001) VCaP, a cell-based model system of human prostate cancer. *In Vivo* **15**:163–168
11. Thompson, J., Lepikhova, T., Teixido-Travesa, N., Whitehead, M., Palvimo, J.J., Jänne, O.A. (2006) Small carboxyl-terminal domain phosphatase attenuates androgen receptor-dependent transcription. *EMBO J* **25**:2757–2767
12. Paakinaho, V., Makkonen, H., Jääskeläinen, T., Palvimo, J.J. (2010) Glucocorticoid receptor activates poised FKBP51 locus through long-distance interactions. *Mol Endocrinol* **24**:511–525

distribution of DNA-binding activities, affinity as well as association and dissociation rates. The AR can be present in cell or tissue extracts or can be produced in eukaryotic or prokaryotic expression systems. In our experience, the DNA binding by the AR is independent of its source; the concentration of the AR is much more important in determining whether binding can be demonstrated or not. The specificity of the interactions can be demonstrated by competition with specific and non-specific (cold) DNA and by inducing a super shift with specific antibodies against AR. Here, we describe EMSA for isolated AR-DBD as well as for full-length AR in extracts from transfected cells ectopically expressing AR or AR-positive prostate cell lines.

3.1. In Silico Search for AREs

A position-specific probability matrix was developed based on aligned known AREs (**Fig. 6.1**). To search for candidate AREs in any given genomic sequence, this sequence is entered in FASTA format into the “matrix scan” application on the ”regulatory sequence analysis tools” at http://rsat.ulb.ac.be/rsat/ (11). This application will provide a table of candidate AREs with *P*-values indicating the resemblance to the matrix (*see* **Note 8**).

3.2. Functional Analysis of AREs

3.2.1. Transient Transfection Experiments

Transient transfections are best performed at least three times independently and in triplicate on cells plated in 96-well plates at 10^4 cells per well. To each well approximately 120 ng DNA in total is added: 100 ng luciferase reporter, 10 ng beta-galactosidase expression vector, and 1–10 ng of an AR expression vector (*see* **Note 9**). The relative amount of each of the plasmids in the mix can be critical, and within one experiment, each well should receive the same amount of reporter and expression plasmids. After transfection and stimulation, the luciferase activity measured in the extracts is divided by the beta-galactosidase value which is used as internal control to monitor the transfection efficiency:

1. Day 1: The cells are seeded into 96-wells at a density of 10^4 cells/well in the appropriate medium supplemented with 5% DCC.

	−7	−6	−5	−4	−3	−2	−1	0	1	2	3	4	5	6	7
A	**77**	0	**80**	**100**	0	**73**	12	23	31	42	4	38	38	12	27
C	0	0	0	0	**100**	4	15	39	15	19	11	12	8	**73**	27
G	19	**100**	8	0	0	0	42	19	23	4	**77**	0	31	0	4
T	4	0	12	0	0	23	31	19	31	35	8	50	23	15	42

Fig. 6.1. Position-Specific Probability Matrix of AREs. This matrix is based on 26 AREs for which in vitro binding in EMSA and functionality in transient transfections was available from literature or our own work (6). Positions are relative to the central base of the spacer. For each position the percentage of occurrence is calculated for each base. Percentages above 70 are highlighted in *bold*.

2. Day 2: Cells are transfected with a mixture of 100 ng reporter plasmid and 1–10 ng expression plasmids for the AR and 5–10 ng pCMV-beta-galactosidase expression plasmid per well. Prepare the DNA mixes for transfection of *n* wells in a small volume (typically 10 μL). Per microgram of DNA mix, 1.5 μL GeneJuice transfection reagent is added to serum-free medium (calculated at 5 μL per well) and left at 25°C for 5 min. Then, $n \times 5$ μL of GeneJuice medium mix is added to the DNA mixes and left at room temperature for 15–30 min before delivering $1/n$ of the mix to each of targeted wells.
3. Day 3: The medium is replaced by medium with or without 10 nM of the synthetic androgen methyltrienolone.
4. Day 4: The medium is aspirated and cells are lysed in 25 μL 1× passive lysis buffer. From each well, 2.5 μL is transported to a white 96-well pit. For the luciferase assay, each well is injected with 10 μL of luciferase assay reagent and measured for 500 ms in the Luminoskan Ascent luminometer. For the beta-galactosidase assay, 20 μL galacton + 1/100 diluted in Reaction Buffer Diluent is added per 2.5 μL cell lysate. After 1 h at room temperature in the dark, 30 μL Accelerator II is added. Luminescence is measured for 500 ms.
5. Relative luciferase activities are calculated as luciferase luminescence divided by the beta-galactosidase luminescence activity. Induction factors indicate the hormone responsiveness and are calculated by dividing the relative luciferase activity from extracts of cells stimulated by hormone by the relative luciferase activity in extracts from cells that were not stimulated.

3.2.2. Making Stable Cell Lines

Transient transfections have been criticized for the fact that transient reporter DNA does not reflect the chromatin status of the endogenous genes, and this could affect the activity of an ARE. To circumvent this problem, one can make integrated reporter constructs.

HEK 293-derived cell lines were made using the Flp-In T-REx system of Invitrogen. The HEK 293 host cell line contains a genomically integrated FRT site which serves as a binding and cleavage site for the FLP recombinase. This FRT site has been inserted upstream of an ATG initiation codon of the LacZ/Zeocin gene whose expression is controlled by the SV40 early promoter. A hygromycin resistance gene with a FRT site embedded in the 5′ coding region was inserted in the reporter plasmids. Cotransfection of the reporter vectors with an expression vector for the FLP recombinase enables a homologous recombination between the genomic and the plasmid FRT sites. After this recombination the SV40 promoter and ATG initiation codon are in frame with the hygromycin resistance gene with a concomitant loss of the beta-galactosidase expression and zeocin resistance. This allows efficient selection of stable cell lines:

1. Before transfection, determine the optimal concentration hygromycin B for selection of the Flp-In T-REx cell line (normally between 10 and 400 μg/mL).
2. Day 1: cells are seeded into 6-wells at a density of 3×10^5 cells/well in DMEM medium supplemented with 10% FCS and 15 μg/mL blasticidin.
3. Day 2: cells are transfected with the reporter construct and the Flp recombinase in a 9:1 ratio using the GeneJuice transfection reagent (see **Section 3.2**). The cells are kept overnight at 30°C, since this is the optimal temperature for the recombinase.
4. Day 3: medium is replaced by DMEM 10% FCS supplemented with 15 μg/mL blasticidin. From this moment on, cells are always incubated at 37°C.
5. Day 4: Split the cells, targeting 25% confluency in 6-well plates.
6. Day 5: cells will be attached to the culture dish and medium is removed and replaced by selective medium containing blasticidin and the pre-determined concentration of hygromycin (see point 1).
7. Next days: replace the medium every 3–4 days until foci of cell growth can be identified.
8. Mark the position of the colonies on the bottom of the culture dish. Aspirate the medium from the culture disk. Take 1–10 μL of medium and pipet up and down above the indicated colonies. Select at least 10 colonies for further development.
9. Replate the cells first to 24-well plates, later to 6 cm plates when the cells are starting to become confluent, replacing the medium every 3–4 days.
10. Screen for colonies which have lost the beta-galactosidase activity. This indicates that the reporter gene is inserted at the FRT site. Up to half of the colonies can result from insertions elsewhere.

3.2.3. Testing Androgen Responsiveness of Stable Cell Lines

HEK-293 cells do not express the endogenous AR. Therefore, the cells need to be transfected with an expression plasmid for the human AR (see **Section 3.1** for transfection protocol). The amount of transfected expression vector should be optimized. Cells are plated in 96-well plates at 10^4/well, transfected with an AR expression plasmid and the next day stimulated or not with 10 nM R1881. After 24 h stimulation, cell extracts can be tested for luciferase assay and induction factors can be calculated as in **Section 2.1**.

Alternatively, we made stable cell lines based on an in-house made Flp-In T-REx cell line with an integrated hAR expression cassette. This ensures that all cells are AR positive, express the protein to the same level, and avoid the need for transient transfections.

3.3. Preparation of Protein Extracts from COS-7 or VCaP Cells

COS-7 cells are extremely useful for the preparation of ectopic protein expression since they are easy to transfect and since they express the SV40 large T antigen which enables amplification of expression plasmids with an SV40 ORI. This results in high expression levels of proteins like the human AR, enabling EMSA with whole cell or nuclear extracts. Alternatively, EMSA can be performed with protein extracts of, e.g., VCaP cells expressing high levels of endogenous AR:

1. Day 1: 5×10^6 COS-7 or VCaP cells are plated in a 15 cm petri dish.
2. Day 2: COS-7 cells are transfected with 7 μg of AR expression plasmid; VCaP cells are not transfected.
3. Day 3: 10 nM R1881 is added to the medium (*see* **Note 10**). After 1 h incubation with R1881, wash the cells twice with 5 mL ice cold PBS. From this step on, work at 4°C.
4. Collect the cells in 5 mL PBS by scraping the plate with a rubber policeman.
5. Centrifuge for 1 min at 3000×*g* to collect the cells in the pellet.
6. Proceed to either whole cell protein extraction or nuclear extraction.

3.3.1. Preparation of Whole Cell Extracts for EMSA

1. Resuspend the cell pellet in 100 μL extraction buffer supplemented with PMSF, DDT, and protease inhibitor cocktail.
2. Freeze and thaw the solution three times by shifting between liquid nitrogen and ice. Vortex between each cycle.
3. Centrifuge for 10 min at 20,000×*g* to pellet the cellular debris.
4. Aliquot the supernatant in 20 μL aliquots and store at –80°C.
5. The protein concentration of the solution is measured with Coomassie Brilliant Blue protein assay (Thermo Scientific).
6. Protein expression and quality should be verified by SDS-PAGE and Western blot for the AR.

3.3.2. Preparation of Nuclear Extracts for EMSA

1. Resuspend the cell pellet in 500 μL buffer A.
2. Leave on ice for 10 min and then vortex for 30 s.

3. Centrifuge 1 min at 8000×*g* at 4°C.
4. Discard the supernatant and resuspend the nuclear pellet in 100 μL of buffer C.
5. Leave on ice for 20 min for high-salt extraction of the proteins.
6. Vortex for 30 s.
7. Centrifuge for 2 min at 20,000×*g* to remove cellular and nuclear debris.
8. Aliquot the supernatant in 20 μL aliquots and snap freeze in liquid nitrogen. Store at –80°C.
9. Protein concentrations are measured with Coomassie Brilliant Blue protein assay.
10. Protein expression and receptor integrity should also be verified by SDS-PAGE and Western blotting with anti-AR antibody.

3.4. Prokaryotic Expression of AR-DBD

The cDNA fragment encoding the DNA-binding domain of the AR was cloned in the pGEX2TK vector (Promega) in frame with the GST coding cassette, under control of the lac promoter (12). This vector is transformed in BL21 *E. coli* strain. The expression of the fusion protein is induced with IPTG, cells are lysed and the protein purified on a glutathione column. Finally, the thrombin site between the GST and DBD is cleaved, so that the DBD can be eluted.

1. Day 1: inoculate 5 mL of LB-medium supplemented with ampicillin.
2. Day 2: transfer the culture to 50 mL of the same medium and grow at 37°C until an OD_{600} of 0.5 is reached. Add IPTG up to a final concentration of 0.5 mM.
3. After 2 h of growth at 20°C with shaking, collect the cells by centrifugation at 5000×*g* for 10 min. From here on perform all steps at 4°C or on ice.
4. Resuspend the pellet in 3 mL PBS supplemented with protease inhibitor mix, and lyse the cells by sonication in a Bioruptor (Diagenode, Luik, Belgium) with 10 cycles of 40 s at amplitude 40 and 20 s rest on ice between cycles.
5. Centrifuge the solution at 20,000×*g* for 10 min.
6. Add glutathione beads (prewashed in PBS) to the supernatant and incubate for 1 h.
7. Centrifuge to collect the column in the tip of the tube.
8. Wash the beads five times with TBS and resuspend in 300 μL TBS. This mix can be kept at –20°C, provided glycerol is added to a final concentration of 50%.

9. To separate the DBD from the GST on the column, it is incubated for 30 min with 1 U of thrombin dissolved in PBS.
10. Spin the tube, so that the supernatant containing the DBD can be collected. Wash the column with 100 μL TBS and pool the supernatant with the earlier fraction.
11. The quality of the DBDs can be verified by SDS-PAGE, the quantity can be determined with the Coomassie Brilliant Blue protein assay.

3.5. Labeling Oligos

The double-stranded oligonucleotides containing the ARE sequences are labeled with [α-^{32}P]dNTP by a fill-in reaction with the Klenow fragment of DNA-polymerase I. N depends on the first nucleotide that can be filled by the Klenow fragment:

1. Mix the complementary oligonucleotides at a 1:1 molar ratio at 1 pmol/μL in annealing buffer.
2. The oligos are annealed by incubation of the samples at 94°C for 2 min, followed by slowly cooling to room temperature.
3. Five picomole of the annealed oligos is incubated with 5 units of the Klenow fragment and 3 μL of [α-^{32}P]dNTP (3000 Ci/mmol, Perkin Elmer) in a final volume of 20 μL. The reaction mix is incubated for 30 min at 37°C, cold dNTPs are added, and the reaction is allowed to continue for another 30 min.
4. The radioactive probe is separated from free dNTPs on a sephadex column as described in Maniatis et al. (13).
5. Determine the radioactivity in 3 μL of the sample in a β-counter.
6. Dilute the radioactive probe to 20,000 counts per minute (cpm) per microliters.

3.6. Electrophoretic Mobility Shift Assay

In the EMSA, the proteins are incubated with DNA to allow binding, and subsequently the free probe is separated from the bound probe by electrophoresis. Specificity of AR binding is demonstrated by the induction of a super shift by an AR-specific antibody:

1. Prepare a 5% AA/BA 29/1 mix supplemented with 0.25x TBE and 0.05% Triton X-100. Just before pouring the gel, add APS to 0.0625% and TEMED to 0.3%. Mix briefly and pour between the glass plates to polymerize.
2. The protein extracts or purified DBDs are incubated for 20 min on ice with radiolabeled probe (20,000 cpm) in D100 buffer, supplemented with 0.05% Triton X-100, 1 mM DTT, and poly (dIdC) (*see* **Note 11**).

a. The gel is pre-run for 20 min at 120 V (= 10 V/cm).

b. Protein-bound probe and unbound probe are resolved by electrophoresis at 10 V/cm for 2 h.

c. The presence of the AR in the shifted band can be demonstrated by a super shift induced by adding 1 μL of a threefold dilution of an anti-AR antibody (in-house made rabbit antiserum against the first 20 amino acids of the AR) (*see* **Note 9**).

d. Gels are dried using a vacuum dryer for 1.5 h at 70°C.

e. The radioactive signal is visualized and quantified when necessary with a STORM 840 PhosphorImager device (Molecular Dynamics Inc.) (*see* **Note 12**).

4. Notes

1. In our hands, the source of AR does not change its affinity or selectivity for DNA. The main objective is to obtain a protein extract that contains sufficient AR to enable demonstration of DNA binding in EMSA.
2. The pGL4 version of the luciferase vector (Promega) was also used. Although it gave comparable luciferase values, the sequence of this ORF is richer in CpG motifs and hence may be more sensitive to DNA methylation effects.
3. At least 6 bp flanking sequence is needed to enable demonstration of AR and AR-DBD binding (10). Therefore, we always clone ARE sequences including these flanking bases. Cloning two or more copies of a candidate ARE upstream of a reporter gene will increase the responsiveness, while a single copy of a weak ARE will result in very low induction factors.
4. Methyltrienolone, also known as R1881, is used because testosterone and dihydrotestosterone can sometimes give less reliable results. This is probably due to metabolism of the natural ligands.
5. PMSF precipitates should be dissolved before use, by vortexing at room temp for a time as short as possible.
6. Acrylamide is a neurotoxin and handling should always be performed wearing gloves.
7. Different versions of the luciferase coding cassette are in use. For this purpose, we prefer the pGL3 series from Promega which encode a luciferase that is not targeted to lysosomes and putative transcription factor binding sites in

the coding sequence are modified. The pGL4 cassette has been modified further to reduce the number of putative transcription factor binding sites, but has an increased CpG content, which might interfere with gene expression.

8. In a leave-one-out analysis, this approach was able to predict approximately 75% of proven AREs (6).
9. We use human AR cloned into the vector pSG5 either as a wild-type sequence or as a N-terminally Flag-tagged version. The latter enables detection by anti-Flag antibodies, which is an advantage when different mutants or receptors are compared (14). All plasmids are available by e-mail request to frank.claessens@med.kuleuven.be.
10. The ligand will activate and stabilize the AR and induce its nuclear translocation. As a consequence, it is not possible to study the effect of hormone on DNA binding by EMSA.
11. In case extracts are used, add 50 ng/μL poly(dIdC), and in case purified AR-DBD is used, add 2.5 ng/μL poly(dIdC).
12. Quantifying EMSA signals enables the determination of affinity constants as has been described (13), but this has only been possible for the DBDs, mainly because the concentration of the AR in nuclear or whole cell extracts is too low and because many other DNA-binding proteins are present.

References

1. Shaffer, P.L., Jivan, A., Dollins, D.E., Claessens, F., and Gewirth, D.T. (2004) Structural basis of androgen receptor binding to selective androgen response elements. *Proc. Natl. Acad. Sci. USA* **101**, 4758–63.
2. Claessens., F., Alen, P., Devos, A., Peeters, B., Verhoeven, G., and Rombauts, W. (1996) The androgen-specific probasin response element 2 interacts differentially with androgen and glucocorticoid receptors. *J. Biol. Chem.* **271**, 19013–6.
3. Claessens, F., Rushmere, N.K., Davies, P., Celis, L., Peeters, B., and Rombauts, W.A. (1990) Sequence-specific binding of androgen-receptor complexes to prostatic binding protein genes. *Mol. Cell. Endocrinol.* **74**, 203–12.
4. Claessens, F., Celis, L., Peeters, B., Heyns, W., Verhoeven, G., and Rombauts W. (1989) Functional characterization of an androgen response element in the first intron of the C3(1) gene of prostatic binding protein. *Biochem. Biophys. Res. Commun.* **164**, 833–40.
5. Wang, Q., Li, W., Liu, X.S., Carroll, J.S., Jänne, O.A., Keeton, E.K., Chinnaiyan, A.M., Pienta, K.J., and Brown M. (2007) A hierarchical network of transcription factors governs androgen receptor-dependent prostate cancer growth. *Mol. Cell* **27**, 380–92.
6. Denayer, S., Helsen, C., Thorrez, L., Haelens, A., and Claessens, F. (2010) The rules of DNA recognition by the androgen receptor. *Mol. Endocrinol.* **24**, 898–913.
7. Haelens, A., Verrijdt, G., Callewaert, L., Peeters, B., Rombauts, W., and Claessens F. (2001) Androgen-receptor-specific DNA binding to an element in the first exon of the human secretory component gene. *Biochem. J.* **353**, 611–20.
8. Alen, P., Claessens, F., Verhoeven, G., Rombauts, W., and Peeters, B. (1999) The androgen receptor amino-terminal domain plays a key role in p160 coactivator-stimulated gene transcription. *Mol. Cell. Biol.* **19**, 6085–97.
9. Verrijdt, G., Schauwaers, K., Haelens, A., Rombauts, W., and Claessens, F. (2002)

Functional interplay between two response elements with distinct binding characteristics dictates androgen specificity of the mouse sex-limited protein enhancer. *J. Biol. Chem.* **277**, 35191–201.

10. Moehren, U., Denayer, S., Podvinec, M., Verrijdt, G., and Claessens, F. (2008) Identification of androgen-selective androgen-response elements in the human aquaporin-5 and Rad9 genes. *Biochem. J.* **411**, 679–86.
11. Turatsinze, J.V., Thomas-Chollier, M., Derance, M., and van Helden, J. (2008) Using RSAT to scan genome sequences for transcription factor binding sites and cis-regulatory modules. *Nat. Protoc.* **3**, 1578–88
12. Schoenmakers, E., Alen, P., Verrijdt, G., Peeters, B., Verhoeven, G., Rombauts, W., and Claessens, F. (1999) Differential DNA binding by the androgen and glucocorticoid receptors involves the second Zn-finger and a C-terminal extension of the DNA-binding domains. *Biochem. J.* **341**, 515–21.
13. Maniatis, T., Fritsch, E.F., and Sambrook, J. (1982) Molecular cloning: A laboratory manual, Cold Spring Harbor Laboratory Press, New York, NY.
14. Schauwaers, K., De Gendt, K., Saunders, P.T., Atanassova, N., Haelens, A., Callewaert, L., Moehren, U., Swinnen, J.V., Verhoeven, G., Verrijdt, G., and Claessens, F. (2007) Loss of androgen receptor binding to selective androgen response elements causes a reproductive phenotype in a knockin mouse model. *Proc. Natl. Acad. Sci. USA* **104,** 4961–6.

Chapter 7

In Vitro and In Vivo Silencing of the Androgen Receptor

Helen Cheng, Eric Leblanc, and Paul S. Rennie

Abstract

The androgen receptor (AR) plays a pivotal role in the progression of prostate cancer from the androgen-dependent to the castration-resistant state, making it a potential target for therapy. In this chapter, we describe the preparation and use of sublines of LNCaP and C4-2 human prostate cancer cells which have been engineered to stably express a doxycycline (DOX)-inducible AR shRNA in order to study the in vitro and in vivo effects of AR knockdown.

Key words: Androgen receptor, RNA interference, doxycycline-inducible short hairpin RNA (shRNA), AR knockdown, prostate cancer, androgen dependence, castration resistance.

1. Introduction

The lethal end stage of prostate cancer occurs after progression of the disease from androgen dependence to castration resistance, upon failure of various forms of androgen withdrawal therapy. Currently, there are no curative treatment options for castration-resistant prostate cancer (CRPC), which has a median survival of approximately 18 months (1). Gene expression profiles have shown that the androgen receptor (AR) gene is the only one that is consistently upregulated when prostate xenograft tumors become castration resistant, suggesting a pivotal role of the androgen receptor in CRPC (2). Cross talk between AR-signaling pathways and other non-steroidal signaling pathways can lead to the activation of the AR by agents other than androgens (3–5), which consequently results in prostate cancer growth despite medical or surgical castration. Accordingly, the central role played by the AR makes it an attractive therapeutic

F. Saatcioglu (ed.), *Androgen Action*, Methods in Molecular Biology 776,
DOI 10.1007/978-1-61779-243-4_7, © Springer Science+Business Media, LLC 2011

target in the treatment of CRPC. Indeed, it has been shown that knockdown of the AR using an RNA interference (RNAi) approach in the androgen-responsive LNCaP or in the castration-resistant C4-2 human prostate cancer xenograft models results in impeded tumor growth, decreased serum prostate-specific antigen (PSA), delayed tumor progression to castration resistance (6), and, in some cases, complete tumor regression (7).

RNA interference (RNAi) is a powerful tool designed to inhibit expression of specific target genes (8). RNAi can be used in the form of a synthetic 21-nucleotide-long, double-stranded short interfering RNA (siRNA) or a plasmid-based short hairpin RNA (shRNA) (9). While the action of the siRNA is transient, a lentiviral vector-mediated shRNA can be expressed constitutively in a cell and thereby achieve long-term silencing of target genes. In this chapter, we describe the design and construction of lentiviral vectors carrying shRNA-targeting AR and subsequent stable expression of the shRNA in LNCaP prostate cancer cells. We also describe the creation of a line of LNCaP and C4-2 cells containing doxycycline (DOX)-inducible AR shRNA, in which the shRNA can be turned on when these cells or their xenograft tumors are exposed to the antibiotic DOX.

2. Materials

2.1. Software Analysis

1. Xeragon software (QIAGEN): http://www1.qiagen.com/Products/GeneSilencing/CustomSiRna/SiRnaDesigner. Interagon
2. BLAST analysis: http://blast.ncbi.nlm.nih.gov/Blast.cgi

2.2. Cloning of the AR and Scrm shRNAs into the Shuttle Vector pSHAG-1-tet

1. Plasmid pSHAG-1 was obtained from Dr Greg Hannon (Cold Spring Harbor Laboratory). *See* **Section 3.1** for details.
2. Standard buffers, restriction enzymes *Bse*RI, *Bam*HI, *Hin*dIII.
3. Ultrapure agarose.
4. QIAquick PCR Purification Kit (QIAGEN) (*see* **Note 1**).
5. Calf intestinal phosphatase.
6. QIAEX II Gel Extraction Kit (QIAGEN).
7. Oligonucleotides are synthesized by Integrated DNA Technologies (IDT).
8. Oligonucleotide annealing buffer: 10 mM Tris, pH 7.5, 100 mM NaCl (*see* **Note 2**).

9. 5× Ligase buffer and T4 DNA ligase.
10. Chemically competent *Escherichia coli* DH5α.
11. Luria-Bertani medium (LB) and Bacto-agar plates supplemented with 50 μg/mL of kanamycin.
12. QIAprep Spin Miniprep Kit (QIAGEN).
13. QIAfilter Plasmid Maxi Kit (QIAGEN).

2.3. Screening of AR shRNAs for Knockdown of AR in Transactivation Assays

1. LNCaP and C4-2 prostate cancer cell lines are maintained in RPMI 1640 (Invitrogen) supplemented with 5% fetal bovine serum (FBS) (Invitrogen) and penicillin/streptomycin (P/S) antibiotic (Invitrogen). The cells are cultured in a humidified 37°C incubator with 5% (V/V) CO_2.
2. 175-cm Tissue culture flasks, 10-cm dishes, 6-well and 96-well plates.
3. Lipofectin reagent (Invitrogen).
4. Charcoal-stripped fetal bovine serum (CSS).
5. Synthetic androgen methyltrienolone (R1881): 10^{-5} M in 100% ethanol (Perkin Elmer).
6. Luciferase Assay System (Promega).
7. BCA Protein Quantification Kit (Pierce).
8. Orion L microplate luminometer.

2.4. Generation of LNCaP and C4-2 Cell Lines Expressing DOX-Inducible AR and Scrm shRNAs

1. Gateway LR Clonase II Enzyme Kit (Invitrogen).
2. TE buffer: 10 mM Tris, pH 8.0, 1 mM EDTA.
3. One Shot Stbl3 chemically competent *E. coli* cells.
4. 293T human embryo kidney cells are maintained in Dulbecco's modified Eagle's medium (DMEM) supplemented with 5% FBS and P/S.
5. Poly-L-lysine solution (Sigma).
6. ProFection Mammalian Transfection System (Promega).
7. Vironostika HIV-1 Antigen Microelisa System (bio Mérieux).
8. 0.45-μm Microfilters.
9. Trypsin (0.25%) and ethylenediaminetetraacetic acid (EDTA).
10. Puromycin: 10 mg/mL in sterile ddH_2O. Dimethyl sulfoxide (DMSO) (*see* **Note 3**).
11. Dimethyl sulfoxide.
12. Blasticidin: 10 mg/mL in ddH_2O (Invitrogen).

2.5. Efficiency of AR Knockdown In Vitro

1. Doxycycline (DOX): 1 mg/mL in ddH_2O. Filter sterilized and stored in small aliquots at –20°C (BD Biosciences) (*see* **Note 4**).
2. TRIzol reagent (Invitrogen).
3. RIPA buffer: 0.5% Sodium deoxycholate, 0.1% SDS, and 1% IGEPAL. Dilute 1:1 with phosphate buffered saline (PBS).
4. Complete protease inhibitor cocktail tablets: Dissolve 1 tablet in 1 mL of sterile water (50×). Rotate at 4°C until dissolved (Roche).
5. Primary antibodies include AR (1:200) (AR441 or N20), PSA (1:200) (C-19), and vinculin (1:500) (Santa Cruz).
6. CellTiter 96 Aqueous Non-Radioactive Cell Proliferation Assay (MTS) (Promega).

2.6. Efficiency of AR Knockdown In Vivo

1. BD Matrigel (BD Biosciences).
2. Male athymic nude mice (Harlan).
3. Micro-hematocrit capillary tube (Fisher Scientific).
4. Prostate Specific Antigen Enzyme Immunoassay Test Kit (PSA ELISA).
5. Doxycycline hyclate: 20 mg/mL in ddH_2O (Sigma). Filter sterilized and stored at –20°C in small aliquots.
6. Buffered formalin (Fisher Scientific).
7. Antibodies for immunohistochemistry include AR (1:100) (AR N20) (Santa Cruz) and Ki67 (1:500) (Lab Vision Corporation).

3. Methods

Vector-based shRNAs have been used effectively to suppress expression of target genes via RNAi. The shRNA cassette (containing the promoter for RNA polymerase III (U6) and the shRNA) is cloned into a retrovirus vector (e.g. lentivirus) which is then used to produce viral particles that can infect recipient mammalian cells, generating cell lines that stably express the shRNA. However, continual constitutive expression of shRNA may have deleterious effects on cells, whereas an inducible system provides more controlled RNAi exposure. For this strategy, the shRNA cassette is modified to incorporate two tetracycline operator 2 ($TetO_2$) sites upstream of the U6 promoter for tetracycline-regulated expression of the shRNA. Two stable cell lines are generated: (1) a cell line that expresses the tetracycline repressor (TR) constitutively and (2) a TR-expressing cell line that expresses the shRNA of interest. Binding of the TR to the $TetO_2$ sites represses transcription of the shRNA and this repression is abrogated in the presence of the antibiotic tetracycline (or its more potent derivative, DOX), resulting in the expression of the shRNA. To generate a stable cell line that can be induced to express shRNAs, Invitrogen has recently developed a kit, BLOCK-iT Inducible H1 RNAi Entry Vector Kit (*see* **Note 5**).

To design shRNA sequences for the AR, it is necessary to use an algorithm such as the Xeragon software provided by QIAGEN. It is estimated that the probability of obtaining an effective shRNA is 1 out of 3, hence in our study we have designed and tested four AR shRNAs and one scrambled control and cloned them into the shuttle vector pSHAG-1-tet. Before going further, the most effective shRNA plasmid should be determined by its ability to suppress AR expression in transient transfection experiments, measured by AR transcriptional activity in transactivation assays (*see* **Section 3.4**). This shRNA is then used in the generation of an LNCaP cell line stably expressing the shRNA of interest. The effect of AR knockdown on AR expression and cell proliferation in vitro is then investigated to confirm its effectiveness.

3.1. Design of the Shuttle Vector for AR and Scrm shRNA Cloning

pSHAG-1 vector contains the Gateway pENTR/D-TOPO backbone (Invitrogen) with *att*L sites for easy transfer of the RNAi cassette into other suitable Gateway destination vectors. It also contains the human U6 promoter which is recognized by RNA polymerase III for efficient expression of the shRNA. The presence of *Bse*RI and *Bam*HI restriction sites enables directional cloning of the shRNA. pSHAG-1 was further modified (Dr Alice Mui, University of BC) to contain two tetracycline operator 2 ($TetO_2$) sites in tandem upstream of the U6 promoter for tetracycline-regulated expression of the shRNA (pSHAG-1-tet).

3.2. Design of the AR and Scrm shRNA Oligos

1. The sequence of the AR shRNA is designed using an algorithm provided by Xeragon. The sequence for the scrambled shRNA is selected randomly and all sequences are subjected to BLAST search to ensure that there is no sequence homology to any known genes. Core sequences for the four AR shRNAs and their positions in the coding region of the *AR* gene are listed in **Table 7.1** (*AR* accession no. NM_000044).
2. We designed the shRNA target sequence to be inserted into pSHAG-1-tet vector using *BseR1/BamH1* restriction sites (*see* **Table 7.1** for specific sequences). In the following steps, N represents any nucleotides and X represents their complement. In addition, the number (N_1, N_2 or X_1, X_2, etc.) shows the nucleotide position in the sequence.

Table 7.1
Sequences of the AR and scrambled shRNAs

AR shRNA sequences	Position in the coding region of the *AR* gene
AR shRNA 1: 5′-AAGCTCAAGGATGGAA GTGCAGTTAGGGCTG-3′	1106–1136
AR shRNA 2: 5′-AGCAGCAGGAAGC AGTATCCGAAGGCAGCAG-3′	1705–1735
AR shRNA 3: 5′-CACAGCCGAAGAAG GCCAGTTGTATGGACCG-3′	2432–2462
AR shRNA 4: 5′-GAAAGCACTGCTA CTCTTCAGCATTATTCCA-3′	3539–3569

Scrambled shRNA Sequence 5′-CCGTACCTACACGC AGCGCTGACAACAGT TT-3′

(a) Core sequence of 31 nucleotides obtained from Xeragon:
5′-N_1---N_{31}-3′ (*see* **Note 6**)

(b) Insert a "C" before N_{30} (*see* **Note 7**) to obtain the following sequence:
5′-N_1---N_{29} **C**N_{30} N_{31}-3′

(c) Get reverse complement of (b) to get 5′-$X_{31}X_{30}$**G**X_{29}---X_1-3′

(d) Remove $X_{31}X_{30}$ to get
5′-**G**X_{29}---X_1-3′

(e) Add 5′-GAAGCTTG-3′ to the 3′-end of (d) to get
5′-**G**X_{29}---X_1GAAGCTTG-3′ (*see* **Note 8**)

(f) To the end of (e), add (b) to get
5′-**G**X_{29}---X_1GAAGCTTGN_1---N_{29}**C**$N_{30}N_{31}$-3′

(g) Add RNA polymerase III terminator TTTTTT to (f) to get
5′-**G**X_{29}---X_1GAAGCTTGN_1---N_{29}**C**$N_{30}N_{31}$TTTTT T-3′ = oligo A

(h) Drop G from oligo A to get
5′-X_{29}---X_1GAAGCTTGN_1---N_{29}**C**$N_{30}N_{31}$TTTT TT-3′

(i) Get reverse complement of (h) to get
5′-AAAAAA$X_{31}X_{30}$**G**X_{29}---X_1CAAGCTTCN_1---N_{29}-3′

(j) To (i), add GATC to 5′-end to get
5′-GATCAAAAAA$X_{31}X_{30}$**G**X_{29}---X_1CAAGCTTCN_1---N_{29}-3′

(k) To (j), add CG to the 3′-end to get 5′-GATCAAAAAA$X_{31}X_{30}$**G**X_{29}---X_1CAAGCTTCN_1---N_{29}CG-3′=oligo B (*see* **Note 9**).

3.3. Cloning of the Four AR shRNAs into the Shuttle Vector pSHAG-1-tet

1. Digest pSHAG-1-tet with *Bse*RI and *Bam*HI:

	Volume (μL)
pSHAG-1-tet (10 μg)	10
10× React 2 (Invitrogen)	5
10× BSA	5
Sterile ddH_2O	25

Add 5 μL of *Bse*RI restriction enzyme and incubate for 1–2 h at 37°C. Heat denature the enzyme at 65°C for 20 min. Check to see if the vector is linearized by running 2 μL of digested product on an agarose gel. Purify the DNA using the QIAquick PCR Purification Kit and set up a second digest with *Bam*HI:

	Volume (μL)
*Bse*RI-digested pSHAG-1-tet	50
10× React 3 (Invitrogen)	6

Add 4 μL of *Bam*HI restriction enzyme and incubate for 1–2 h at 37°C. At the end of digestion, add 1 μL of calf intestinal phosphatase (CIP) and incubate at 50°C for 1 h. Add 0.5 μL of CIP and incubate for a further 30 min. Purify the DNA on a 1% agarose gel. Excise the DNA band and extract using the QIAEXII Gel Extraction Kit.

2. Anneal the upper and lower strands (oligos A and B) of the shRNA to make double-stranded oligos. Resuspend lyophilized oligos A and B in annealing buffer and dilute them to 250 ng/μL:

	Volume (μL)
Oligo A	1
Oligo B	1
Oligo annealing buffer	8

Heat to 95°C for 4 min in a thermal cycler and cool at the ramp rate of 0.1°C/s to 22°C.

3. Ligate double-stranded oligos into pre-digested pSHAG-1 vector at a molar ratio of 10:1 (insert:vector), taking into account the sizes of the oligos and the vector:

	Volume (μL)
Digested pSHAG-1-tet (100 ng/μL)	1
Annealed ds oligos (20 ng/μL)	1
5× Ligase buffer	2
Sterile ddH_2O	5.5
T4 DNA ligase	0.5

Incubate overnight at 16°C. Transform 5 μL of ligated product into chemically competent *E. coli* DH5α and plate onto Luria-Bertani (LB) plates containing 50 μg/mL kanamycin. On the following day, pick 10 kanamycin-resistant colonies and set up 5 mL overnight cultures in LB supplemented with 50 μg of kanamycin. Isolate plasmid DNA using QIAprep Spin Miniprep Kit. Perform diagnostic restriction enzyme digest with *Hin*dIII which cuts inside the hairpin to generate a linearized plasmid. Generate sequencing of 2–3 positive transformants to confirm the presence and correct orientation of the double-stranded oligonucleotide insert and its sequence integrity (*see* **Note 10**). Once the correct clone has been identified, make glycerol stock of the bacteria and store at −80°C for long-term storage.

4. Grow large-scale amounts (200 mL) of the plasmid DNA using the QIAfilter Plasmid Maxi Kit (*see* **Note 11**).

3.4. Screening of the AR shRNA for Effectiveness in Knocking Down the AR in a Transactivation Assay

1. Seed LNCaP cells at a density of 3 × 10^5 cells/well in a six-well plate.
2. Co-transfect the cells with the AR shRNA-expressing plasmid and the –6.1-kb PSA-luciferase reporter when cells are at 60% confluence (usually takes 1 or 2 days).
3. For each well, use 1 μg of scrm or AR shRNA in pSHAG-1-tet and 1 μg of–6.1-kb PSA-luciferase reporter. For each microgram of plasmid DNA, use 3 μL of lipofectin reagent.
4. To prepare a lipofectin:DNA mixture for a total of six wells, add 36 μL of lipofectin reagent to 360 μL of serum-free RPMI (no antibiotic) (ratio of lipofectin to medium is 1:10). Incubate the mixture for 40 min at room temperature.
5. In a 15-mL tube, mix DNA with medium at a ratio of 1:100, i.e., for 12 μg of DNA, add 1.2 mL of medium.
6. Add lipofectin reagent to DNA dropwise, with mixing in-between. Incubate for 20 min at room temperature.

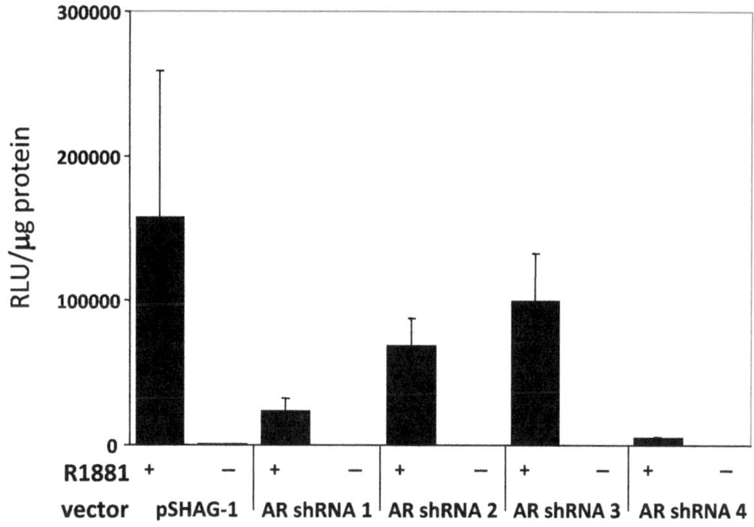

Fig. 7.1. Transactivation assay. LNCaP cells were seeded at a density of 3×10^5 cells in six-well plates, allowed to grow for 24 h, and then transiently co-transfected with −6.1-kb PSA-luciferase (1 µg/well) along with one of the four AR shRNA plasmids or empty vector (1 µg/well). Media was replaced on the following day with fresh media containing 5% CSS supplemented with 1 nM R1881 or with vehicle. Cells were harvested after 48 h of treatment and luciferase activity was measured in a luminometer. Results are presented as relative luciferase units (RLUs) per microgram protein.

7. Add 6 mL of RPMI to the lipofectin/DNA mix. Remove medium from cells and add 1 mL of the mix to each well. Incubate the cells for a further 16 h.
8. Add 1 mL of RPMI + 10% CSS with or without 2 nM R1881 to each well. Harvest the cells after 24–48 h.
9. Remove medium and add 250 µL of luciferase lysis buffer to each well. Incubate with gentle shaking for 15 min at room temperature.
10. Remove 5 µL for protein determination (BCA).
11. Remove 20 µL for luciferase assay.
12. Relative luciferase units (RLUs) are normalized to protein concentrations and are plotted (*see* **Fig. 7.1**).

3.5. Transfer of the AR shRNA Cassette Containing the $TetO_2$ Sites, U6 Promoter, and shRNA from pSHAG-1-tet to pLenti6/BLOCK-iT-DEST by LR Recombination

	Volume (µL)
shRNA in pSHAG-1-tet (150 ng/µL)	1
pLenti6/BLOCK-iT-DEST (300 ng/µL)	1
5× LR clonase reaction buffer	2
TE buffer	4
LR clonase enzyme	2

Incubate at room temperature for 60 min. Add 1 μL of proteinase K and incubate at 37°C for 10 min. Transform 5 μL of the reaction mixture into *E. coli* Stbl3 (*see* **Note 12**).

3.6. Generation of Stable Cell Line Expressing the Tet Repressor

1. Plate 293T cells in 5% DMEM on poly-L-lysine-coated (*see* **Note 13**), 10-cm tissue culture dishes at a density of 1.5 × 10^6 cells/dish. Grow the cells to 50–60% confluency. This usually takes 1–2 days.
2. Two hours before transfection, replace cell culture medium with 6 mL of fresh growth medium.
3. For each 10-cm dish, prepare two 1.5-mL sterile Eppendorf tubes.
4. Thaw all transfection reagents. Warm to room temperature and mix thoroughly:

   Tube 1

   Plasmid DNA

pMX (TR expression plasmid)	10 μg
pGP3 (Gag/Pol expression plasmid)	10 μg
pVSV-G (lentiviral envelope expression plasmid)	5 μg
2 M $CaCl_2$	37 μL
Sterile ddH_2O to 250 μL	

5. In a tissue culture hood, gently vortex the 2× HBS. Aliquot 250 μL into tube 2. Slowly add the DNA solution in tube 1 dropwise to the HBS while vortexing.
6. Incubate the combined solution at room temperature for 30 min.
7. Vortex again, then immediately add the solution, dropwise, to cells. Swirl plate gently to mix. Incubate at 37°C for 12–16 h.
8. Replace with 8 mL of fresh growth medium. Incubate for a further 24 h.
9. Collect the medium containing the lentivirus every day for a total of 3 days, replenishing with 8 mL of medium each time. Medium collected from day 1 will have the highest titer of viruses (*see* **Note 14**).
10. Virus-containing medium is filtered through 0.45-mm filter to remove cellular debris and can now be used to transduce recipient LNCaP or C4-2 cells or can be stored at –80°C (*see* **Note 15**).
11. To assay for lentiviral titer, 293T cells are transfected as described above, replacing the TR plasmid with 10 μg pHR-CMV-EGFP. Virus-containing medium is again

collected, filtered, and assayed for viral P24 antigen using the Vironostika HIV-1 Antigen Microelisa System as per manufacturer's instruction. p24 antigen (1 ng) is estimated to be equivalent to 10^6 pfu of virus. A known number of viral particles are then added to a known number of LNCaP cells and, 72 h post-infection, the number of GFP-positive cells are counted using UV microscopy. Multiplicity of infection (MOI) is calculated by dividing the number of viral particles by the number of GFP-positive cells.

12. LNCaP or C4-2 cells are seeded at a density of 3×10^6 in 10-cm dishes in RPMI + 5% FBS until they are about 70% confluent (*see* **Fig. 7.2**). Medium is removed and the cells are infected with the virus collected in Step 10 at a multiplicity of infection (MOI) of 10–60.
13. After 24 h, replace with 10 mL of fresh growth medium. Incubate overnight at 37°C.
14. Trypsinize and seed cells (1:2 split) in medium containing 2 μg/mL of puromycin (*see* **Note 16**). Approximately every 3–4 days, aspirate the medium and replace with fresh medium containing puromycin at 2 μg/ml, until puromycin-resistant colonies can be identified (generally 10–14 days after selection).
15. Expand the cells and make frozen stocks in RPMI +15% FBS + 10% DMSO for long-term storage.

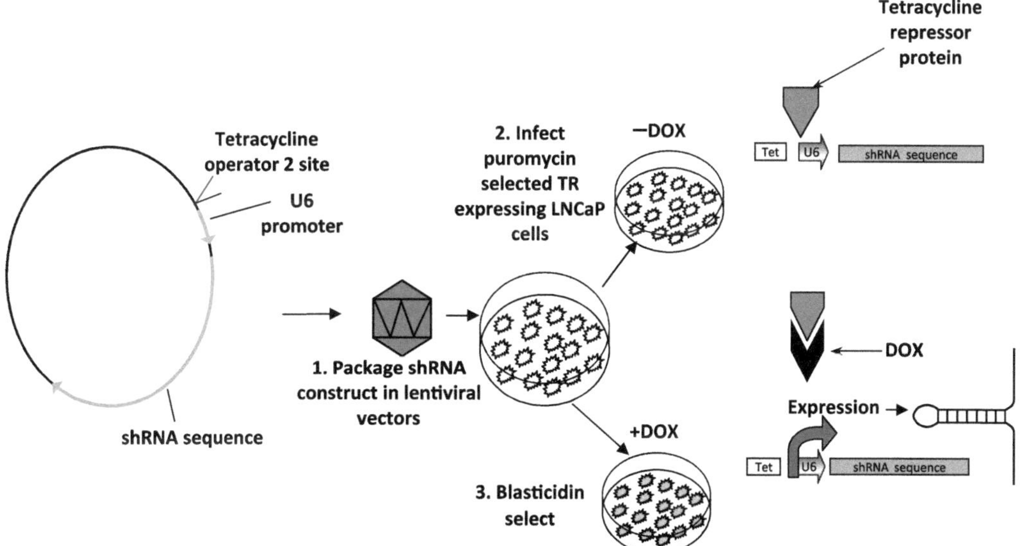

Fig. 7.2. Generation of DOX-inducible, AR shRNA-expressing cell lines. The AR shRNA cassette which contains the $TetO_2$ sites, human U6 promoter, and the shRNA was cloned in the Gateway pLenti6/BLOCK-iT-DEST and, together with lentiviral packaging plasmids, was co-transfected into the 293T cells. Viral particles were collected and used to transduce TR-expressing LNCaP or C4-2 cells and, after selection with blasticidin, the resistant cells were expanded.

3.7. Generation of DOX-Inducible, AR shRNA-Expressing LNCaP and C4-2 Cell Lines

1. Repeat Step 2 of **Section 3.4**, replacing pMX with AR shRNA or scrambled shRNA in pLenti6/BLOCK-iT-DEST and pGP3 with the packaging plasmid pCMVΔR8.2.
2. Select with 5 μg/mL of blasticidin and 1 μg/mL of puromycin.
3. Expand blasticidin-resistant cells and make freezer cultures.

3.8. Effectiveness of AR Knockdown In Vitro – Northern Blot Analysis

Plate scrm or AR shRNA-expressing LNCaP cells at a density of 2×10^6 cells in a 10-cm dish (*see* **Fig. 7.3a**). On the following day, add DOX at a concentration of 1 μg/mL and incubate the cells for a further 48 h (*see* **Note 17**). Remove the medium from the cells and add 1 mL of ice-cold TRIzol to extract RNA, following the protocol provided by the manufacturer. Analyze 20 μg of total RNA by Northern blot following the methods described in *Methods in Molecular Biology*, v 31, part V. The human AR probe used to hybridize the RNA is a 980-bp fragment

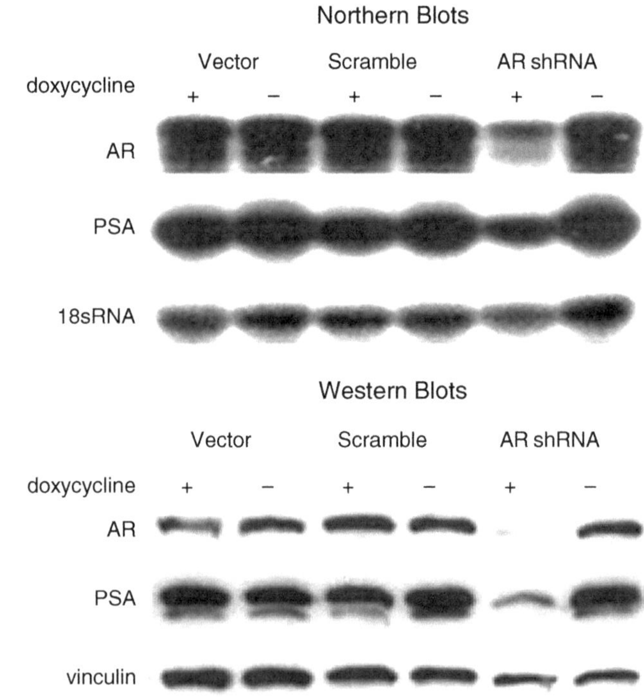

Fig. 7.3. Northern and Western blot analyses of LNCaP cell lines containing DOX-inducible shRNA. LNCaP cells stably expressing the empty vector, DOX-inducible scrambled shRNA, or DOX-inducible AR shRNA were exposed to 1 μg/mL of DOX or vehicle for 2 days. Cells were harvested and total RNA (Northern blot, **a**) or proteins (Western blot, **b**) were isolated, then migrated by electrophoresis, blotted, and probed for either AR or PSA. 18S rRNA or the protein vinculin was used for normalization.

(+1,059 to +2039) and the PSA probe used is the full-length cDNA. 18S ribosomal RNA is used as loading control.

3.9. Effectiveness of AR Knockdown In Vitro – Western Blot Analysis

1. Plate scrm or AR shRNA-expressing LNCaP cells at a density of 5×10^5 cells per well in a six-well plate (*see* **Fig. 7.3b**).
2. On the following day, add DOX at a concentration of 1 μg/mL and incubate the cells for 48 h.
3. Remove the medium and add 250 μL of cold RIPA buffer containing protease inhibitors. Incubate with shaking for 1 h at 4°C.
4. Centrifuge at 18,000×*g* for 30 min at 4°C.
5. For Western blot analysis, use 50 μg of protein as determined by the BCA assay.

3.10. Cell Growth Assay

1. Seed 3000 LNCaP cells stably expressing scrm or AR shRNA (750 cells for C4-2) in a volume of 100 μL in each well of a 96-well plate. A total of five plates are prepared. Allow the cells to attach and grow for 24 h.
2. On the following day, add 100 μL of medium containing 0.2 nM R1881 or vehicle (±2 μg/mL of DOX). Cells are replenished with fresh medium every 3 days. Every 2 days, one plate is removed and cell numbers are measured using CellTiter 96 Aqueous Non-Radioactive Cell Proliferation Assay (MTS) according to manufacturer's instructions. For the remaining plates, aspirate the medium and replace with fresh medium containing R1881 or DOX at the appropriate concentration.
3. Read the absorbance at 490 nm using an ELISA plate reader. Subtract the blank (medium alone) and plot the A_{490} values against the number of days (*see* **Fig. 7.4**).

3.11. Effectiveness of AR Knockdown In Vivo: LNCaP Xenograft Model

1. All animal procedures are carried out according to the guidelines of the Canadian Council of Animal Care with appropriate institutional certification.
2. Grow LNCaP cells expressing scrm or AR shRNA in antibiotic-free medium in 175-cm tissue culture flasks until it reaches a confluence of 90%. One flask contains enough cells to inject two mice at two sites (about 2.5×10^6 cells per site).
3. Add 3 mL of trypsin–EDTA to each flask and incubate at 37°C for 5 min.
4. Add 7 mL of RPMI + 5% FBS to neutralize the trypsin.
5. Combine all the cells and centrifuge at 150×*g* for 4 min.

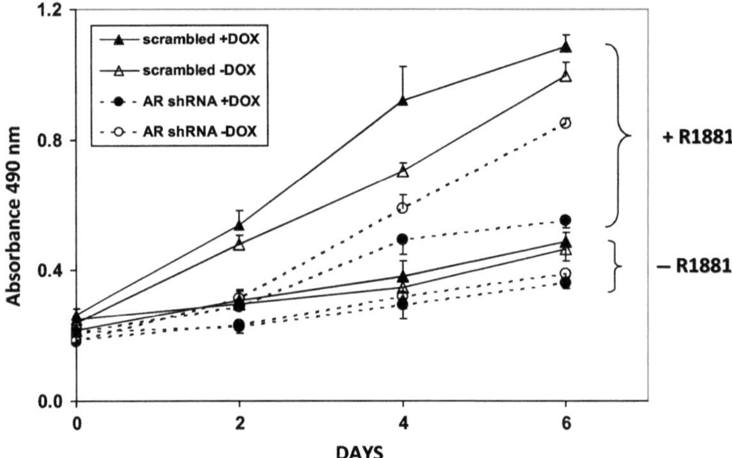

Fig. 7.4. Effects of AR shRNA and scrambled shRNA expression on LNCaP cell growth. LNCaP cells stably expressing DOX-inducible AR shRNA (*dashed lines*) or the DOX-inducible scrambled control (*solid lines*) were plated in 96-well plates (3000 cells/well) and incubated for 1 day in medium containing 5% CSS. Cells were then treated with or without DOX in the presence or the absence of 0.1 nM R1881. Cell number was then evaluated every 2 days using the MTS assay. *Black symbols* represent +DOX treatment and open symbols represent −DOX treatment.

6. Resuspend cells in serum-free, antibiotic-free RPMI to 5×10^6 cells/100 μL.
7. Combine 100 μL of cells with 100 μL of Matrigel and inoculate subcutaneously into the two flank regions of male athymic nude mice.
8. Monitor the mice for tumor growth and when the tumors become palpable, measure their volumes weekly using calipers. Tumor size is calculated using the formula: length × width × depth × 0.5236.
9. Collect 50 μL of blood from the tail vein using a micro-hematocrit capillary tube. Centrifuge in a horizontal micro-centrifuge for 10 min and collect the serum. PSA is measured by ELISA, according to manufacturer's protocol.
10. When PSA values reach 50–75 ng/mL, half of the mice are castrated via scrotal incision and all of the mice receive DOX-treated water. Six mice are used per treatment group. Water is changed every 3 days (*see* **Fig. 7.5a**).
11. When the size of the tumor exceeds 10% of the body weight, the mice are euthanized.
12. Tumor volume and PSA values are expressed as percentage of pretreatment levels and plotted against days post-treatment (*see* **Fig. 7.5b**).

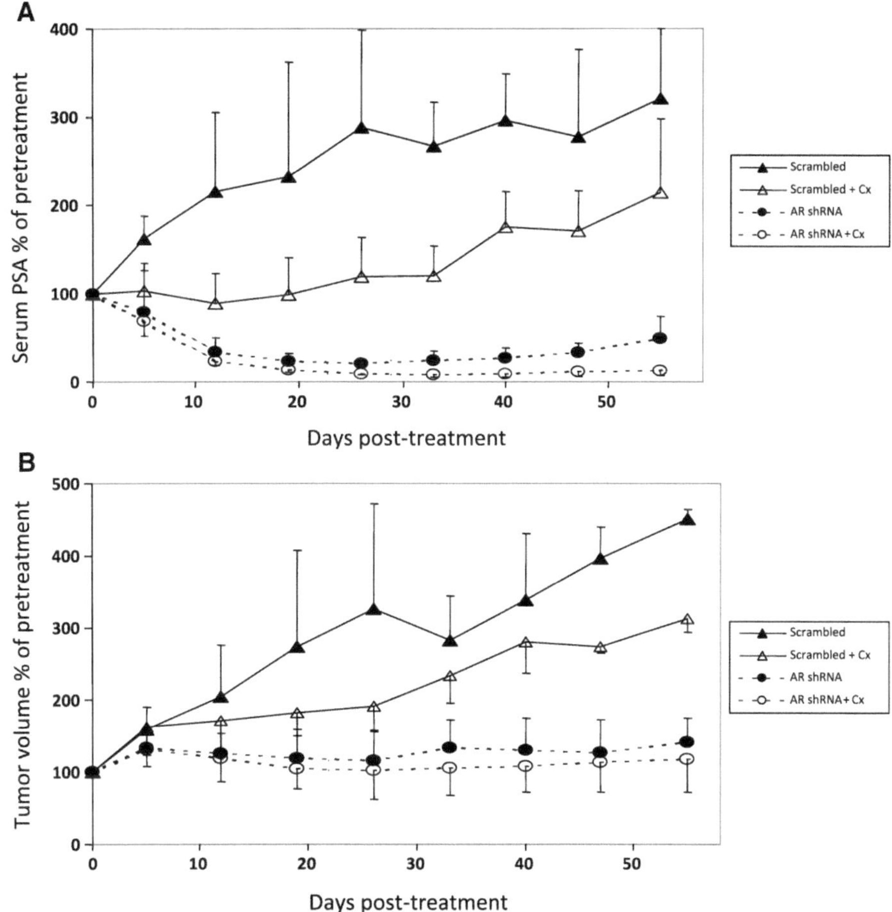

Fig. 7.5. Effect of induced AR shRNA and scrambled shRNA expression on serum PSA and growth of LNCaP xenograft tumors. LNCaP cell lines having either DOX-inducible AR shRNA or DOX-inducible scrambled shRNA were subcutaneously injected into male nude mice. Serum PSA was monitored weekly until it reached 50–75 ng/mL. Half the mice in each tumor group were then castrated and given DOX in their drinking water, whereas the other half were given only DOX in their drinking water. Serum PSA (**a**) and tumor volume (**b**) were measured weekly. Data is expressed as percentage of pretreatment values. The *dashed lines* and *circles* are values from the AR shRNA xenograft group and the *solid lines* and *triangles* are those obtained from the scrambled shRNA group. *Black symbols* indicate values in non-castrated hosts and *open symbols* indicate those measured in castrated animals.

13. In the LNCaP xenograft model, time to progression is defined as the length of time required after castration for serum PSA to return to or increase above precastrate levels.
14. Student's *t-test* is used to evaluate statistically significant differences.

4. Notes

1. All the kits are used according to manufacturer's instructions and no details will be given.
2. All solutions used are made with Milli-Q ddH_2O with a resistivity of 18.2 MΩ cm. Water is sterilized in an autoclave.
3. Antibiotics including kanamycin, puromycin, blasticidin, and DOX are made in sterile ddH_2O, filtered sterilized, and stored at –20°C as small aliquots.
4. DOX is light sensitive and should be shielded from light. It is toxic and a mask should be worn while weighing out the powder.
5. The kit provided by Invitrogen requires the use of zeocin and blasticidin as selection agents. LNCaP cells are sensitive to the combination and should not be exposed for a long period of time. We recommend using high and low doses alternately and making freezer cultures of the surviving cells as soon as they become available.
6. The shRNA core sequence provided by Xeragon is a 21-mer and nucleotides spanning both sides of the 21-mer are added to achieve the length of 31 bp. Avoid stretches of more than three of the same nucleotide.
7. The oligo must end in a "C" so that RNA polymerase III, which initiates at a "G" in the U6 promoter, will initiate precisely at the first base of the antisense strand.
8. The hairpin sequence contains the *Hin*dIII digest site which enables easy identification of the shRNA insert.
9. Synthesize oligonucleotides with 5′ phosphorylated ends.
10. There may be difficulty in sequencing the shRNA insert because the hairpin sequence is an inverted repeat that can form a secondary structure during sequencing. This can be improved by using high-quality, purified plasmid DNA, by adding DMSO to a final concentration of 5% in the sequencing mix, and increasing the amount of template. The use of modified BigDye chemistries and DNA relaxing agents is also useful (10).
11. It is important to have good quality plasmid DNA and kits from QIAGEN are used routinely to extract DNA.
12. *Escherichia coli* Stbl3 cells are ideal for cloning unstable inserts such as lentiviral DNA containing direct repeats. Transformed cells are grown in LB supplemented with 100 μg/mL of ampicillin and 15–30 μg/mL of chloramphenicol at 30°C.

13. To prevent cells from lifting off with a change of medium, tissue culture plates are rinsed with poly-L-lysine solution. After removing the solution, plates are dried with the fan in a tissue culture hood.
14. At this point, the medium contains infectious viruses and care should be taken in the handling and disposal of the samples. Perform all manipulations in a certified biosafety cabinet. All solutions and plasticwares containing viruses should be rinsed with bleach and disposed of as biohazardous waste.
15. Virus-containing medium can be stored at –80°C for up to 6 months without loss of titer. Aliquot in small volumes to avoid repetitive freezing and thawing of the same tube. After long-term storage, they should be re-titered before use.
16. The working puromycin concentration for mammalian cell lines ranges from 1 to 10 μg/ml. First step is to determine the minimum concentration that is required to kill non-transduced cells within 10–14 days of puromycin selection and then use this concentration for selection.
17. Typically, 1 μg/mL of DOX is used to induce expression of the shRNA; however, doses as low as 5 ng/mL are found to be effective in knocking down the AR.

Acknowledgments

The authors acknowledge the work of Dr Ladan Fazli, Mary Bowden, and Fariba Ghaidi in the development of these techniques in our lab. We also thank Drs Greg Hannon, Alice Mui, and Rob Kay for providing the plasmids. This work was supported by grants from the Prostate Cancer Foundation and the Canadian Institutes of Health Research.

References

1. Petrylak, D. P., Tangen, C. M., Hussain, M. H., Lara, P. N., Jr., Jones, J. A., Taplin, M. E., Burch, P. A., Berry, D., Moinpour, C., Kohli, M., Benson, M. C., Small, E. J., Raghavan, D., and Crawford, E. D. (2004) Docetaxel and estramustine compared with mitoxantrone and prednisone for advanced refractory prostate cancer, *N Engl J Med* **351**, 1513–1520.
2. Chen, C. D., Welsbie, D. S., Tran, C., Baek, S. H., Chen, R., Vessella, R., Rosenfeld, M. G., and Sawyers, C. L. (2004) Molecular determinants of resistance to antiandrogen therapy, *Nat Med* **10**, 33–39.
3. Wang, Q., Li, W., Zhang, Y., Yuan, X., Xu, K., Yu, J., Chen, Z., Beroukhim, R., Wang, H., Lupien, M., Wu, T., Regan, M. M., Meyer, C. A., Carroll, J. S., Manrai, A.

K., Janne, O. A., Balk, S. P., Mehra, R., Han, B., Chinnaiyan, A. M., Rubin, M. A., True, L., Fiorentino, M., Fiore, C., Loda, M., Kantoff, P. W., Liu, X. S., and Brown, M. (2009) Androgen receptor regulates a distinct transcription program in androgen-independent prostate cancer, *Cell* **138**, 245–256.

4. Gregory, C. W., Fei, X., Ponguta, L. A., He, B., Bill, H. M., French, F. S., and Wilson, E. M. (2004) Epidermal growth factor increases coactivation of the androgen receptor in recurrent prostate cancer, *J Biol Chem* **279**, 7119–7130.
5. Sadar, M. D. (1999) Androgen-independent induction of prostate-specific antigen gene expression via cross-talk between the androgen receptor and protein kinase A signal transduction pathways, *J Biol Chem* **274**, 7777–7783.
6. Cheng, H., Snoek, R., Ghaidi, F., Cox, M. E., and Rennie, P. S. (2006) Short hairpin RNA knockdown of the androgen receptor attenuates ligand-independent activation and delays tumor progression, *Cancer Res* **66**, 10613–10620.
7. Snoek, R., Cheng, H., Margiotti, K., Wafa, L. A., Wong, C. A., Wong, E. C., Fazli, L., Nelson, C. C., Gleave, M. E., and Rennie, P. S. (2009) In vivo knockdown of the androgen receptor results in growth inhibition and regression of well-established, castration-resistant prostate tumors, *Clin Cancer Res* **15**, 39–47.
8. Downward, J. (2004) Use of RNA interference libraries to investigate oncogenic signalling in mammalian cells, *Oncogene* **23**, 8376–8383.
9. Hannon, G. J., and Conklin, D. S. (2004) RNA interference by short hairpin RNAs expressed in vertebrate cells, *Methods Mol Biol* **257**, 255–266.
10. Taxman, D. J., Livingstone, L. R., Zhang, J., Conti, B. J., Iocca, H. A., Williams, K. L., Lich, J. D., Ting, J. P., and Reed, W. (2006) Criteria for effective design, construction, and gene knockdown by shRNA vectors, *BMC Biotechnol* **6**, 7.

Chapter 8

Analysis of Interdomain Interactions of the Androgen Receptor

Elizabeth M. Wilson

Abstract

High-affinity binding of testosterone or dihydrotestosterone to the androgen receptor (AR) triggers the androgen-dependent AR NH_2- and carboxyl-terminal (N/C) interaction between the AR NH_2-terminal F*XX*LF motif and the activation function 2 (AF2) hydrophobic binding surface in the ligand-binding domain. The functional importance of the AR N/C interaction is supported by naturally occurring loss-of-function AR AF2 mutations where AR retains high-affinity androgen binding but is defective in AR F*XX*LF motif binding. Ligands with agonist activity in vivo such as testosterone, dihydrotestosterone, and the synthetic anabolic steroids induce the AR N/C interaction and increase AR transcriptional activity in part by slowing the dissociation rate of bound ligand and stabilizing AR against degradation. AR ligand-binding domain competitive antagonists inhibit the agonist-dependent AR N/C interaction. Although the human AR N/C interaction is important for transcriptional activity, it has an inhibitory effect on transcriptional activity from AF2 by competing for p160 coactivator L*XX*LL motif binding. The primate-specific AR coregulatory protein, melanoma antigen gene protein-A11 (MAGE-A11), modulates the AR N/C interaction through a direct interaction with the AR F*XX*LF motif. Inhibition of AF2 transcriptional activity by the AR N/C interaction is relieved by AR F*XX*LF motif binding to the F-box region of MAGE-11. Described here are methods to measure the androgen-dependent AR N/C interdomain interaction and the influence of transcriptional coregulators.

Key words: Androgen receptor, steroid receptor, N/C interaction, mammalian two-hybrid assay, MAGE-11, MAGE-A11.

1. Introduction

The androgen receptor (AR) is a ligand-dependent transcription factor essential for male sex development and a critical factor in prostate cancer. AR binds the two biologically active androgens, testosterone and dihydrotestosterone (DHT), with similar

F. Saatcioglu (ed.), *Androgen Action*, Methods in Molecular Biology 776,
DOI 10.1007/978-1-61779-243-4_8, © Springer Science+Business Media, LLC 2011

high affinity. Androgen binding causes AR to translocate to the nucleus, bind to DNA-response elements, and interact with coregulatory proteins to promote the transcription of androgen-dependent genes required for male sex development and reproductive function. Early studies investigating the kinetics of androgen binding demonstrated that DHT dissociates from AR more slowly than does testosterone, a property that contributes to the greater physiological potency of DHT (1, 2). The greater effectiveness of DHT during development is evident from the incomplete masculinization of genetic males deficient in DHT due to naturally occurring gene mutations in the 5α-reductase enzyme that converts testosterone to DHT (3).

Both testosterone and DHT dissociate with slower kinetics from full-length AR than from an AR carboxyl-terminal, ligand-binding domain (LBD) fragment that lacks the NH_2-termimal region (4). In the presence of androgen, an AR NH_2-terminal fragment interacts with an AR DNA and ligand-binding domain fragment, and the complex binds DNA and activates the prostate-specific antigen (PSA) enhancer/promoter (5). These observations, together with the results of mammalian two-hybrid interaction assays (6, 7), provided the first evidence for an androgen-dependent AR NH_2- and carboxyl-terminal (N/C) interaction.

The importance of the AR N/C interaction in male reproductive physiology is suggested by naturally occurring AR gene mutations in the LBD that disrupt the AR N/C interaction without altering equilibrium androgen-binding affinity (7–13). These LBD mutations cause the androgen insensitivity syndrome that results in partial or complete failure of masculinization in 46XY genetic males by disrupting the AR N/C interaction and p160 coactivator binding (6–8). Ligands that induce the AR N/C interaction display complete agonist activity in vivo, which indicates that the mammalian two-hybrid N/C interaction assay is a useful method to identify ligands that function as active androgens (14) (*see* **Note 1**). The N/C interaction contributes to agonist potency in part by slowing the ligand dissociation rate and stabilizing AR (6). The AR N/C interaction assay is also useful in conjunction with transcription assays to identify AR antagonists (14).

The androgen-dependent AR N/C interaction is mediated by the AR NH_2-terminal F*XX*LF motif ^{23}FQNLF^{27} binding to a hydrophobic cleft in the LBD surface known as activation function 2 (AF2) (9, 15, 16). Both the AR F*XX*LF motif and AF2 are flanked by complementary charged amino acid residues that facilitate their interaction (17). Assays of the AR N/C interaction require the expression of an AR NH_2-terminal fragment that contains the AR-20–30 NH_2-terminal region with ^{23}FQNLF^{27} sequence (17). The coexpressed interacting LBD fragment must contain AR LBD residues 658–919, an AR fragment that retains high-affinity androgen binding but displays rapid androgen dissociation kinetics when expressed alone (2). The two-hybrid AR N/C interaction assay is performed in mammalian cells (*see* **Note 2**) using GAL4 DNA-binding domain and VP16 activation domain fusion proteins, and a GAL4–luciferase reporter gene. Deletion of the hinge region from the carboxyl-terminal LBD fragment minimizes an inhibitory effect (*see* **Note 3**). The assay can be performed using AR NH_2-terminal and carboxyl-terminal fragments with an androgen-responsive luciferase reporter vector such as PSA-Enh-Luc (*see* **Note 4**).

The androgen-dependent AR N/C interaction promotes the expression of AR target genes (18, 19). However, not all androgen-responsive enhancer/promoter regions require the AR N/C interaction. Most notably, the mouse mammary tumor virus (MMTV) and sex-limited gene enhancer/promoters do not require the AR N/C interaction that was required for maximal induction of the PSA and other androgen-responsive genes (18).

The human AR N/C interaction inhibits p160 coactivator L*XX*LL motif binding to AF2, which decreases overall AR transcriptional activity derived from AF2 (20, 21). Competitive inhibition at the AF2 site in the LBD occurs between the AR F*XX*LF motif and the p160 coactivator L*XX*LL motifs that bind the same AF2 hydrophobic cleft on the LBD surface (22). The ~10 fold higher binding affinity for the AR F*XX*LF motif relative to p160 coactivator L*XX*LL motifs imposes an inhibitory effect on AF2 activity. Through this mechanism, the AR N/C interaction shifts the dominant activation region from AF2 in the LBD to activation function 1 (AF1) in the human AR NH_2-terminal region. In normal physiology, AF1 is androgen-dependent because of the inhibitory effect of the unliganded LBD. An AR NH_2-terminal and DNA-binding domain fragment that lacks the LBD is constitutively active (23).

The AR N/C interaction is modulated by AR coregulatory proteins (*see* **Note 5**). One recently characterized AR coregulator that influences the AR N/C interaction is melanoma antigen gene protein-A11 (MAGE-11). MAGE-11 is a member of the *MAGE* gene family of cancer-testis antigens that evolved in a species-specific manner. MAGE-11 is expressed only in humans and other primates and is not expressed in rodents. MAGE-11 binds the AR F*XX*LF motif and directly recruits p160 coactivators to the AR NH_2-terminal region (24, 25). MAGE-11 binds the AR F*XX*LF motif through a MAGE-11 F-box that is posttranslationally modified. MAGE-11 is phosphorylated at Thr-360 within the F-box (amino acid residues 329–369) by cell cycle checkpoint kinase Chk1 in response to epidermal growth factor (EGF). MAGE-11 is phosphorylated at Ser-174 outside the F-box by MAP kinase in response to serum stimulation (25). In contrast to the minimal AR-20–30 amino acid region required to bind AF2 in the AR N/C interaction, a longer AR-16–36 region that contains the F*XX*LF motif is required to bind MAGE-11. Thus, overlapping AR F*XX*LF motif regions mediate interactions with AR AF2 in the AR N/C interaction and with MAGE-11. This suggests different flanking sequence requirements for AR F*XX*LF motif interactions. MAGE-11 interaction with AR also depends on the EGF-stimulated monoubiquitinylation of MAGE-11 outside the F-box that is triggered by phosphorylation at Thr-360 within the F-box (26). The recent evolutionary appearance of MAGE-11 as an AR coregulator in primates provides important and novel regulatory control on AR function.

Thus, the AR N/C interaction regulates AR function in response to agonists, is inhibited by AR antagonists, and modulated by coactivators. Two-hybrid assays of the AR N/C interaction in mammalian cells provide a measure of ligand potency for agonists or antagonists, and the functional effects of AR coregulators and naturally occurring and targeted AR mutations.

2. Materials

2.1. AR N/C Interaction Assay Reagents

1. HeLa epithelial cells derived from a human cervix adenocarcinoma (CCL-2; American Type Culture Collection, Rockville, MD) (*see* **Note 2**).
2. HeLa cell medium: Minimum essential medium with or without phenol red contains 10% fetal bovine serum (FBS); 2 mM L-glutamine (5.5 ml of 200 mM 100× L-glutamine, added to 500 ml media); penicillin; and streptomycin (5.5 ml of 10,000 IU/ml 100× penicillin and streptomycin, added to 500 ml media).
3. Eukaryotic expression and reporter vectors:
   GAL-AR-658–919 is a fusion protein of the GAL4 DNA-binding domain amino acid residues 1–147 and human AR carboxyl-terminal amino acid residues 658–919 that constitutes the AR LBD (2, 27).

   VP-AR-1–660 is a fusion protein of VP16 transactivation domain amino acid residues 411–456 and human AR NH_2-terminal amino acid residues 1–660 that includes the DNA-binding domain (7).

   5XGAL4Luc3 is a luciferase reporter vector with five copies of the GAL4 upstream enhancer element (17, 28).
4. Fugene 6 Transfection Reagent (Roche Diagnostics, Indianapolis, IN) is stored at 4°C.
5. Phosphate buffered saline (PBS).

6. Testosterone, DHT (Steraloids, Inc., Newport, RI) and the synthetic androgen, methyltrienolone (R1881) (PerkinElmer, Waltham, MA) 10 mg/ml stocks, are prepared fresh each month in 100% ethanol and stored at −20°C. Dilutions of steroid stocks are prepared fresh each week in 100% ethanol.
7. Luciferase lysis buffer: 1% Triton X-100, 2 mM EDTA, and 25 mM Tris-phosphate, pH 7.8.
8. D-Luciferin, monopotassium salt (Thermo Scientific, Rockford, IL); 0.1 ml of 0.1 M D-luciferin is added automatically per well in 96-well assay plates in a luminometer.
9. Luciferase reading buffer: 25 mM Glycylglycine, 15 mM $MgCl_2$, 5 mM ATP, and 0.5 mg/ml bovine serum albumin, pH 7.8; 0.1 ml reading buffer is added automatically per well in 96-well assay plates in a luminometer.
10. Twelve-well treated nonpyrogenic polystyrene tissue culture plates (Corning, Inc., Corning, NY).
11. 96-Well, nontreated, flat-bottomed white polystyrene microtiter plates (Costar; Corning, Inc., Corning, NY).
12. 15-ml Sterile RNase/DNase-free nonpyrogenic polypropylene centrifuge tubes (Corning, Inc., Corning, NY).
13. Automated Lumistar Galaxy multi-well plate luminometer (BMG Labtech, Germany).

3. Methods

3.1. HeLa Cell Culture

HeLa cells (*see* **Note 2**) are propagated in MEM supplemented with 10% FBS, L-glutamine, penicillin, and streptomycin. Cells are passaged twice each week at 1:7 dilution. Cells are harvested by washing T150 flasks with 10 ml PBS, adding 2 ml of a 0.05% trypsin and 0.53 mM EDTA solution/flask, and incubating at 37°C for 5 min to release cell adhesion. MEM containing 10% FBS is added to inactivate trypsin. Cells are counted using a hemocytometer, plated at 5×10^4 cells/well in 12-well plates, and transfected using FuGene 6 Transfection Reagent.

3.2. Experimental Design

The AR N/C interaction is performed as a mammalian two-hybrid assay. Cotransfection of an AR NH_2-terminal F*XX*LF motif-containing fragment with an AR carboxyl-terminal fragment that contains the LBD and AF2 binding surface increases reporter gene activity in the presence of an AR agonist. The AR N/C interaction requires the addition of an active androgen such as testosterone, DHT, or synthetic androgen R1881

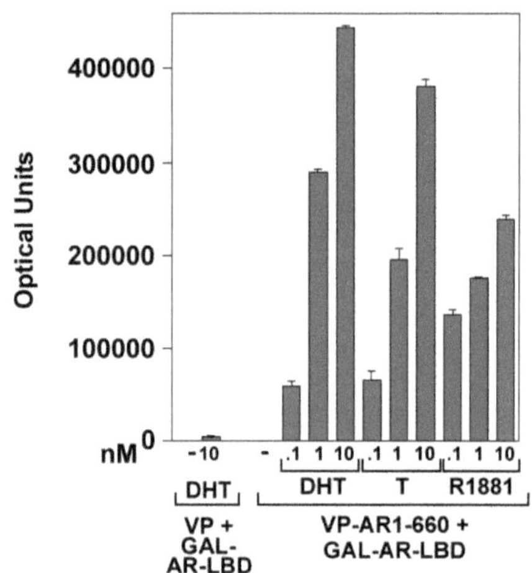

Fig. 8.1. Androgen-dependent AR N/C interaction. HeLa cells (5 $\times$ 10^4/well of 12-well plates) were transfected using Fugene 6 with (per well) 0.1 μg 5XGAL4Luc3 reporter and 50 ng GAL-AR-658–919 (GAL-AR-ligand-binding domain (-LBD)) in the presence of 50 ng pVP16 empty vector (VP) or 50 ng VP-AR-1–660 that codes for the AR NH_2-terminal and DNA-binding domains. The day after transfection, the medium was exchanged with phenol red-free, serum-free medium in the absence and the presence of 0.1, 1, and 10 nM dihydrotestosterone (DHT), testosterone (T), or the synthetic androgen methyltrienolone (R1881), respectively. Cells were incubated overnight at 37°C and luciferase activity was determined. The data are representative of the androgen-dependent mammalian two-hybrid AR N/C interaction assay.

(**Fig. 8.1**), mibolerone, or anabolic steroid (*see* **Note 1**). The concentration of steroid that increases luciferase light units by greater than fivefold is indicative of ligand potency and a ligand-dependent AR intramolecular and/or intermolecular interaction (*see* **Note 6**). The AR N/C interaction is not induced by antagonists such as hydroxyflutamide or Casodex (bicalutamide). These antagonists inhibit the agonist-induced AR N/C interaction in a dose-dependent manner (**Fig. 8.2**). The effects of naturally occurring AR mutations identified in patients with the androgen insensitivity syndrome, somatic mutations in prostate cancer tissue (21), or targeted mutations designed to establish the AR sequence requirements for the N/C interaction can be tested in the two-hybrid assay when mutations are introduced into the AR NH_2-terminal or LBD fragments. Assays are set in duplicate or triplicate with an agonist dose response range between 0.01 and 10 nM and an antagonist dose response range between 50 nM and 1 μM. The optimal dose response concentration for the AR N/C interaction in the two-hybrid assay is 10 nM testosterone, DHT, or synthetic androgen (**Fig. 8.1**). The N/C interaction assay has also been demonstrated for other steroid receptors using NH_2- and carboxyl-terminal fragments (*see* **Note 7**).

When the AR N/C interaction is performed using GAL4 and VP16 fusion vectors, a 5XGAL4Luc reporter vector is used. The

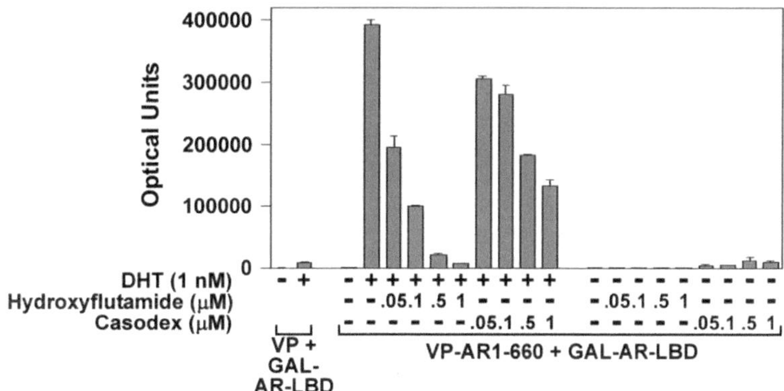

Fig. 8.2. Inhibition of the DHT-induced AR N/C interaction by AR antagonists. HeLa cells were transfected with 5XGAL4Luc3 reporter vector and GAL-AR-658–919 (GAL-AR-LBD) with pVP16 empty vector (VP) or VP-AR-1–660 as described in **Fig. 8.1**. The next day, cells were incubated in phenol red-free, serum-free media in the absence and the presence of 1 nM dihydrotestosterone (DHT) with and without 0.05, 0.1, 0.5, and 1 μM hydroxyflutamide or Casodex (bicalutamide), and the same increasing concentrations of hydroxyflutamide or Casodex alone. The data show that the DHT-induced AR N/C interaction measured in a mammalian two-hybrid assay is inhibited in a dose-dependent manner by increasing concentrations of AR antagonists, hydroxyflutamide and Casodex, and that these antagonists alone do not induce the AR N/C interaction.

AR N/C interaction may also be performed using the androgen-responsive luciferase reporter vector PSA-Enh-Luc, or the less active MMTV-Luc, transfected into HeLa cells with the AR DNA and ligand-binding domain fragment AR-507–919 and AR NH_2-terminal fragment AR-1–503 that lacks the DNA-binding domain (*see* **Note 4**). Experimental procedures for both assays are otherwise identical.

3.3. Transfection of HeLa Cells Using FuGene 6

DNAs are aliquoted into microfuge tubes and stored for not more than 1 week at –20°C using per well:

50 ng Gal-AR-658–919 (*see* **Note 3**)

50 ng VP-AR-1–660

0.1 μg 5XGAL4Luc3

Coregulator expression plasmid DNA (25–50 ng/well of 12-well plates) can be added to test for effects on the AR N/C interaction by including equivalent amounts of empty vector DNA added to controls (*see* **Note 5**).

Day 1: Plate 5 × 10^4 HeLa cells/well in 12-well plates with 1 ml HeLa cell medium containing phenol red and incubate overnight in a 5% CO_2 cell culture incubator.

Day 2:

1. Aspirate the medium and add 0.75 ml/well fresh HeLa cell medium containing phenol red using a sterile repeat pipette and return the plates to the cell culture incubator.

2. Calculate 43 μl × number of wells (include four extra wells in the calculation) of serum-free, phenol red-free medium and add to a 15-ml centrifuge tube.
3. Briefly warm the Fugene reagent and vortex for 1 s.
4. Add 0.6 μl Fugene reagent × total number of wells directly to the aliquoted serum-free media avoiding contact with the plastic tube.
5. Vortex for 1 s and incubate for 5 min at room temperature.
6. Thaw the previously aliquoted expression plasmid and luciferase reporter DNA.
7. Add sufficient Fugene cell medium solution for 43 μl/well to the aliquoted DNA.
8. Vortex each tube for 1 s and incubate for 15 min at room temperature.
9. Vortex briefly and add 40 μl of DNA–Fugene cell medium solution to each well.
10. Return the plates to the 37°C cell culture incubator and incubate overnight.

Day 3

1. Aspirate the medium and replace with 1 ml serum-free, phenol red-free HeLa cell medium with and without ligand.
2. Return the plates to the 37°C cell culture incubator and incubate overnight.

Day 4

1. Aspirate the media and wash each well with 1 ml PBS.
2. Aspirate PBS twice to dryness and add 0.25 ml luciferase lysis buffer using a repipetter.
3. Gently rock the plates on a platform shaker for 30 min at room temperature.
4. Aliquot 0.1 ml from each well into a 96-well microtiter assay plates and measure luciferase activity using an automated luminometer.

4. Notes

1. *N/C interaction screen for AR agonists and antagonists:* Androgens that induce the AR N/C interaction have agonist activity in vivo. The AR N/C interaction is inhibited by classical AR antagonists such as hydroxyflutamide

and Casodex that bind the AR LBD with moderate affinity (14). Examples of AR agonists that induce the AR N/C interaction are testosterone, DHT, and the anabolic steroids oxandrolone and fluoxymesterone which bind with lower affinity than testosterone or DHT but are potent agonists in vivo (14). These findings suggest that the AR N/C two-hybrid interaction assay can be used to identify agonists that increase AR transcriptional activity in vivo when performed in cells with suitable ligand uptake and lack of metabolism (*see* **Note 2**). The AR N/C interaction has been used to screen AR agonists and antagonists, establish ligand dependence and motif-binding specificity, and investigate *AR* gene mutations that cause the androgen insensitivity syndrome (10, 29). While AR antagonists competitively inhibit the agonist-induced N/C interaction (**Fig. 8.2**), high concentrations of an AR antagonist, e.g., 10 μM hydroxyflutamide, may have agonist activity with wild-type AR in transient transfection experiments, but only weakly induce the AR N/C interaction (6, 14, 30). Somatic AR mutations in prostate cancer can enhance the ability of antagonists to induce the AR N/C interaction and increase AR transcriptional activity.

2. *AR N/C interaction assay in other cell lines:* The AR N/C interaction has been performed in Chinese hamster ovary (CHO) (6, 7, 9, 15, 31), human hepatocellular carcinoma HepG2 (17, 32), and HeLa cells (2, 18, 21). HeLa cells are advantageous because they contain low levels of steroid-metabolizing enzymes. This is evident by the similar activities of the naturally occurring androgens testosterone and DHT, and the synthetic androgen methyltrienolone (R1881), which is less susceptible to metabolism than are naturally occurring androgens. HepG2 cells derive from a human hepatocellular carcinoma and thus may express liver-derived, steroid-metabolizing enzymes. This is supported by the weaker activities of testosterone and DHT compared to synthetic androgens when assayed in HepG2 cells. The AR N/C interaction has also been performed in yeast (33). In this case, consideration should be given to ligand uptake and metabolism. While a variety of cell lines can be used to perform the AR N/C interaction assay, possible complications to be considered are ligand uptake and metabolism, and the influence of endogenous transcriptional coregulators that can differ between cell lines.

3. *Inhibition of the AR N/C interaction by the hinge region:* Human AR has 919 amino acids that include the NH_2-terminal residues 1–558, DNA-binding domain residues 559–624, hinge region residues 625–676, and LBD residues

677–919 (34). However, the number of amino acid residues in human AR varies between individuals depending on the length of the polymorphic NH_2-terminal glutamine repeat that begins at amino acid residue 58. Length of the AR CAG-encoded glutamine repeat influences AR transcriptional capacity through mechanisms that are not completely understood (35–37).

Initial experiments that identified and characterized the androgen-dependent AR N/C interaction made use of the AR LBD fragment AR-624–919, which contains the LBD and the entire hinge region (6, 7). It was shown subsequently that human AR hinge region residues 628–646, 624–658, or 629–636 have an inhibitory effect on the AR N/C interaction and AR transcriptional activity (2, 38, 39). The inhibitory region contains part of the bipartite AR nuclear-targeting signal at residues 617–633 (39, 40). However, inhibition may be independent of a detrimental effect on AR nuclear transport, since inhibition was also observed using a GAL4 DNA-binding domain–AR LBD fusion protein that has an independent nuclear-targeting signal (2). The inhibitory effect of the AR hinge region appears to be mediated through the AR AF2 site (2). A naturally occurring AR-R629W hinge mutation that disrupts the AR N/C interaction caused severe androgen insensitivity without altering androgen-binding affinity and nuclear localization (12). These findings support a negative influence of the hinge region on the AR N/C interaction and AR transcriptional activity in vivo. The phenotypic expression of decreased AR transcriptional activity in the androgen insensitivity syndrome resulting from mutations that disrupt the AR N/C interaction supports the functional importance of the AR N/C interaction in vivo.

The inhibitory effect of the hinge region does not appear to result from structural artifacts associated with expression of the AR NH_2-terminal and LBD fragments, since inhibition was also observed in full-length AR. While the precise mechanism for AR hinge region inhibition of AF2 is not known, the inhibitory effect decreases AF2 binding of the AR F*XX*LF and p160 coactivator L*XX*LL motifs (2). This suggests that AR AF2 may have greater inherent transcriptional capacity than previously recognized, especially when p160 coactivator levels are increased as in castration-recurrent prostate cancer (41). In HeLa cells in the absence of overexpressed p160 coactivators, AR AF2 transcriptional activity was almost undetectable when GAL-AR-LBD was expressed in the presence of 10 nM DHT in the absence of an interacting AR NH_2-terminal fragment (**Fig. 8.1**). This reflects in part the genetic changes in the AR LBD AF2 site during evolution that have weakened the binding affinity for p160 coactivator L*XX*LL motifs and improved binding of the F*XX*LF motifs (22). However, in prostate cancer cells, such as CWR-R1 cells that have higher levels of coactivators, expression of GAL-AR-LBD alone can have significant activity (2).

Thus, the contribution of AF2 to AR transcriptional activity relative to AR NH_2-terminal AF1 is limited by the inhibitory effect of the AR N/C interaction, by genetic changes in AF2 that weaken p160 coactivator L*XX*LL motif binding relative to the F*XX*LF motif, through inhibitory mechanisms involving the hinge region, and through the complement of cell-specific coregulatory proteins.

4. *Promoter specificity of the AR N/C interaction*: Androgen-responsive enhancer/promoters differ in the requirement for the AR N/C interaction. For example, transcriptional activation of the PSA enhancer is increased through mechanisms that are increased by the AR N/C interaction (18, 19), whereas some androgen-responsive enhancers/promoters do not require the AR N/C interaction. The most notable example is MMTV which is activated to a similar extent in transient transfection assays whether or not the AR N/C interaction is disrupted by mutations in the AR F*XX*LF motif (18). The extent of activation of PSA-Enh-Luc or MMTV-Luc by the coexpression of AR-507–919 and AR-1–503 is influenced by the concentration of endogenous coregulators which differs between cell lines. The full complement of transcriptional regulators that influence the AR N/C interaction and AR transcriptional activity remains to be defined.

5. *AR FXXLF motif and influence of coregulators*: The effect of coregulatory proteins on the AR N/C interaction can be assessed in the mammalian two-hybrid assay. However, nonspecific inhibitory effects associated with the addition of increasing amounts of coactivator DNA require appropriate controls. Expression vector DNA concentrations should be kept to a minimum (100 ng or less/well in 12-well plates) to avoid nonspecific inhibition. Transient transfections may be performed in monkey kidney CV1 cells using calcium phosphate precipitation of DNA when assessing the effects of coregulatory proteins on AR transcriptional activity (24).

   The AR N/C interaction assay measures the androgen-dependent AR NH_2-terminal F*XX*LF motif ^{23}FQNLF^{27} interaction with the AF2 site in the LBD. The AR F*XX*LF motif also serves as the principal interaction site for MAGE-11, an AR coregulator that increases AR transcriptional activity by relieving inhibition of p160 coactivator L*XX*LL motif binding at AF2 and through direct interactions with transcriptional coregulators (20, 25). The

AR F*XX*LF motif was also implicated as a site involved in AR degradation by the proteasome (42). An NH_2-terminal W*XX*LF (^{433}WHTLF^{437}) contributes to ligand-independent AR transcriptional activity and to the AR N/C interaction through mechanisms that remain to be established (15, 18, 43).

F*XX*LF-like motifs that mediate androgen-dependent interactions with AR AF2 are present in the AR coregulatory proteins ARA54, ARA55, and ARA70 (32, 44–46). While the affinity of these F*XX*LF-related motifs has not been reported, their inability to slow the dissociation rate of bound androgen when inserted to replace the AR F*XX*LF motif suggests that AR AF2 affinity is weaker for these coregulator F*XX*LF motifs than for the AR F*XX*LF motif (32). Furthermore, MAGE-11 does not interact with F*XX*LF-related motifs present in these AR coregulators, indicating specificity for the AR F*XX*LF motif (24). MAGE-11 contains an F*XX*IF motif that serves as part of a recognition sequence for p160 coactivators. Interaction between the MAGE-11 F*XX*IF and an F-box-like region in transcriptional mediator protein 2 (TIF2), and AR F*XX*LF motif binding to the MAGE-11 F-box, suggests a novel protein–protein interaction paradigm of F*XX*LF motif binding to the F-box (25).

A number of AR coregulatory proteins are reported to influence the AR N/C interaction, some of which involve interactions with F*XX*LF-related motifs. Cyclin D1 binds the AR NH_2-terminal F*XX*LF motif and inhibits the AR N/C interaction and AR transcriptional activity independent of its ability to recruit histone deacetylases (47). p53 contains an α-helical F*XX*LF-like recognition motif in its activation domain and inhibits the AR N/C interaction (48, 49). hRAD9 contains a carboxyl-terminal F*XX*LF motif, inhibits the AR N/C interaction, and represses AR transcriptional activity (50). FoxO1 binding to the AR NH_2-terminal region inhibits the AR N/C interaction (51). Inhibition of the AR N/C interaction is caused by corepressor SMRT that interacts with the AR NH_2-terminal region (52, 53) and by N-CoR that interacts with the AR carboxyl-terminal region (54, 55).

MAGE-11 contains an F*XX*IF motif and increases AR transcriptional activity by competitively inhibiting AR F*XX*LF motif binding to AF2 in the AR N/C interaction, and by directly recruiting p160 coactivators (24). c-Jun, although not reported to contain an F*XX*LF motif, enhanced the AR N/C interaction, AR binding to DNA, and AR transcriptional activity (56). Effects of AR coregulatory proteins can be assessed in two-hybrid interaction assays as described here. Endogenous levels of these and other coregulators yet to be defined influence the AR N/C interaction. Experiments using small inhibitory RNAs to knock down endogenous protein expression can therefore provide additional evidence for the effects of coregulatory proteins in the AR N/C interaction and AR function.

6. *Intramolecular versus intermolecular AR N/C interaction:* The AR N/C interaction may occur as an intramolecular and an intermolecular interaction. Early studies on androgen insensitivity syndrome AR mutations suggest that an intermolecular AR N/C interaction facilitates AR binding to DNA as an anti-parallel dimer (5–7). Structural analysis using fluorescence resonance energy transfer (FRET) suggests an intermolecular AR N/C interaction in the nucleus and an intramolecular interaction in the cytoplasm (57). Additional FRET analysis suggests that the AR N/C interaction occurs when AR is mobile, but not when transiently bound to DNA (58), and an intramolecular N/C interaction for transcriptionally active AR bound to DNA (59). Under some conditions, an intermolecular AR N/C interaction could contribute to runaway domain swapping (60), in which amyloid-like fibrils are associated with the degenerative phenotype that results from glutamine expansion diseases, such as spinal bulbar muscular atrophy which is caused by an expanded AR NH_2-terminal glutamine repeat (61). However, the length of the polymorphic glutamine repeat in the human AR NH_2-terminal region has not been shown to influence the AR N/C interaction (6).

7. *N/C interaction in other steroid receptors*: A ligand-dependent N/C interaction has been reported for the progesterone receptor (PR) (62), estrogen receptor-α (63), and mineralocorticoid receptor (MR) (64), although none has been as extensively characterized as the AR N/C interaction. The glucocorticoid receptor (GR) does not undergo an agonist-induced N/C interaction. However, introducing a p160 coactivator L*XX*LL motif into the NH_2-terminal region to create a GR chimera slows the dissociation rate of dexamethasone ($t_{1/2}$ 168 min), a synthetic glucocorticoid, compared to wild-type GR ($t_{1/2}$ 31 min) (20). This demonstrates the ability of an N/C interaction to slow ligand dissociation and stabilize a receptor.

   The longest PR-B contains an extended NH_2-terminal L*XX*LL-like motif upstream region absent from the shorter PR-A. This unique PR-B upstream region imparts greater transcriptional activity through cooperative interactions involving a PR-B N/C interaction when bound to DNA (65).

The MR N/C interaction displays striking ligand specificity, with strong induction by aldosterone and the synthetic agonist 9α-fludrocortisol, and weak or no interaction with other physiologically relevant ligands such as deoxycorticosterone or cortisol (64, 66). While the precise nature of the MR NH_2-terminal interaction motif has yet to be characterized, evidence suggests that like AR, the MR N/C interaction contributes to MR transcriptional activity by discriminating ligand-specific effects in vivo. An MR corepressor recruited by agonist-bound MR inhibited the MR N/C interaction (67).

References

1. Wilson, E. M., and French, F. S. (1976) Binding properties of androgen receptors: evidence for identical receptors in rat testis, epididymis, and prostate. *J. Biol. Chem.* **251**, 5620–5629.
2. Askew, E. B., Gampe, R. T., Stanley, T. B., Faggart, J. L., and Wilson, E. M. (2007) Modulation of androgen receptor activation function 2 by testosterone and dihydrotestosterone. *J. Biol. Chem.* **282**, 25801–25816.
3. Imperato-McGinley, J., Guerrero, L., Gautier, T., and Peterson, R. E. (1974) Steroid 5-alpha-reductase deficiency in man: an inherited form of male pseudohermaphroditism. *Science* **186**, 1213–1215.
4. Zhou, Z. X., Lane, M. V., Kemppainen, J. A., French, F. S., and Wilson, E. M. (1995) Specificity of ligand-dependent androgen receptor stabilization: receptor domain interactions influence ligand dissociation and receptor stability. *Mol. Endocrinol.* **9**, 208–218.
5. Wong, C. I., Zhou, Z. X., Sar, M., and Wilson, E. M. (1993) Steroid requirement for androgen receptor dimerization and DNA binding. Modulation by intramolecular interactions between the NH_2-terminal and steroid-binding domains. *J. Biol. Chem.* **268**, 19004–19012.
6. Langley, E., Zhou, Z. X., and Wilson, E. M. (1995) Evidence for an antiparallel orientation of the ligand activated human androgen receptor dimer. *J. Biol. Chem.* **270**, 29983–29990.
7. Langley, E., Kemppainen, J. A., and Wilson, E. M. (1998) Intermolecular NH_2-/carboxyl-terminal interactions in androgen receptor dimerization revealed by mutations that cause androgen insensitivity. *J. Biol. Chem.* **273**, 92–101.
8. Quigley, C. A., Tan, J. A., He, B., Zhou, Z. X., Mebarki, F., Morel, Y., Forest, M., Chatelain, P., Ritzen, E. M., French, F. S., and Wilson, E. M. (2004) Partial androgen insensitivity with phenotypic variation caused by androgen receptor mutations that disrupt activation function 2 and the NH_2- and carboxyl-terminal interaction. *Mech. Ageing Dev.* **125**, 683–695.
9. He, B., Kemppainen, J. A., Voegel, J. J., Gronemeyer, H., and Wilson, E. M. (1999) Activation function 2 in the human androgen receptor ligand binding domain mediates interdomain communication with the NH_2-terminal domain. *J. Biol. Chem.* **274**, 37219–37225.
10. Ghali, S. A., Gottlieb, B., Lumbroso, R., Beitel, L. K., Elhaji, Y., Wu, J., Pinsky, L., and Trifiro, M. A. (2003) The use of androgen receptor amino/carboxyl-terminal interaction assays to investigate androgen receptor gene mutations in subjects with varying degrees of androgen insensitivity. *J. Clin. Endocrinol. Metab.* **88**, 2185–2193.
11. Thompson, J., Saatcioglu, F., Jänne, O. A., and Palvimo, J. J. (2001) Disrupted amino- and carboxyl-terminal interactions of the androgen receptor are linked to androgen insensitivity. *Mol. Endocrinol.* **15**, 923–935.
12. Deeb, A., Jääskeläinen, J., Dattani, M., Whitaker, H. C., Costigan, C., and Hughes, I. A. (2008) A novel mutation in the human androgen receptor suggests a regulatory role for the hinge region in amino-terminal and carboxy-terminal interactions. *J. Clin. Endocrinol. Metab.* **93**, 3691–3696.
13. Jääskeläinen, J., Deeb, A., Schwabe, J. W., Mongan, N. P., Martin, H., and Hughes, I. A. (2006) Human androgen receptor gene ligand-binding-domain mutations leading to disrupted interaction between the N- and C-terminal domains. *J. Mol. Endocrinol.* **36**, 361–368.

14. Kemppainen, J. A., Langley, E., Wong, C. I., Bobseine, K., Kelce, W. R., and Wilson, E. M. (1999) Distinguishing androgen receptor agonists and antagonists: distinct mechanisms of activation by medroxyprogesterone acetate and dihydrotestosterone. *Mol. Endocrinol.* **13**, 440–454.
15. He, B., Kemppainen, J. A., and Wilson, E. M. (2000) FXXLF and WXXLF sequences mediate the NH_2-terminal interaction with the ligand binding domain of the androgen receptor. *J. Biol. Chem.* **275**, 22986–22994.
16. He, B., and Wilson, E. M. (2002) The NH_2-terminal and carboxyl-terminal interaction in the human androgen receptor. *Mol. Gen. Metab.* **75**, 293–298.
17. He, B., and Wilson, E. M. (2003) Electrostatic modulation of steroid receptor recruitment of the LXXLL and FXXLF motifs. *Mol. Cell. Biol.* **23**, 2135–2150.
18. He, B., Lee, L. W., Minges, J. T., and Wilson, E. M. (2002) Dependence of selective gene activation on the androgen receptor NH_2- and carboxyl-terminal interaction. *J. Biol. Chem.* **277**, 25631–25639.
19. Callewaert, L., Verrijdt, G., Christiaens, V., Haelens, A., and Claessens, F. (2003) Dual function of an amino-terminal amphipatic helix in androgen receptor-mediated transactivation through specific and nonspecific response elements. *J. Biol. Chem.* **278**, 8212–8218.
20. He, B., Bowen, N. T., Minges, J. T., and Wilson, E. M. (2001) Androgen-induced NH_2- and carboxyl-terminal interaction inhibits p160 coactivator recruitment by activation function 2. *J. Biol. Chem.* **276**, 42293–42301.
21. He, B., Gampe, R. T., Hnat, A. T., Faggart, J. L., Minges, J. T., French, F. S., and Wilson, E. M. (2006) Probing the functional link between androgen receptor coactivator and ligand binding sites in prostate cancer and androgen insensitivity. *J. Biol. Chem.* **281**, 6648–6663.
22. He, B., Gampe, R. T., Kole, A. J., Hnat, A. T., Stanley, T. B., An, G., Stewart, E. L., Kalman, R. I., Minges, J. T., and Wilson, E. M. (2004) Structural basis for androgen receptor interdomain and coactivator interactions suggests a transition in nuclear receptor activation function dominance. *Mol. Cell* **16**, 425–438.
23. Simental, J. A., Sar, M., Lane, M. V., French, F. S., and Wilson, E. M. (1991) Transcriptional activation and nuclear targeting signals of the human androgen receptor. *J. Biol. Chem.* **266**, 510–518.
24. Bai, S., He, B., and Wilson, E. M. (2005) Melanoma antigen gene protein MAGE-11 regulates androgen receptor function by modulating the interdomain interaction. *Mol. Cell. Biol.* **25**, 1238–1257.
25. Askew, E. B., Bai, S., Hnat, A. T., Minges, J. T., and Wilson, E. M. (2009) Melanoma antigen gene protein-A11 (MAGE-11) F-box links the androgen receptor NH_2-terminal transactivation domain to p160 coactivators. *J. Biol. Chem.* **284**, 34793–34808.
26. Bai, S., and Wilson, E. M. (2008) Epidermal growth factor-dependent phosphorylation and ubiquitinylation of MAGE-11 regulates its interaction with the androgen receptor. *Mol. Cell. Biol.* **28**, 1947–1963.
27. Finkel, T., Duc, J., Fearon, E. R., Dang, C. V., and Tomaselli, G. F. (1993) Detection and modulation in vivo of helix–loop–helix protein–protein interactions. *J. Biol. Chem.* **268**, 5–8.
28. Wagner, B. L., Norris. J. D., Knotts, T. A., Weigel, N. L., and McDonnell, D. P. (1998) The nuclear corepressors NCoR and SMRT are key regulators of both ligand- and 8-bromo-cyclic AMP-dependent transcriptional activity of the human progesterone receptor. *Mol. Cell. Biol.* **18**, 1369–1378.
29. Raivio, T., Palvimo, J. J., Dunkel, L., Wickman, S., and Jänne, O. A. (2001) Novel assay for determination of androgen bioactivity in human serum. *J. Clin. Endocrinol. Metab.* **86**, 1539–1544.
30. Kemppainen, J. A., and Wilson, E. M. (1996) Agonist and antagonist activities of hydroxyflutamide and Casodex relate to androgen receptor stabilization. *Urology* **48**, 157–163.
31. Berrevoets, C. A., Doesburg, P., Steketee, K., Trapman, J., and Brinkmann, A. O. (1998) Functional interactions of the AF-2 activation domain core region of the human androgen receptor with the amino-terminal domain and with the transcriptional coactivator TIF2 (transcriptional intermediary factor2). *Mol. Endocrinol.* **12**, 1172–1183.
32. He, B., Minges, J. T., Lee, L. W., and Wilson, E. M. (2002) The FXXLF motif mediates androgen receptor-specific interactions with coregulators. *J. Biol. Chem.* **277**, 10226–10235.
33. Doesburg, P., Kuil, C. W., Berrevoets, C. A., Steketee, K., Faber, P. W., Mulder, E., Brinkmann, A. O., and Trapman, J. (1997) Functional in vivo interaction between the amino-terminal, transactivation domain and the ligand binding domain of the androgen receptor. *Biochemistry* **36**, 1052–1064.

34. Lubahn, D. B., Joseph, D. R., Sar, M., Tan, J., Higgs, H. N., Larson, R. E., French, F. S., and Wilson, E. M. (1988) The human androgen receptor: complementary deoxyribonucleic acid cloning, sequence analysis and gene expression in prostate. *Mol. Endocrinol.* **2**, 1265–1275.
35. Choong, C. S., Kemppainen, J. A., Zhou, Z. X., and Wilson, E. M. (1996) Reduced androgen receptor gene expression with first exon CAG repeat expansion. *Mol. Endocrinol.* **10**, 1527–1535.
36. Choong, C. S., Kemppainen, J. A., and Wilson, E. M. (1998) Evolution of the primate androgen receptor: a structural basis for disease. *J. Mol. Evol.* **47**, 334–342.
37. Choong, C. S., and Wilson, E. M. (1998) Trinucleotide repeats in the human androgen receptor: a molecular basis for disease. *J. Mol. Endocrinol.* **21**, 235–257.
38. Wang, Q., Lu, J., and Yong E. L. (2001) Ligand- and coactivator-mediated transactivation function (AF2) of the androgen receptor ligand-binding domain is inhibited by the cognate hinge region. *J. Biol. Chem.* **276**, 7493–7499.
39. Haelens, A., Tanner, T., Denayer, S., Callewaert, L., and Claessens, F. (2007) The hinge region regulates DNA binding, nuclear translocation, and transactivation of the androgen receptor. *Cancer Res.* **67**, 4514–4523.
40. Zhou, Z. X., Sar, M., Simental, J. A., Lane, M. V., and Wilson, E. M. (1994) A ligand-dependent bipartite nuclear targeting signal in the human androgen receptor. Requirement for the DNA-binding domain and modulation by NH_2-terminal and carboxyl-terminal sequences. *J. Biol. Chem.* **269**, 13115–13123.
41. Gregory, C. W., He, B., Johnson, R. T., Ford, O. H., Mohler, J. L., French, F. S., and Wilson, E. M. (2001) A mechanism for androgen receptor-mediated prostate cancer recurrence after androgen deprivation therapy. *Cancer Res.* **61**, 4315–4319.
42. Chandra, S., Shao, J., Li, J. X., Li, M., Longo, F. M., and Diamond, M. I. (2008) A common motif targets huntingtin and the androgen receptor to the proteasome. *J. Biol. Chem.* **283**, 23950–23955.
43. Dehm, S. M., Regan, K. M., Schmidt, L. J., and Tindall, D. J. (2007) Selective role of an NH_2-terminal WxxLF motif for aberrant androgen receptor activation in androgen depletion independent prostate cancer cells. *Cancer Res.* **67**, 10067–10077.
44. Hsu, C. L., Chen, Y. L., Yeh, S., Ting, H. J., Hu, Y. C., Lin, H., Wang, X., and Chang, C. (2003) The use of phage display technique for the isolation of androgen receptor interacting peptides with (F/W)XXL(F/W) and FXXLY new signature motifs. *J. Biol. Chem.* **278**, 23691–23698.
45. van de Wijngaart, D. J., Dubbink, H. J., Molier, M., de Vos, C., Trapman, J., and Jenster, G. (2009) Functional screening of FxxLF-like peptide motifs identifies SMARCD1/BAF60a as an androgen receptor cofactor that modulates TMPRSS2 expression. *Mol. Endocrinol.* **23**, 1776–1786.
46. van de Wijngaart, D. J., van Royen, M. E., Hersmus, R., Pike, A. C., Houtsmuller, A. B., Jenster, G., Trapman, J., and Dubbink, H. J. (2006) Novel FXXFF and FXXMF motifs in androgen receptor cofactors mediate high affinity and specific interactions with the ligand-binding domain. *J. Biol. Chem.* **281**, 19407–19416.
47. Burd, C. J., Petre, C. E., Moghadam, H., Wilson, E. M., and Knudsen, K. E. (2005) Cyclin D1 binding to the androgen receptor NH_2-terminal domain inhibits AF2 association and reveals dual roles for AR corepression. *Mol. Endocrinol.* **19**, 607–620.
48. Shenk, J. L., Fisher, C. J., Chen, S. Y., Zhou, X. F., Tillman, K., and Shemshedini, L. (2001) p53 represses androgen-induced transactivation of prostate-specific antigen by disrupting hAR amino- to carboxyl-terminal interaction. *J. Biol. Chem.* **276**, 38472–38479.
49. Uesugi, M., and Verdine, G. L. (1999) The alpha-helical FXXPhiPhi motif in p53: TAF interaction and discrimination by MDM2. *Proc. Natl. Acad. Sci. USA* **96**, 14801–14806.
50. Wang, L., Hsu, C. L., Ni, J., Wang, P. H., Yeh, S., Keng, P., and Chang, C. (2004) Human checkpoint protein hRad9 functions as a negative coregulator to repress androgen receptor transactivation in prostate cancer cells. *Mol. Cell. Biol.* **24**, 2202–2213.
51. Ma, Q., Fu, W., Li, P., Nicosia, S. V., Jenster, G., Zhang, X., and Bai, W. (2009) FoxO1 mediates PTEN suppression of androgen receptor N- and C-terminal interactions and coactivator recruitment. *Mol. Endocrinol.* **23**, 213–225.
52. Dotzlaw, H., Moehren, U., Mink, S., Cato, A. C., Iñiguez Lluhí, J. A., and Baniahmad, A. (2002) The amino terminus of the human AR is target for corepressor action and antihormone agonism. *Mol. Endocrinol.* **16**, 661–673.
53. Liao, G., Chen, L. Y., Zhang, A., Godavarthy, A., Xia, F., Ghosh, J. C., Li, H., and Chen, J. D. (2003) Regulation of andro-

gen receptor activity by the nuclear receptor corepressor SMRT. *J. Biol. Chem.* **278**, 5052–5061.

54. Cheng, S., Brzostek, S., Lee, S. R., Hollenberg, A. N., and Balk, S. P. (2002) Inhibition of the dihydrotestosterone-activated androgen receptor by nuclear receptor corepressor. *Mol. Endocrinol.* **16**, 1492–1501.
55. Wu, Y., Kawate, H., Ohnaka, K., Nawata, H., and Takayanagi, R. (2006) Nuclear compartmentalization of N-CoR and its interactions with steroid receptors. *Mol. Cell. Biol.* **26**, 6633–6655.
56. Bubulya, A., Chen, S. Y., Fisher, C. J., Zheng, Z., Shen, X. Q., and Shemshedini, L. (2001) c-Jun potentiates the functional interaction between the amino and carboxyl termini of the androgen receptor. *J. Biol. Chem.* **276**, 44704–44711.
57. Schaufele, F., Carbonell, X., Guerbadot, M., Borngraeber, S., Chapman, M. S., Ma, A. A., Miner, J. N., and Diamond, M. I. (2005) The structural basis of androgen receptor activation: intramolecular and intermolecular amino-carboxy interactions. *Proc. Natl. Acad. Sci. USA* **102**, 9802–9807.
58. van Royen, M. E., Cunha, S. M., Brink, M. C., Mattern, K. A., Nigg, A. L., Dubbink, H. J., Verschure, P. J., Trapman, J., and Houtsmuller, A. B. (2007) Compartmentalization of androgen receptor protein-protein interactions in living cells. *J. Cell Biol.* **177**, 63–72.
59. Klokk, T. I., Kurys, P., Elbi, C., Nagaich, A. K., Hendarwanto, A., Slagsvold, T., Chang, C. Y., Hager, G. L., and Saatcioglu, F. (2007) Ligand-specific dynamics of the androgen receptor at its response element in living cells. *Mol. Cell. Biol.* **27**, 1823–1843.
60. Guo, Z., and Eisenberg, D. (2006) Runaway domain swapping in amyloid-like fibrils of T7 endonuclease I. *Proc. Natl. Acad. Sci. USA* **103**, 8042–8047.
61. La Spada, A. R., Wilson, E. M., Lubahn, D. B., Harding, A. E., and Fischbeck, K. H. (1991) Androgen receptor gene mutations in X-linked spinal and bulbar muscular atrophy. *Nature* **352**, 77–79.
62. Tetel, M. J., Giangrande, P. H., Leonhardt, S. A., McDonnell, D. P., and Edwards, D. P. (1999) Hormone-dependent interaction between the amino- and carboxyl-terminal domains of progesterone receptor in vitro and in vivo. *Mol. Endocrinol.* **13**, 910–924.
63. Kraus, W. L., McInerney, E. M., and Katzenellenbogen, B. S. (1995) Ligand-dependent, transcriptionally productive association of the amino- and carboxyl-terminal regions of a steroid hormone nuclear receptor. *Proc. Natl. Acad. Sci. USA* **92**, 12314–12318.
64. Rogerson, F. M. and Fuller, P. J. (2003) Interdomain interactions in the mineralocorticoid receptor. *Mol. Cell. Endocrinol.* **200**, 45–55.
65. Tung, L., Abdel-Hafiz, H., Shen, T., Harvell, D. M., Nitao, L. K., Richer, J. K., Sartorius, C. A., Takimoto, G. S., and Horwitz, K. B. (2006) Progesterone receptors (PR)-B and -A regulate transcription by different mechanisms: AF-3 exerts regulatory control over coactivator binding to PR-B. *Mol. Endocrinol.* **20**, 2656–2670.
66. Pippal, J. B., Yao, Y., Rogerson, F. M., and Fuller, P. J. (2009) Structural and functional characterization of the interdomain interaction in the mineralocorticoid receptor. *Mol. Endocrinol.* **23**, 1360–1370.
67. Murai-Takeda, A., Shibata, H., Kurihara, I., Kobayashi, S., Yokota, K., Suda, N., Mitsuishi, Y., Jo, R., Kitagawa, H., Kato, S., Saruta, T., and Itoh, H. (2010) NF-YC functions as a corepressor of agonist-bound mineralocorticoid receptor. *J. Biol. Chem.* **285**, 8084–8093.

Chapter 9

Methods to Study Dynamic Interaction of Androgen Receptor with Chromatin in Living Cells

Hatice Zeynep Kirli and Fahri Saatcioglu

Abstract

Androgen receptor (AR) is a ligand-dependent transcription factor that belongs to the nuclear receptor superfamily. In the presence of its specific ligands, AR translocates into the nucleus, interacts with chromatin at hormone response elements (HREs) and recruits a variety of coregulators and basal transcription factors to regulate transcription. Using green fluorescent protein (GFP) labelling and the tandem gene array system of mouse mammary tumour virus (MMTV), the interaction of AR with HREs can be visualized and studied in live cells. The MMTV array in nuclei can be specifically detected by DNA fluorescence in situ hybridization (DNA FISH) and thereby specific binding of GFP-AR to the array can be confirmed in the presence of specific ligands. The transcriptional activity of GFP-AR at the MMTV array can be visualized by RNA FISH in combination with interactions of GFP-AR or its cofactors, or different components of the transcriptional initiation complex, by indirect immunofluorescence (IF). Finally, using fluorescence recovery after photobleaching (FRAP), dynamic interactions of GFP-AR with the chromatin template can be studied. Methods to carry out these experiments are described herein.

Key words: Androgen receptor, chromatin, transcription, confocal microscopy, DNA FISH, RNA FISH, FRAP.

1. Introduction

Androgen receptor (AR) mediates the action of androgens through interaction with chromatin and regulation of target gene expression. Upon ligand binding, AR changes conformation, translocates to the nucleus, binds to its specific hormone response elements (HREs) in the vicinity of target genes and recruits coregulators which results in the assembly of the general transcription machinery and regulation of transcription (1, 2). The classical

F. Saatcioglu (ed.), *Androgen Action*, Methods in Molecular Biology 776,
DOI 10.1007/978-1-61779-243-4_9, © Springer Science+Business Media, LLC 2011

view of nuclear receptor (NR) action suggests that, similar to other transcription factors, NRs are stably associated with their target sites in chromatin in the presence of ligand, serving as a platform for sequential recruitment of protein complexes that results in the regulation of transcription (3, 4). Advances in green fluorescent protein (GFP) technology and quantitative live cell microscopy have resulted in an alternative view, called the "hit-and-run" model, where ligand-bound NRs interact only transiently with the chromatin template, recruit other factors and are dynamically displaced from the target site on a timescale of seconds (5, 6). The importance of this short-term binding dynamics in the context of long-term cyclical behaviour of chromatin decondensation and transcription binding events has previously been discussed (7).

Using the murine mammary adenocarcinoma cell line 3134 that contains the tandem gene array of mouse mammary tumour virus long terminal repeat (MMTV LTR), the dynamic interactions between steroid receptors and their HREs in chromatin can be visualized in living cells (6). 3134 cells contain 200 copies of MMTV-Harvey viral ras (v-Ha-ras) gene stably integrated into chromosome 4. The derivative cell line, 3108, stably expresses GFP-AR under the control of a tetracycline repressible promoter (8). MMTV DNA can be visualized by DNA FISH, confirming the specific recruitment of ligand-bound receptor to the array (9, 10). FRAP analysis revealed that GFP-AR transiently interacts at the MMTV array and its recovery kinetics is strongly ligand dependent, as agonist-bound GFP-AR showed reduced recovery compared to unliganded or antagonist-bound GFP-AR (8). There are also differences in the dynamic interactions of wild-type GFP-AR with chromatin compared with a transcriptionally deficient mutant (7). RNA FISH confirmed GFP-AR-mediated activation of transcription at the MMTV array in a ligand-dependent manner (8). These methods are described herein.

2. Materials

2.1. Cell Culture

1. 3108 cell line as previously described (8) (*see* **Note 1**).
2. Dulbecco's modified Eagle's medium (DMEM) supplemented with 4.5 g/l D-glucose, 2 mM L-glutamine, 5 mg/ml penicillin–streptomycin, 10 μg/ml tetracycline and 10% foetal bovine serum (FBS) (*see* **Note 2**).
3. Trypsin/EDTA.

4. Phosphate buffered saline (PBS): 150 mM Sodium chloride (NaCl), 6.7 mM disodium hydrogen phosphate dihydrate, 1.85 mM sodium dihydrogen phosphate monohydrate in water (pH 7.4) (*see* **Note 3**); sterilized by autoclaving.
5. Charcoal-stripped FBS is prepared by treating with activated charcoal (Sigma Aldrich) and then sterile filtering in sequence through filters of 0.45 and 0.2 μm pore sizes (*see* **Note 4**).
6. Synthetic androgen methyltrienolone (R1881) (DuPont NEN Research Products) is dissolved in 100% ethanol and stored at –20°C. 1000× Stock solutions are prepared by dilution at 10 μM in 100% ethanol.
7. Hydroxyflutamide (OHF) (Schering-Plough Research Institute, Kenilworth, NJ, USA) is dissolved in 100% ethanol and stored at –20°C. 1000× Stock solutions are prepared by dilution at 10 mM in 100% ethanol, also kept at –20°C. Make new dilution once a month.

2.2. Fluorescence Recovery After Photobleaching (FRAP)

1. MatTek 35-mm glass bottom microwell dishes (MatTek Corporation).
2. DMEM.
3. DMEM without phenol red (BioWhittaker, Cambrex Bio Science) as in Step 2, but supplemented with 10% charcoal-stripped FBS.

2.3. DNA FISH in Combination with Indirect Immunofluorescence (IF)

1. Six-well tissue culture dishes and glass slides (sterilized, 0.96–1.06 mm thickness). Microscope coverslips (22 mm × 22 mm).
2. DMEM with phenol red as in Step 2, but supplemented with 10% charcoal-stripped FBS.
3. Paraformaldehyde (Electron Microscopy Sciences, Hatfield, PA): Prepare 3% (w/v) solution in PBS. Add the paraformaldehyde powder into water, add a drop of 4 M NaOH and dissolve by heating the solution to 65°C on a hot plate while stirring. After cooling down to room temperature, add 10× PBS to get a final solution of 1× PBS (adjust pH to 7.4 with HCl). Store in single-use aliquots at –20°C.
4. Permeabilization solution: 0.5% Triton X-100 in PBS.
5. RNase solution: Buffer E1 of the JETSTAR Plasmid Purification Kit (Genomed, GmbH) with 100 μg/ml RNase.
6. DIG-Nick Translation Mix and anti-digoxigenin–rhodamine, Fab fragments (Roche Applied Science).

7. 20 mg/ml Glycogen and 3 M sodium acetate, pH 5.2, used to precipitate labelled DNA probe.
8. Saline–sodium citrate (SSC) buffer (pH 7.0): Prepare 20× stock with 3 M sodium chloride and 0.3 M trisodium citrate dihydrate in water (adjust pH to 7.0 with HCl). Prepare working solution by dilution of one part with nine parts water (2×) or with four parts water (4×).
9. Deionized formamide (minimum 99.5% GC) and tRNA from Baker's yeast (10 mg/ml stock solution in 10 mM Tris–HCl, pH 7.4, 1 mM EDTA) (Sigma Aldrich); mouse cot-1 DNA (1 mg/ml stock solution in 10 mM Tris–HCl, pH 7.4, 1 mM EDTA diluted 1:9 in water) and salmon sperm DNA (Invitrogen) (10 mg/ml stock solution in water).
10. DNA denaturation solution: 50% Formamide in 2× SSC.
11. Dehydration solutions: 70% Ethanol, 90% ethanol, 100% ethanol.
12. 2.5× Hybridization buffer: 2.5× SSC, 25% dextran sulphate and 2.5 mg/ml tRNA. Make stock (e.g. 2 ml) and store at –20°C in aliquots.
13. Hybridization mixture: 1× Hybridization buffer, 25% formamide, 200 ng labelled DNA probe.
14. Humid chamber: Dark, plastic foil-covered box with wet paper tissue inside.
15. Sealing solution for coverslips: Prepare 2% agarose in water, heat in a microwave oven to melt and then cool to seal coverslips to the glass slide.
16. Antibody dilution buffer for primary antibody in immunofluorescence: PBS.
17. Primary antibody for immunofluorescence: Anti-GFP rabbit IgG fraction (Invitrogen) (*see* **Note 5**).
18. Secondary antibody for immunofluorescence: Anti-rabbit Alexa Fluor 488 (*see* **Note 5**).
19. Detection solution for the hybridized probe: 1.5 μg/ml Anti-digoxigenin–rhodamine in 4× SSC, 0.1% BSA and 0.01% Tween 20.
20. Mounting medium: ProLong Gold Antifade Reagent (Invitrogen).

2.4. RNA FISH in Combination with IF

1. Six-well tissue culture dishes and glass slides as in **Section 2.3**, Step 1.
2. DMEM and supplements as in **Section 2.3**, Step 2.
3. Paraformaldehyde: Prepare 3 and 5% (w/v) solution in PBS (*see* **Section 2.3**, Step 3).

4. Permeabilization solution: As in **Section 2.3**, Step 4. Antibody dilution buffer for immunofluorescence: PBS.
5. Primary antibody for immunofluorescence: Anti-GFP rabbit IgG fraction (Invitrogen), anti-CREB-binding protein (CBP) NT rabbit polyclonal antibody (Upstate, Millipore) (*see* **Note 5**).
6. Secondary antibody for immunofluorescence: Anti-rabbit Alexa Fluor 488, anti-rabbit Alexa Fluor 647 (Invitrogen).
7. DIG-Nick Translation Mix and anti-digoxigenin–rhodamine, Fab fragments.
8. 20 mg/ml Glycogen and 3 M sodium acetate, pH 5.2.
9. SSC buffer (pH 7.0) as described in **Section 2.3**, Step 8.
10. Deionized formamide, tRNA from Baker's yeast, mouse cot-1 DNA and salmon sperm DNA (as in **Section 2.3**, Step 9).
11. Hybridization buffer (2.5×) as described in **Section 2.3**.
12. Hybridization mixture as in **Section 2.3**.
13. Humid chamber.
14. Sealing solution for coverslips, as in **Section 2.3**, Step 15.
15. Detection solution for the hybridized probe, as in **Section 2.3**, Step 19.
16. Nuclear stain: 1000× DAPI (4,6-diamidino-2-phenylindole) stock (5 μg/ml) in water. Prepare working solution by dilution of one part in thousand parts of PBS (*see* **Note 6**).
17. Mounting medium, as in **Section 2.3**, Step 20.

3. Methods

In order to allow GFP-AR expression, cells are grown in medium without tetracycline for 3 days prior to the experiment, including 2 days in phenol red-free medium with 10% charcoal-stripped serum. To study the effects of steroid hormones in vitro and in vivo, endogenous hormones, growth factors and cytokines are removed by charcoal stripping of serum. Ligand-specific dynamics of AR at its response elements can be studied in living cells by time-lapse microscopy which facilitates visualization of GFP-tagged AR at the MMTV array against the background of a large number of unbound molecules in the nucleus (8). GFP-AR is observed as a bright spot at the MMTV array in the nucleus within the background of the single copy binding sites. By FRAP,

GFP-AR at the array is bleached for a short time and images of the array are acquired before and after the bleach pulse. Fluorescence intensities in the regions of interest are analysed in order to generate the FRAP recovery curve. This allows to quantitatively measure the kinetics of GFP-AR binding to the MMTV array (8).

The tandem repeat organization of the MMTV in the nucleus allows following of transcription from the MMTV promoter in single cells. Using RNA FISH, a labelled probe which includes the target gene reporter v-ras is hybridized with the ras transcript in the tandem array in fixed samples and subsequently the labelled probe is detected by confocal microscopy. Analysis of fluorescence intensity of the probe is proportional to the magnitude of ligand-mediated transcription at the MMTV array (8). In combination with IF, recruitment of chromatin-associated proteins, such as cofactors and modified histones, can be observed on the array.

The tandem gene repeat of MMTV can be localized in the nuclei of the cells by performing DNA FISH. This method follows a similar procedure as RNA FISH with the addition of the RNase treatment and denaturation of the DNA in order to hybridize the probe to the ras reporter gene in the tandem array. Colocalization of ligand-bound GFP-AR with the DNA FISH signal of MMTV confirms the specific recruitment of the nuclear receptor to the array. Furthermore, DNA FISH signal can be used for possible condensation/decondensation events that would indicate chromatin transitions (10).

3.1. FRAP

1. The cells are passaged when approaching 80% confluence with trypsin/EDTA to provide maintenance cultures on 100-mm tissue culture dishes and experimental cultures on 35-mm glass bottom microwell dishes (MatTek dishes). Seventy thousand cells are plated per MatTek dish in growth medium without tetracycline and incubated for 24 h (*see* **Note 7**).
2. Aspirate off the medium the next day and wash once with 2 ml PBS. Feed cells with phenol red-free growth medium without tetracycline and incubate for 48 h.
3. Add R1881 (to a final concentration of 10 nM) or OHF (to a final concentration of 1 μM) directly to the culture medium from the stock solutions. Incubate cells for 1 and 1.5 h, respectively (*see* **Note 8**).
4. FRAP analysis is carried out on an Olympus FluoView 1000 confocal laser scanning microscope with an incubator maintained at 37°C. Images are captured with a 60×/1.4-numerical aperture oil immersion objective and 50-mW argon laser. Five single prebleach images are acquired followed by a brief bleach pulse of 100 ms using 405-nm laser line at 100% laser power (laser output, 10%) without

attenuation. Single optical sections are acquired at 500 ms intervals by using a 488-nm laser line with laser power attenuated to 10% (*see* **Note 9**). Representative images are shown in **Fig. 9.1a**.

5. The bleached area, the nucleus and an area outside of the cell are marked as regions of interest and fluorescence intensities are analysed by Olympus FV10-ASW 1.7b software and Microsoft Excel. Average fluorescence intensities of the bleached area and the nucleus are determined in each single optical section by subtracting the average intensity in the area outside of the monitored cell. Fluorescence intensity in the bleached region is normalized to that in the whole nucleus for each time point. The average fluorescence intensity of the bleached area in prebleach images is then determined. Fluorescence intensity of the bleached region is normalized to the average prebleach intensity in each single image. FRAP recovery curves are generated using Microsoft Excel (*see* **Note 10**) as presented in **Fig. 9.1b**.

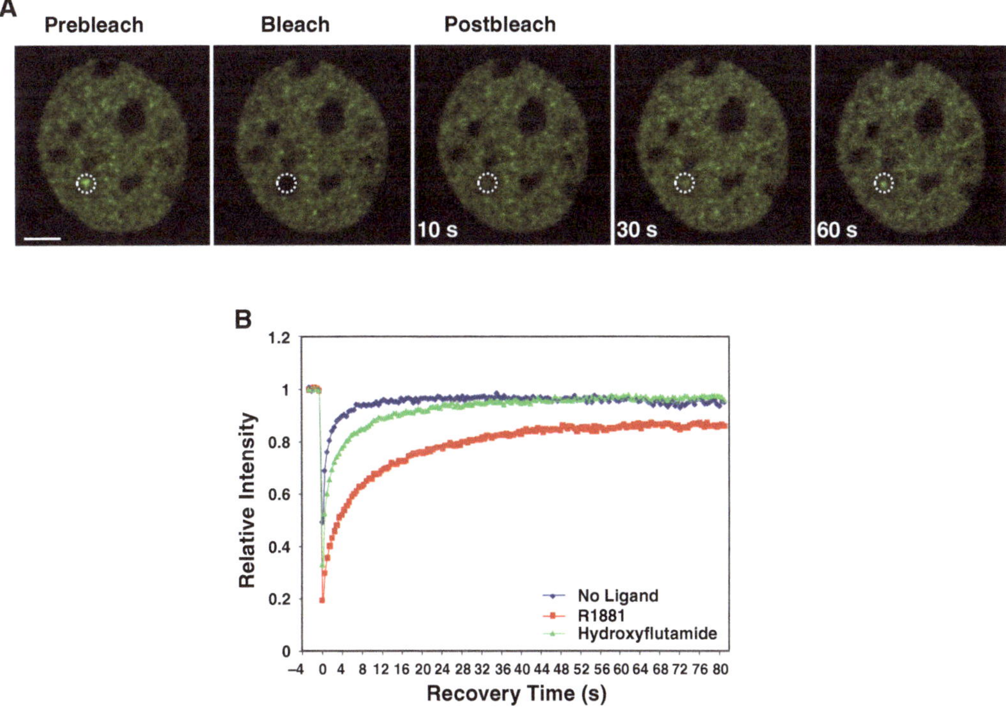

Fig. 9.1. Analysis of AR dynamics at the MMTV array by FRAP. (**a**) 3108 cells were treated with R1881 (10 nM) for 1 h. The cells were imaged before and during recovery after bleaching of GFP-AR at the array. Images taken at 10, 30 and 60 s are shown indicating the rapid recovery kinetics of GFP-AR at the MMTV array. Bar, 4 μm. (**b**) Cells were either left untreated, treated with R1881 (10 nM) for 1 h or with OHF (1 μM) for 1.5 h. For each condition, approximately 25 cells were analysed. Quantitative FRAP analysis reveals significantly slower recovery kinetics of GFP-AR in the presence of agonist R1881 compared with unliganded or antagonist-bound GFP-AR (8).

3.2. DNA FISH

3.2.1. Labelling Reaction with DIG-dUTP

1. Add 16 μl water containing 1 μg DNA template (pM18 plasmid which includes the MMTV LTR, vras) (for five reactions) to a reaction tube (*see* **Note 11**).
2. Add 4 μl of DIG-Nick Translation Mix.
3. Mix gently, spin down and incubate at 15°C for 105 min.
4. Run 0.2 μg of the DNA template on a 1% agarose gel to check the probe size after the 105 min incubation. During gel electrophoresis, the rest of the sample can be kept on ice. The fragment size should be less than 500 bp. Otherwise, incubate the DNA template for longer time at 15°C.
5. Precipitate the DNA by addition of 3 μl of 20 mg/ml glycogen, 1/10 volume 3 M sodium acetate, pH 5.2, and 3 volumes of ice-cold 100% ethanol.
6. Mix well and incubate at –80°C for 30 min.
7. Centrifuge at 15,800×*g* (13,000 rpm on a benchtop centrifuge) for 30 min at 4°C.
8. Aspirate the supernatant and air-dry the pellet.
9. Resuspend the pellet in 16 μl water.
10. Vortex until the pellet is dissolved and incubate at 37°C for 30 min. Store at –20°C.

3.2.2. Preparation of Cells

1. The cells are passaged as described above (*see* **Section 3.1**), except that for the experiments they are grown on coverslips in six-well dishes. Coverslips are sterilized by washing with 70% ethanol in a tube. They are then placed into the six-well dishes, one coverslip per well, and washed twice with 2 ml PBS in order to remove any traces of ethanol. Cells are then grown on the coverslips in growth medium without tetracycline and incubated for 24 h.
2. Aspirate off the medium the next day and wash once with 2 ml PBS. Add growth medium without tetracycline and incubate for 48 h.
3. Add R1881 (to a final concentration of 10 nM) directly to the culture medium. Place cells in the incubator for 4 h (*see* **Note 8**).

3.2.3. Preparation of the Labelled DNA Probe

1. Per reaction, mix 200 ng of DNA probe (4 μl), 5 μl of tRNA (10 mg/ml), 4 μl of mouse cot-1 DNA (0.1 mg/ml) and 5 μl of salmon sperm DNA (10 mg/ml), and mix well (*see* **Note 12**).
2. Add 1/10 volume of 3 M sodium acetate, pH 5.2, and 3 volumes of ice-cold 100% ethanol.
3. Mix well by vortexing and incubate at –80°C for 30 min.

4. Centrifuge at 15,800×*g* (13,000 rpm on a benchtop centrifuge) for 30 min at 4°C.
5. Aspirate the supernatant and air-dry the pellet.
6. Resuspend the pellet in 6 μl formamide and vortex for 5 min.
7. Spin down and incubate at 37°C for 10 min. Vortex again and spin down.
8. Repeat Step 7 twice (*see* **Note 13**).
9. Add 8.4 μl water, vortex and incubate at 37°C for 10 min.
10. Denature the probe at 95°C for 8 min and then keep on ice until use.
11. Just before the hybridization, add 9.6 μl of 2.5× hybridization buffer to the probe and mix by pipetting.

3.2.4. Hybridization of DNA Probe

1. Aspirate off the medium after 4 h treatment of cells with R1881.
2. Fix cells in 3% paraformaldehyde at room temperature for 20 min (*see* **Note 14**).
3. Wash cells three times with PBS (2 ml) each for 5 min.
4. Add 2 ml of permeabilization solution and incubate at room temperature for 10 min (*see* **Note 15**).
5. Treat cells with RNase solution (2 ml) at room temperature for 15 min.
6. Place a clean glass slide on a hot plate heated to 95°C and pipette 100 μl of DNA denaturation solution onto it. Dry excess of liquid from the coverslip and place it on the top of the denaturation solution (cells down).
7. Incubate for 5 min at 95°C.
8. Place the glass slide on a metal block kept on ice and incubate for 5 min.
9. Place coverslip back into the six-well dish and dehydrate cells sequentially in 70, 90 and 100% ethanol, each for 5 min on ice.
10. Remove the coverslip from the six-well dish and air-dry with cell side facing up.
11. Pipette 24 μl of probe (*see* **Section 3.2.3**) onto a clean glass slide and place on top of the solution.
12. Seal the coverslip with molten agarose (2%) and let air-dry.
13. Incubate at 37°C in humid chamber for 18–24 h.
14. Remove agarose from the coverslip and add 2× SSC to loosen it from the slide. Place it in a clean six-well dish to continue with the procedure.

15. Wash once with 2× SSC and then once with 2× SSC containing 0.05% Triton X-100 at room temperature each for 15 min. Shake the dish every 3–4 min.
16. Wash once with 2× SSC for 15 min and once with 4× SSC for 5 min at room temperature (*see* **Note 16**).
17. For immunofluorescence detection, wash twice with PBS, each for 10 min at room temperature.
18. Dilute the primary antibody in PBS (anti-GFP, 1:400) and apply 75–100 μl on a piece of parafilm (*see* **Note 17**).
19. Incubate at room temperature for 1.5 h in the dark.
20. Wash three times with PBS, each for 5 min at room temperature.
21. Pipette 75–100 μl of detection solution (*see* **Section 2.4**, Step 16) containing the secondary antibody (anti-rabbit Alexa Fluor 488 1:200). Dry excess of liquid from the coverslip and place it on the top of the detection solution (cells down).
22. Incubate at 37°C in a humid chamber for 1 h.
23. Place coverslip back to six-well dish and wash once with 4× SSC containing 0.05% Triton X-100 for 10 min at room temperature.
24. Wash twice with 4× SSC, each for 10 min and then once with 2× SSC for 5 min at room temperature.
25. Wash twice with PBS, each for 5 min at room temperature.
26. Mount the coverslip by inverting onto a drop of 10 μl of mounting medium on a clean glass slide. Avoid bubbles and let dry at room temperature in the dark for 24 h.
27. The slide is viewed under a confocal microscope. Excitation at 488 nm induces GFP fluorescence (green emission) for GFP-AR. Rhodamine fluorescence (red emission) is induced by excitation at 555 nm for the DNA FISH signal. Olympus FV10-ASW 1.7b software can be used to overlay fluorescence images as shown in **Fig. 9.2**.

3.3. RNA FISH

3.3.1. Labelling Reaction with DIG-dUTP

As described in **Section 3.2.1**.

3.3.2. Preparation of Cells

As described in **Section 3.2.2**.

3.3.3. Preparation of the Labelled DNA Probe

As described in **Section 3.2.3**.

3.3.4. Indirect Immunofluorescence and Hybridization of DNA Probe

1. Aspirate off the medium after 4 h treatment with R1881.
2. Fix cells in 3% paraformaldehyde at room temperature for 20 min (*see* **Note 14**).

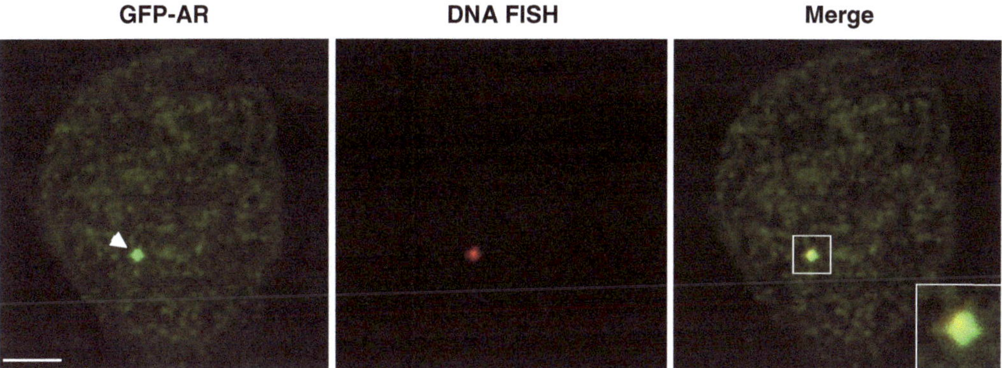

Fig. 9.2. MMTV LTR detected by DNA FISH. 3108 cells were treated with R1881 (10 nM) for 4 h and processed for DNA FISH. GFP-AR was detected by indirect immunofluorescence using an anti-GFP antibody and MMTV DNA was detected by DNA FISH using a probe specific to the entire MMTV LTR array. Confocal images were obtained on an Olympus FluoView 1000 confocal laser scanning microscope using a 60× oil immersion lens. Ligand-bound GFP-AR was colocalized with the DNA FISH signal confirming the specific interaction of GFP-AR with MMTV chromatin. *Arrow* points to GFP-AR on the MMTV array. Bar, 4 μm.

3. Wash cells three times with PBS (2 ml), each for 5 min.
4. Add 2 ml of permeabilization solution and incubate at room temperature for 10 min (*see* **Note 15**).
5. Dilute the primary antibody in PBS (anti-GFP 1:400; anti-CBP NT 1:100) and apply 75–100 μl on a piece of parafilm (*see* **Note 17**).
6. Dry off excess PBS from the coverslip and place it on top of the antibody solution (cells facing down).
7. Incubate at room temperature for 1 h in the dark.
8. Wash cells three times with PBS (2 ml) each for 5 min.
9. Dilute the secondary antibody in PBS (anti-rabbit Alexa Fluor 488 1:200; anti-rabbit Alexa Fluor 647 1:200) and apply it as described for the primary antibody in Steps 5 and 6.
10. Incubate at room temperature for 1 h in the dark.
11. Wash three times with PBS each for 10 min.
12. Fix with 5% paraformaldehyde at room temperature for 10 min (*see* **Note 18**).
13. Wash two times with PBS each for 8 min and once with 2× SSC for 5 min.
14. Pipette 24 μl of probe onto a clean glass slide.
15. Mount the coverslip on the probe and seal with molten agarose (2%) and let air-dry.
16. Incubate at 37°C in a humid chamber for 18–24 h.

17. Remove agarose from the coverslip and add 2× SSC to loosen it from the slide. Place it in a fresh six-well dish to continue with the experimental procedure.
18. Wash once with 2× SSC and then once with 2× SSC containing 0.05% Triton X-100 at room temperature each for 15 min. Shake the dish every 3–4 min.
19. Wash once with 2× SSC for 15 min and once with 4× SSC for 5 min at room temperature.
20. Pipette 75–100 μl of detection solution for the hybridized probe onto a clean glass slide. Dry excess liquid from the coverslip and place it on top of the detection solution (cells down).
21. Incubate at 37°C in a humid chamber for 1 h.
22. Place the coverslip back to the six-well dish and wash it three times with PBS each for 5 min.
23. DAPI stock solution is diluted in PBS at 1:1000 and added directly to cells for 5 min at room temperature to stain the DNA and to visualize the nuclei.

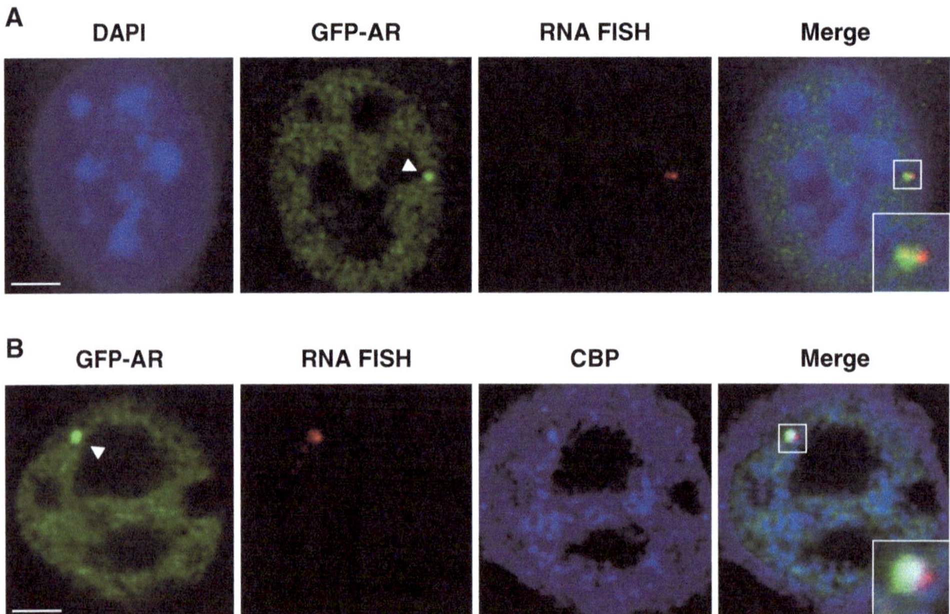

Fig. 9.3. Detection of GFP-AR-dependent transcription at the MMTV array by RNA FISH and recruitment of cofactor CBP to the array. (**a**) 3108 cells were treated with R1881 (10 nM) for 4 h and processed for RNA FISH. GFP-AR was detected by indirect immunofluorescence using an anti-GFP antibody. GFP-AR colocalizes with nascent MMTV transcripts. Bar, 4 μm. (**b**) Three thousand one hundred and eight cells were treated with R1881 (10 nM) for 4 h and subjected to RNA FISH and indirect immunofluorescence microscopy for CBP. Endogenous CBP colocalizes with nascent MMTV transcripts and GFP-AR. Confocal images were obtained on an Olympus FluoView 1000 confocal laser scanning microscope using a 60× oil immersion lens. Image for CBP is pseudocoloured with blue as staining with Alexa Fluor 647 emits far-red fluorescence. *Arrows* point to GFP-AR. Bar, 4 μm.

24. Wash the sample twice with PBS each for 5 min at room temperature.
25. Mount the coverslip by inverting onto a drop of 10 μl mounting medium on a clean glass slide. Avoid bubbles and let dry at room temperature in the dark for 24 h.
26. View the slide by confocal microscopy. Excitation at 488 nm induces Alexa Fluor 488 fluorescence (green emission) for GFP-AR. Rhodamine fluorescence (red emission) is induced by excitation at 555 nm for the RNA FISH signal. DAPI fluorescence (blue emission) and excitation at 633 nm induces for CBP stained with Alexa Fluor 647 (far-red emission). Olympus FV10-ASW 1.7b software can be used to overlay fluorescence images and analyse fluorescence intensities as shown in **Fig. 9.3a**, **b**.

4. Notes

1. There are currently no other cell lines with arrayed HREs that can be used to study AR–chromatin interactions, as opposed to general nucleoplasmic interactions of AR in living cells (11). However, such lines can be generated, for example, by using sequential cloning steps in *E. coli* (12) or using a plasmid that contains a mammalian replication initiation origin and a matrix attachment region from the Chinese hamster *dhfr* gene (13).
2. Tetracycline stock is prepared at 5 mg/ml in 100% ethanol and stored in the dark at –20°C. It is added to the medium each time the cells are passaged.
3. Unless stated otherwise, all solutions are prepared in ultrapure water from the Milli-Q system (Millipore) that is sterilized by autoclaving (20 min, 121°C).
4. 3 g of active charcoal powder is added to 45 ml of FBS in a 50-ml sterile centrifuge tube and incubated for 18 h at 4°C by rotation. The tube is centrifuged at 3500×*g* for 20 min at 4°C and the liquid is transferred into a new tube containing a new batch of 3 g active charcoal powder. The tube is incubated for 2 h at 4°C by rotation and centrifuged at 3500×*g* for 20 min at 4°C. All liquid is filtered first through a sterile filter with a pore size of 0.45 μm and then one with a pore size of 0.2 μm. Charcoal-stripped serum aliquots are stored at –20°C until use.

5. The immunofluorescence step is not required but highly recommended for visualization of GFP-AR as the GFP fluorescence signal decreases during the FISH procedure due to several washing steps. The GFP signal is enhanced by IF with anti-GFP as primary antibody and Alexa Fluor 488 as secondary antibody to visualize GFP-AR (green emission). If two or more primary antibodies are going to be used on the same coverslip, antibodies with different host animals need to be chosen in order to avoid cross-reactivity of the secondary antibodies.
6. DAPI stock is stored in the dark at –20°C.
7. The cells should be at 20–30% confluence the day after plating and at 80% confluence the day of the FRAP.
8. The incubation time needed with agonists and antagonists depends on the experimental setup. FRAP recovery curves with different incubation times should be used to find the optimum time point.
9. The microscope and imaging settings are different for each microscope, cell line and the protein of interest. They need to be adjusted for optimum results.
10. All cells and arrays that go out of focus during imaging are discarded from the analysis, as well as cells that have varying fluorescence intensity in the bleached area. After determination of FRAP recovery curve for each single cell, the data from all cells that have 10–30% fluorescence intensity left at the array after bleaching are pooled together for generating the average FRAP recovery curve.
11. Larger batches of the labelled probe can be made and stored at –20°C for 3 months.
12. The reaction mix should be prepared fresh for each experiment.
13. The pellet should completely dissolve during the last incubation at 37°C.
14. Cells can also be fixed with 3.5 or 4% paraformaldehyde at room temperature for 15 min.
15. Cells can also be permeabilized for 5 min on ice instead of at room temperature.
16. After the last wash (*see* Step 1), DNA FISH can be carried out with an additional indirect immunofluorescence step for visualization of GFP-AR as the denaturation step results in the loss of GFP fluorescence, cofactors or general transcription factors. If DNA FISH is performed without

immunofluorescence, the steps in RNA FISH after the last wash are followed (*see* **Section 3.3.4**, Steps 20–26).

17. The dilution of primary and secondary antibodies needs to be optimized for each cell line and protein of interest.
18. If RNA FISH is carried out without the indirect immunofluorescence step, cells are fixed, permeabilized with Triton X-100 and washed with PBS (*see* **Section 3.2.4**, Steps 1–4). Cells are then rinsed in 2× SSC for 5 min at room temperature and directly hybridized with the DNA probe (*see* **Section 3.3.4**, Step 14).

Acknowledgements

This work was supported by grants from the Norwegian Cancer Society and the Norwegian Research Council.

References

1. Dilworth, F. J., and Chambon, P. (2001) Nuclear receptors coordinate the activities of chromatin remodeling complexes and coactivators to facilitate initiation of transcription, *Oncogene 20*, 3047–3054.
2. Hager, G. L. (2001) Understanding nuclear receptor function: from DNA to chromatin to the interphase nucleus, *Prog Nucleic Acid Res Mol Biol 66*, 279–305.
3. McKenna, N. J., and O'Malley, B. W. (2002) Combinatorial control of gene expression by nuclear receptors and coregulators, *Cell 108*, 465–474.
4. Shang, Y., Myers, M., and Brown, M. (2002) Formation of the androgen receptor transcription complex, *Mol Cell 9*, 601–610.
5. Hager, G. L., Elbi, C., Johnson, T. A., Voss, T., Nagaich, A. K., Schiltz, R. L., Qiu, Y., and John, S. (2006) Chromatin dynamics and the evolution of alternate promoter states, *Chromosome Res 14*, 107–116.
6. McNally, J. G., Muller, W. G., Walker, D., Wolford, R., and Hager, G. L. (2000) The glucocorticoid receptor: rapid exchange with regulatory sites in living cells, *Science 287*, 1262–1265.
7. Hager, G. L., McNally, J. G., and Misteli, T. (2009) Transcription dynamics, *Mol Cell 35*, 741–753.
8. Klokk, T. I., Kurys, P., Elbi, C., Nagaich, A. K., Hendarwanto, A., Slagsvold, T., Chang, C. Y., Hager, G. L., and Saatcioglu, F. (2007) Ligand-specific dynamics of the androgen receptor at its response element in living cells, *Mol Cell Biol 27*, 1823–1843.
9. Johnson, T. A., Elbi, C., Parekh, B. S., Hager, G. L., and John, S. (2008) Chromatin remodeling complexes interact dynamically with a glucocorticoid receptor-regulated promoter, *Mol Biol Cell 19*, 3308–3322.
10. Muller, W. G., Walker, D., Hager, G. L., and McNally, J. G. (2001) Large-scale chromatin decondensation and recondensation regulated by transcription from a natural promoter, *J Cell Biol 154*, 33–48.
11. Farla, P., Hersmus, R., Trapman, J., and Houtsmuller, A. B. (2005) Antiandrogens prevent stable DNA-binding of the androgen receptor, *J Cell Sci 118*, 4187–4198.
12. Robinett, C. C., Straight, A., Li, G., Willhelm, C., Sudlow, G., Murray, A., and Belmont, A. S. (1996) In vivo localization of DNA sequences and visualization of large-scale chromatin organization using lac operator/repressor recognition, *J Cell Biol 135*, 1685–1700.
13. Shimizu, N., Miura, Y., Sakamoto, Y., and Tsutsui, K. (2001) Plasmids with a mammalian replication origin and a matrix attachment region initiate the event similar to gene amplification, *Cancer Res 61*, 6987–6990.

Chapter 10

FRET Analysis of Androgen Receptor Structure and Biochemistry in Living Cells

Fred Schaufele

Abstract

The androgen receptor (AR) is the central component of a dynamic conformational and interaction cascade initiated by androgenic hormones. AR function can be modified by cellular inputs not examined in test tube studies of AR action. Thus, there is a need to measure AR conformation and biochemistry directly within the cell where the intracellular locations, levels and availability of the hormone, AR, AR-interacting factors, DNA-binding sites, enzymes that modify those components of AR action, and factors that compete for the formation of functional AR–cofactor complexes may affect AR action. The dynamic nature of the AR functional cycle itself may introduce temporal fluctuations in factor status and location to affect AR output in the intact cell. This chapter focuses on the method of Förster resonance energy transfer which uniquely has the resolving power and ability to directly measure the conformation and biochemistry of AR signaling in living cells.

Key words: Androgen receptor, fluorescence microscopy, conformation, cellular biochemistry, Förster resonance energy transfer.

1. Introduction

Fluorescence microscopy can track, directly in living cells, any proteins expressed as fusions with fluorescent proteins (FPs) (1–4). FP tagging enables the amounts and locations of the FP to be measured as a surrogate of androgen receptor (AR) amount and location (*see* Chapters 8 and 9) (5, 6). To visualize the AR–FP fusion, the FP is excited into an activated state by high-energy (shorter wavelength) light (**Fig. 10.1a**, lightning bolt). The captured energy exists as a transient energy field around the fluorophore (second panel) which, as the FP returns to its

F. Saatcioglu (ed.), *Androgen Action*, Methods in Molecular Biology 776,
DOI 10.1007/978-1-61779-243-4_10,

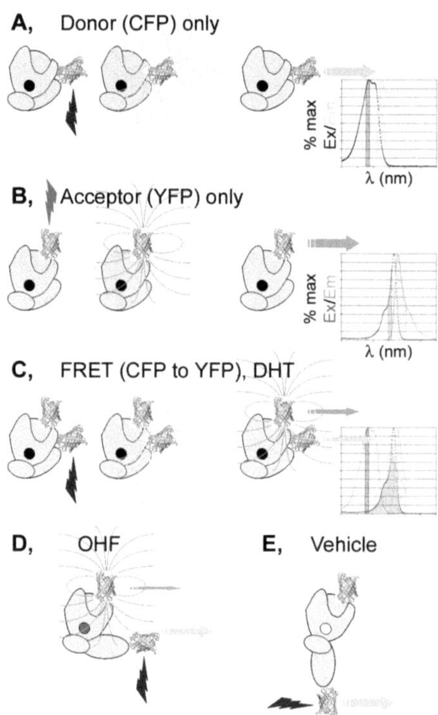

Fig. 10.1. Förster resonance energy transfer. (**a**) Principles of fluorescence: a fluorophore (CFP), attached to the amino terminus of CFP, is excited by light of a high-energy wavelength (lightning bolt). CFP will transiently capture (middle panel, "web") and release (*arrow*) that energy. Graph (*right panel*) shows the extent by which CFP is excited by 350–650-nm light (*left curve*). The corresponding emission curve is shown (*right curve*). Specific excitation and emission wavelengths to select for CFP in a typical CFP–YFP FRET study are shown as *shaded areas* under the curves. The rightmost emission filter depicted is that used to capture YFP emissions upon FRET; it is evident that a large amount of CFP fluorescence is captured in that same FRET channel. (**b**) A similar depiction of the excitation and emission of YFP, and the excitation/collection of YFP in the FRET channel, when YFP is attached to a different domain (the carboxy terminus) of AR. (**c**) When bound with agonist ligand (DHT, *black circle*), the N-terminal and C-terminal domains of AR move CFP and YFP into specific three-dimensional positions that provide a level of energy transfer characteristic of agonist binding. That energy transfer from CFP to YFP is quantified, following corrections for the independent contributions of CFP and YFP to the FRET channel, as the diminution in the amount of CFP fluorescence upon CFP excitation and the corresponding increase in the amount of acceptor fluorescence. The amounts of energy transfer are different when the AR is bound with (**d**) an antagonist ligand or (**e**) no ligand (6).

ground state, is released partly as lower energy (longer wavelength) light (arrow).

The wavelengths of light that excite, and are emitted by, an FP are broad and characteristic of each FP (**Fig. 10.1a**, **b**, right panels). For example, YFP can be excited by "greenish" wavelengths (**Fig. 10.1b**, *see* narrow shaded area under excitation curve) that excite YFP without exciting CFP. Conversely, CFP can be selectively detected by illuminating with "bluish" wavelengths that excite both CFP and YFP, followed by collecting only "cyan" emissions not emitted by YFP (**Fig. 10.1a**). The resulting CFP and YFP fluorescence thus provides information about the locations and amounts of two different FP-tagged proteins (e.g., AR and AR-interacting protein) or of FPs attached to different parts of the same protein (e.g., AR, **Fig. 10.1c**).

The 25–50 Å size of a typical protein domain contrasts to the >2000 Å (200 nm) resolution limits possible for traditional visible light microscopy. Even with newer ultra-resolution light microscopy methods that resolve >500 Å (7, 8), two proteins with attached FPs will appear co-localized even when quite far apart. By contrast, Förster resonance energy transfer (FRET) provides angstrom-level resolutions in FP position that can be combined with light microscopy to measure protein interactions and conformations within microscopically resolved areas of the cell (9–11).

FRET occurs when the transient energy field excited around the higher energy ("donor") FP is of an energy level sufficient to activate a second, nearby FP (**Fig. 10.1c**). Because this "non-radiative" energy transfer occurs only when the acceptor FP is located within the small donor energy field, FRET microscopy is able to distinguish differences in FP positions that are well below the 2000 Å limit of light microscopy. The level of FRET drops rapidly to the sixth power with the distance between the donor and the acceptor fluorophore dipoles (12–14). If, for example, CFP and YFP are 40 Å apart, 78% of activated CFP will transfer energy to YFP compared to the 5% energy transferred when CFP and YFP are 80 Å apart (15). Because energy does not radiate radially from the dipole (*see* **Fig. 10.1c**), the amount of energy transfer will be affected by the orientation of the donor and acceptor FP dipoles in addition to their distance apart from each other. The quantity of energy transfer thus provides information about the effects of cell environment and ligand binding on AR conformation (6, 16). For example, when the donor and acceptor FPs are attached to the N- and C-termini on the same protein, the amounts of energy transfer are different for the androgen-bound (**Fig. 10.1c**), antagonist-bound (**Fig. 10.1d**), and unliganded (**Fig. 10.1e**) AR (*see* **Note 1**).

Agonist-regulated interactions of a donor-labeled AR with an acceptor-labeled AR (or other protein) also bring the FPs into close enough proximity to allow energy transfer (6). Thus, biochemical issues can also be addressed by FRET. Highly sophisticated applications of these cellular biochemistry techniques allow standard biochemical parameters, such as the equilibrium dissociation constant and the $B_{\max}$, to be measured directly in living cells (17).

In this chapter, we discuss how to measure this energy transfer by "fluorescence intensity" methods that quantify the amounts of the emitted photons hitting the detector. Other fluorescence properties affected by FRET and that can be used to measure FRET include "fluorescence lifetime," which is the delay time between FP excitation and emission (18–20), and the polarity of the emitted light (20, 21). Those methods require instruments not available in most laboratories. Intensity-based techniques described here include "FRET/donor" and "acceptor photobleaching" methods readily attainable by many laboratories. We will also touch on the procedures for defining, from FRET/donor and acceptor photobleaching data, the percentage of donor FP energy transferred to the acceptor (the "efficiency" of energy transfer, *E*).

FRET is accompanied by a decrease in the amount of donor fluorescence and, through that energy transfer, the generation of acceptor fluorescence following donor excitation. For FRET/donor measurement, FRET is calculated after the rapid collection of three images:

- Excite acceptor (YFP)/collect "yellow" acceptor emissions (**Fig. 10.1b**).
- Excite donor (CFP) and acceptor (YFP)/collect only "cyan" donor emissions (**Fig. 10.1a**).
- Excite donor and acceptor/collect "yellow" emissions that arise from a combination of the donor itself, from the acceptor itself, and from the energy transferred from the donor to the acceptor (**Fig. 10.1c**). Thus, the third "FRET" channel contains fluorescence from a mixture of sources. The deconstruction of that mixture into its component parts requires the careful calibration of the microscope which will be described below.
- Acceptor photobleaching involves the collection of a second donor (CFP) image following the photo-silencing of the acceptor (YFP). As the elimination of the acceptor allows all CFPs to emit in the donor channel, the quantity of CFP increased following acceptor photobleaching provides another FRET determination that can be determined in most laboratories.

2. Materials

2.1. Cell Preparation and Growth

1. Microscope slides and No. 1 borosilicate cover glasses. *See* **Note 2** for chamber and plate alternatives to the simple slide procedure described here.
2. Appropriate cell line in which to study AR function.

3. Expression vectors for functionally characterized FP-tagged proteins (*see* **Notes 3** and **4**). The best FPs currently available for FRET are mCerulean, a more recent upgrade of CFP (22), and mVenus, a more recent upgrade of YFP (23).
4. Tissue culture media (*see* **Note 5**) appropriate for the cell line, supplemented with 5% or less newborn calf serum depleted of androgens (and other lipophilic hormones) by three sequential bindings to activated charcoal (Sigma) followed by binding to AG1x8 (Bio-Rad) column resin.
5. Transfection reagent appropriate for cell line (*see* **Note 6**).
6. AR agonist or antagonist ligands.

2.2. Image Collection, Background Subtraction, Segmentation, and Quantification

1. Excitation and emission filters and dichroic mirrors to enable the collection of acceptor, donor, and FRET channels (*see* **Note 7**).
2. Software to control microscope and enable rapid image capture (*see* **Note 8**).
3. Quantitatively linear camera (*see* **Note 9**).
4. Camera and software must be able to collect images at >12-bit depth (0-4095 intensity scale) to accurately subtract ratios at 0.1% accuracies for reproducible FRET measurement.
5. Objectives with higher numerical apertures (NAs) are preferred (*see* **Note 10**).
6. If imaging CFP and YFP for FRET, use a mercury, mercury/xenon combination or metal-halide light source that has a strong line in the 430–440 nm area for optimal excitation of the poorly detected CFP.
7. Image analysis software (*see* **Note 11**) is critical for accurate quantification of fluorescence amounts in a region of interest (ROI) in the acceptor, donor, and FRET channels.

2.3. FRET Determination and Interpretation

1. Database software into which the quantified image data are imported for FRET calculation using simple math. Microsoft Excel suffices for simple FRET determination.
2. Calibration tools include AR labeled only with the donor FP (CFP) to establish donor FP bleed-through into the FRET channel.
3. Acceptor (YFP)-only labeled AR establishes acceptor FP bleed-through into the FRET channel (*see* **Notes 12** and **13**).
4. A set of "CFP–AR–YFP" constructs, containing wild-type and mutant ARs dual labeled with both donor and acceptor, that provide different levels of energy transfer when treated

with different AR ligands. This calibration set establishes how well the transferred energy is detected as an increase in the FRET channel, relative to a decrease in the donor channel. Since each dual-labeled protein contains one molecule of donor FP and one molecule of acceptor FP, the protein set also calibrates how well the same amount of donor and acceptor FPs is detected in the donor and FRET channels when there is no FRET.

5. More complex biochemical analysis requires further analytical software, such as GraphPad Prism, capable of conducting non-linear regression analysis to define curves that fit to the calculated data. Those procedures are detailed in (6) and are not discussed here.

3. Methods

3.1. Cell Preparation and Growth

1. Place 25 mm × 25 mm No. 1 borosilicate cover glasses into six-well tissue culture plates and sterilize by placing, with lid off, in tissue culture hood with UV light on for more than 45 min (*see* **Note 14**).
2. Maintain cells in your media of choice.
3. One day prior to transfection, collect the cells following extensive washing with PBS (5× washes) to remove residual media containing trace amounts of androgens. Re-suspend cells in media containing androgen-stripped media (*see* **Notes 5** and **15**) and plate into six-well dishes on top of cover glasses. Press on cover glass with pipette to remove air bubbles under the cover glass. Plate the number of wells you will need for all transfections.
4. The next day, replace media with fresh androgen-stripped media. Ensure that some wells also have media only in them (no cells) which will be used for background subtraction and flattening irregularities in image illumination.
5. Transfect as per the reagent's manufacturer's recommendations. The transfections must include (a) "donor only" cells transfected with CFP-tagged AR; (b) "acceptor only" cells transfected with YFP-tagged AR; and (c) "test" cells transfected with CFP–AR–YFP (intramolecular study) or co-transfected with CFP–AR and a YFP-tagged target (intermolecular study) (*see* **Note 16**). For the transfection and ligand condition anticipated to provide the brightest cells, set up a duplicate cover glass to be used to establish conditions for image collection.

3.2. Image Collection

1. Set up microscope for collection of acceptor, donor, and FRET channels (*see* **Note 17** for a discussion of the relative merits of wide-field and confocal FRET collection).
2. One day (or more) following transfection, add a ligand to a well. Imaging is conducted at specific times, depending on the question addressed, following ligand addition.
3. Use forceps to remove the "duplicate" cover glass from the well. Place cover glass, cells down on a microscope slide, starting with the media droplet accumulated at the bottom edge of the cover glass. This "wicks" a thin film of media without air bubbles between the cells on the cover glass and the slide. A quick inversion onto a Kimwipe blots excess media from the exterior surface of the cover glass (*see* **Notes 15, 18,** and **19**). Do not over dry!
4. Set integration (exposure) times and camera gain using the duplicate slide transfected in **Section 3.1**, Step 5. Establish times for acceptor, donor, and FRET channel collections necessary to average ~2000 intensity units on the 12-bit scale. Keep in mind how FRET will alter those intensities in other transfection or ligand conditions. *See* **Note 6** about adjusting exposure times to collect the large number of dim cells, not the few abnormal bright cells. *See* also **Note 19** regarding multiple integration times.
5. Once collection conditions are established, they must be held constant for all experiments in the study. Obtain a new slide and collect rapidly the acceptor, donor, and FRET images for multiple fields from each slide. Move to random fields, focus rapidly, and collect rapidly. One should easily be able to collect 15 different fields in 7–8 min (*see* **Note 20**). Do not dawdle to inspect pretty images. Do not collect for more than 10 min as the thin layer of media starts to dry out, which affects outcome.
6. In addition to collecting the acceptor, donor, and FRET images for the test, acceptor-only, and donor-only cells under each ligand condition, collect cover glasses with no cells added (*see* **Note 21**) to establish backgrounds. Bubbles or debris can help ensure you are in focus but avoid capturing those air bubbles or debris in the actual 15–20 background fields that you will collect.

3.3. Image Background Subtraction, Segmentation, and Quantification

1. Image analysis is conducted at a different time than the focused, rapid collection of images. We use image collection and processing software (such as MetaMorph) that allows us to create macros to automate collection and analysis into seamless, rapid procedures.

2. Using your image processing software, create an average background image for each of the acceptor, donor, and FRET channels (*see* **Note 22**).
3. Subtract the background image from your acceptor-only, donor–only, and test images using your image processing software. The use of an average background image also corrects for any minor unevenness in image illumination.
4. Create an ROI for fluorescence quantification in the area of the cell of interest. If a YFP-tagged protein stays within one cellular location, the acceptor (YFP) image alone can define the ROIs by automated image segmentation. Since the AR location changes from cytoplasm to nucleus upon agonist ligand addition, we sometimes use cell lines expressing a marker tagged with a red fluorescent protein for automated creation of ROIs in the nucleus for analysis (*see* **Note 23**). That requires the collection of another distinct fluorescence channel.
5. Transfer the ROIs to all of the original and background-subtracted acceptor, donor, and FRET channels. For each ROI, there are thus six matching images consisting of the three fluorescence channels for each of the background-subtracted and original images.
6. For each ROI, record in a spreadsheet (a) average intensity/pixel for each of the three background-subtracted fluorescence channels; (b) maximum intensity/pixel for each of the three original fluorescence channels; and (c) total area in pixels for each ROI. Later analysis will be easiest if all cells from one unique transfection or ligand condition are placed into one spreadsheet, with a different spreadsheet used for the next condition.
7. The use of a single background image assumes that the background for all cover glasses will be similar. As a quality control for background subtraction, record in a "background" spreadsheet the image intensities of the entire field for each of the 15–20 background fields. Those values provide an indication of the extent to which background in your acceptor-only, donor–only, and test ROIs varies from field to field. If those background values are not constant, the use of a single uniform background subtraction will introduce errors into your FRET measurement (*see* **Note 24**).
8. Normalize the background-subtracted intensity values in each of the three channels (Step 6a) for differences in exposure time. For example, if collecting acceptor, donor, and FRET channels at 400, 400, and 200 ms, respectively, multiply the values obtained in the FRET channel by 2 (*see* **Note 9**, quantitatively linear cameras). This simple

correction is conducted as part of an automated set of equations in the spreadsheet and allows the user to easily track the consistencies or inconsistencies in data output from different experiments collected on the same instrument.

3.4. FRET Determination

1. FRET determination uses the fluorescence amounts transferred to the spreadsheet for each ROI. The database software must be capable of easily conducting the simple data calculations and data sorting needed to define the amounts of energy transfer (*see* **Note 25**).
2. For ROIs from the test cells expressing both donor and acceptor FPs, multiply the amount of fluorescence in the background-subtracted acceptor channel by the "acceptor bleed-throughs" to the FRET and the donor channels. Acceptor FP bleed-throughs are calibrated from the cells expressing only the acceptor (**Section 3.5**, Step 2). This defines the amounts of fluorescence in the donor and FRET channels that actually originated from the acceptor FP (*see* **Note 26**).
3. For each ROI, subtract the acceptor bleed-through amounts from the amounts of fluorescence measured in the background-subtracted donor and FRET channels of the same ROI. The remaining fluorescence therefore is that originating from the donor FP.
4. For each ROI, a FRET/donor ratio is obtained by dividing the acceptor bleed-through-corrected amount of fluorescence in the FRET channel by the acceptor bleed-through-corrected amount of fluorescence in the donor channel.
5. Use the sorting operations available in the spreadsheet (*see* **Note 25**) to remove data calculations for ROIs that do not fit quality control criteria. Eliminate any ROI in which any pixel in any channel is saturated. If conducting quality control based on different integration times (*see* **Note 19**), eliminate any ROI not fitting that quality control criteria. The ROI area may also indicate when an ROI is clearly not the size of the intended cellular region. *See* also methods for the mathematical identification of cellular debris based upon the unusual fluorescence properties of that debris in the acceptor, donor, and FRET channels (24). With these quality controls, we seldom delete more than 2% of our ROIs from the analyses. If eliminating more than 10% of your ROIs, you should be concerned about other collection problems causing apparent inaccuracies in your data. We further eliminate ROIs that the system calibrations (**Section 3.5**) show to have insufficient levels of fluorescence from the donor FP to obtain an accurate FRET/donor ratio.

6. The FRET/donor ratio is an easy method of evaluating energy transfer that most laboratories can do. If there is no energy transfer, the amount of donor fluorescence measured represents the actual amount of donor. Thus, after acceptor bleed-through correction, the amount FRET/donor ratio from the test ROIs will be identical to that obtained from the donor-only controls (**Section 3.5**). If there is energy transfer, the amount of fluorescence in the donor channel decreases as the level of fluorescence in the FRET channel increases (i.e., FRET/donor ratio is elevated).
7. It is also possible to conduct the above steps using the same mathematical steps on the actual images themselves (*see* **Note 27**). This can provide data about variations in FRET at any location throughout the image, not just in the specific ROI analyzed.

3.5. System Calibration

1. For FRET/donor calculation, calibrations are done on data collected from two sources: cells expressing only the donor FP and other cells expressing only the acceptor FP. Background-subtracted fluorescence intensities are collected from each ROI in the acceptor, donor, and FRET channels as described above. Eliminate ROIs with saturated pixels or that are of unusual size. You will have a series of ROIs from acceptor (or donor) FP-transfected cells with variable intensities in the acceptor (or donor) channel. **Note 32** includes transfection tips for obtaining a wide range of intensities.
2. For the acceptor FP cells, create a graph of the acceptor channel intensities in each ROI (x-axis) against the FRET channel intensities (y-axis). With good background subtraction and image quantification, these data should form a perfect line ($r^2 > 0.99$ after linear regression analysis) with a y-intercept = 0 (*see* **Note 28**), and the slope indicating the bleed-through of the acceptor FP into the FRET channel. We typically collect >50 ROIs for each analysis. Note that at low acceptor FP expression levels (e.g., intensity/pixel <150 above background), the data points deviate more substantially from the line. This defines the lowest intensity levels at which the data quantification remains reliable.
3. A total of four such graphs are made for the acceptor FP and donor FP control cells to define acceptor FP bleed-through into the FRET channel (e.g., 0.130 ± 0.002 collected from 10 different experiments), acceptor FP bleed-through into the donor channel (0.001 ± 0.000), donor FP bleed-through into the FRET channel (0.573 ± 0.009), and donor FP bleed-through into the acceptor channel (0.002 ± 0.001). These values are collected for every study as quality controls since they should not vary (*see* **Note 29**).

4. The above values are presented only as an example of the consistency you should expect. Calibration values will vary markedly with your instrument as a function of the excitation and emission wavelengths (filters) used, how well those wavelengths pass through your optics, and the different efficiencies by which your camera detects each wavelength.
5. The FRET/donor ratio measured similarly will be highly specific for your instrument and not comparable to measurements made in other laboratories. Further instrument calibrations can be done to establish the correspondence between the amounts of donor channel decrease and FRET channel increase across a spectrum of FRET standards for which the percentage of donor FP energy transfer to a linked acceptor FP is known (24, 25). These instrument-specific calibrations enable the calculation of these *E* values to compare results in different laboratories. It is also possible to directly calculate *E* if you have access to accurate information about the efficiencies of your filter sets and camera and knowledge of the excitation and emission properties of your FPs (26) (*see* **Note 30**).
6. *E* may also be calculated using the acceptor photobleaching method. Immediately following the collection of your acceptor, donor, and FRET images, selectively photobleach the acceptor (*see* **Note 31**). Immediately following the photobleach, collect a second donor image. Record the pre-bleach and post-bleach donor FP intensities in the spreadsheet. Subtract the pre-bleach donor intensity from the post-bleach donor intensity to obtain the amounts of donor "increased" as a result of the post-bleach inability to transfer energy to the now silent acceptor. Divide that "donor increase" intensity by the actual total intensity of donor (the post-bleach donor amount) to arrive at the proportion of donor FP that had been absent when the acceptor FP was available for energy transfer.

3.6. FRET Interpretation

1. The methods quantify a "FRET" value that must be interpreted in biologically meaningful ways. The simplest FRET study is one in which a "sensor" is constructed consisting of CFP and YFP attached to the same protein. For a CFP–AR–YFP construct, that informs the user of whether the addition of a ligand changes the AR in some way and whether different ligands cause changes that are quantitatively similar or occur in the same compartment. Sensors can also be designed to assess "activities," e.g., inserting a phosphorylation site between CFP and YFP. Activity sensors study the location of that enzymatic activity immediately after kinase activation.

2. Sensors ideally are monomeric. For CFP–AR–YFP, the ligand also causes dimerization or oligomerization, leading to a FRET signal. The CFP–AR–YFP probe therefore is a hybrid sensor for the impact of a drug on both conformation and self-association. Temporal FRET microscopy comparison of CFP–AR–YFP with CFP–AR–CFP/YFP–AR–YFP dimerization showed that the ligand-induced oligomerization is distinct from the conformational change in time and intracellular location (6).
3. For the interaction of a CFP-labeled AR with a YFP-labeled interacting factor, the FRET outputs from each cell will vary enormously. This is not due to measurement error but because the amount of FRET depends on the amount of interaction between the two proteins which is governed by standard biochemical concepts. The law of mass action dictates that the interaction of one protein, held constant in concentration, will vary in predictable ways with the amount of the interacting partner. The levels of FRET measured for interaction of a donor FP-labeled protein with an acceptor FP-labeled protein indeed vary in a fashion consistent with the law of mass action (6, 11, 17, 24, 27, 28). Such an analysis relies on conducting studies in which the donor FP-labeled proteins are expressed at relatively constant levels in different cell types together with highly variable levels of the acceptor FP-labeled protein (*see* **Note 32**). High numbers of data points can then be used to fit a curve representing mathematical parameters of the law of mass action. The curve defined by non-linear regression analysis calculates the maximal level of energy transfer (E_{max}) at saturating amounts of acceptor-labeled protein. E_{max} reflects the position and orientations of the FPs in the fully saturated complex. The amount of YFP fluorescence when half of E_{max} is reached is a very crude reflection of the affinity of the CFP-labeled AR for the YFP-labeled interacting factor.
4. In its most sophisticated form, the biochemical analysis includes further instrument calibrations that convert the CFP and YFP fluorescence values into concentrations of the CFP- and YFP-tagged proteins. *E* is converted into the proportion and concentration of CFP-tagged factor in a complex with the YFP-tagged factor (17). This data is then used to plot the actual parameters needed to correctly describe the law of mass action: the molar concentration of the complex (*y*-axis) in relationship to molar concentration of the unbound, YFP-labeled factor (*x*-axis). This analysis must also account for variables, including the stoichiometry of the complex and the limits on the amount of CFP-tagged factor available to interact in the cell, that can be solved

when conducting the regression analysis and curve fitting (17). Thus, the "cellular biochemistry" FRET analysis not only provides the K_d and B_{max} determinations that would be obtained by traditional test tube biochemistry but also details about the structure of the complex (E_{max}), the limitations that the cell places on the amounts of proteins available to interact, the stoichiometry of the complex, and the intracellular locations of the interacting complexes.

4. Notes

1. The conformations measured by FRET are those averaged within the microscopically determined ROI and may vary within the ROI. When assessing FRET pixel by pixel, movement of cellular structures during collection of the necessary acceptor, donor, and FRET images can affect spatial resolution.
2. The methods described here are for the rapid measurement of cells grown on cover glasses. Cells may also be grown on cover glass-thickness multi-well plates or chambers and imaged, still in media, in an incubated chamber on an inverted microscope. Cover glass thicknesses are necessary for use with high NA objectives (*see* **Note 10**).
3. Know the extent to which FP fusion affects the activity of your protein, as determined in appropriate functional studies comparing the native protein with the fusion protein. Ideally, no effect of fusing the FP to your protein will be observed. Short of creating a knock-in mouse in which the fusion protein replaces the endogenous gene (29), you will never be aware of how FP fusion affects the complete spectrum of functions for that factor.
4. Select FPs with highest overlap of the donor emission with the acceptor excitation, but that still can be distinguished in the donor and acceptor collection channels. The donor FP is preferred to have a high quantum yield (a high likelihood of emitting the absorbed energy) and the acceptor FP is desired to be excited relatively well (high molar extinction coefficient).
5. Cell culture media, serum, and even some tissue culture plates or coatings contain fluorescent molecules that, if using CFP, are very bright in the donor and FRET collection channels. Check media alternatives for lowest intrinsic fluorescence in the channels to be collected (i.e., plate cells in different media and measure background). Use the

lowest amount of serum possible to retain the health and growth of your cells. Red-shifted fluorescent proteins can overcome the background that plagues the use of CFP but, to date, a strong, red-shifted alternative to the CFP/YFP FRET pair has not been identified.

6. Where possible, use a transfection reagent that delivers a reasonably consistent amount of DNA to your cells. To test this, package DNA for a single plasmid encoding an FP-tagged protein using a large number of transfection reagents. One to two days following transfection, image. You will observe that some reagents show consistent fluorescence levels with little fluctuation from cell to cell. Quantify fluorescence levels (total and average intensities) per ROI for your records. When capturing images, avoid the natural tendency to set the exposure times based on the brightest cells; multiply that exposure time by 3–5 to see 5–20-fold more cells transfected at a lower, consistent level. Those cells usually are the cells you are interested in analyzing.
7. These filters are available from a variety of manufacturers including Chroma Technology, Omega Optical, and Semrock. The filters are best accommodated in filter wheels (e.g., Sutter Instrument Company) and used together with a single, multi-bandpass dichroic mirror. The filter wheel/single dichroic configuration minimizes pixel shifts that typically occur when using different filter cubes.
8. Microscope software that automates movement of filters around the single dichroic also enables rapid collection.
9. FRET measurements depend upon highly accurate corrections of the contributions of each FP to each image channel. Those corrections are based upon ratios of emissions of each FP into each channel. Those ratios are physical properties of each FP that do not vary with FP amount. However, many cameras (and photomultiplier tubes for confocal collections) are not quantitatively linear, resulting in ratios that vary with intensity. To test for camera linearity prior to purchase, capture the same images at different exposure times and measure whether the background-subtracted fluorescence intensity (*see* procedures in **Sections 3.2** and **3.3**) is the same ratio as the exposure times. Alternatively, FRET analysis programs have been developed for non-linear detection systems in which the ratios are determined across the dynamic range of measurements and subtracted (26).
10. Higher NA objectives provide increased resolution and higher intensity (i.e., shorter exposure time) than does a

lower NA objective of the same magnification. Equally important, the higher NA objective results in a narrower optical section and thereby increases the signal from the FP in relationship to the background from the media and serum (*see* **Note 5**), especially when the optical section is thicker than the cellular ROI.

11. We use MetaMorph software (Molecular Devices) but other options are available including publicly available freeware such as ImageJ. Some of the packaged FRET analysis tools in commercially available software are based on outdated, erroneous understandings of FRET. It is best to understand the concepts outlined here and to evaluate if those packages comply with accurate procedures.
12. It is acceptable for the acceptor FP to be detected in the donor channel since the acceptor FP is accurately measured and its bleed-through correctly subtracted from the donor channel. However, it is not acceptable for the donor FP to be detected in the acceptor channel since, when FRET is present, the amount of donor FP is not known which prevents the accurate subtraction of the donor contribution to the acceptor channel.
13. Because fluorescence calibration values may vary in different intracellular locations or in different cell types (although they typically do not unless the cell environment of protein modifications affects fluorescence values), it is best that all three calibration measurements described in **Section 2.3** are made in the same subcellular structure as the test experiment.
14. These study formats described here require that the cells grow on cover glasses. Many of our studies are conducted using 96-well or 384-well optical plates (manufacturers include Aurora, BD, Costar, Greiner, Matrical) of which some contain thin-bottomed plastics on which some cell types grow better.
15. Since the methods described in this chapter may allow media drying, we recommend using media without phenol red, which affects imaging in the CFP and YFP channels as the media dries out.
16. If specifically conducting an "intramolecular" study to investigate the repositioning of FP-tagged domains within the same AR, one must include a control transfection for the "intermolecular" interaction in which separate vectors, expressing AR tagged with either CFP or YFP, are co-transfected. This is necessary to define the extent to which any change in FRET signal at any time or location originates with AR dimerization rather than intramolecular

folding. Intermolecular measurements are discussed in **Section 3.6**.

17. FRET measurement with wide-field microscopy is typically more accurate than with confocal microscopy. FRET measurement relies on highly precise quantification of the image intensities in the three matched image channels. Confocal microscopy provides not only sharper image resolution but also optical sections that reduce errors from background inconsistencies (*see* **Note 18**). However, cell movements during confocal imaging result in *z* shifts in the thin optical sections for the acceptor, donor, and FRET images. Thus, the bleed-through corrections are not as accurately applied in confocal microscopy as in wide-field microscopy. If using confocal microscopy, these *z* shifts can be minimized to improve FRET measurement by using "between-line" laser activation to collect, almost simultaneously, the acceptor and FRET–donor images on two PMTs. If using a dual-labeled CFP–AR–YFP reporter, which by definition has a constant ratio of YFP to CFP at each image pixel, one can also determine local variations in FRET by using FRET and donor images only, although one does not know the extent of energy transfer at each pixel.
18. Collecting images from cells in a thin film of media minimizes the undesired contributions of media fluorescence (*see* **Note 5**) relative to the fluorescence of the FP-labeled protein in the cell. The disadvantage of this method is the very short time period for cell collection. If collecting cell images in a deep well of media, the background will be variable and the background subtraction less accurate.
19. If you are experiencing background problems and need to establish confidence in your fluorescence quantification, a quick test for appropriate background correction is to capture two different integration times for each channel. For instance, collection at 100 and 400 ms should provide two images that are precisely fourfold different in intensity following background subtraction. In our experience, you should expect a 0.250 ± 0.002 intensity ratio for a fourfold difference in exposure times. If not, fluorescence quantification is not accurate enough for FRET calculation. This is a nice quality control parameter.
20. We typically set up our collections and ligand additions at 10 min intervals. That is, we add ligand at time 0 min, another ligand to another well at 10 min, then start collecting the first cover glass at time 20 min while adding the next ligand to yet another well. We collect fields for

about 8 min and are ready at 30 min to add ligand to yet another well and take the next cover glass to image. That is, images are collected between 20 and 30 min after ligand addition. To collect different times, we simply extend the period between initial ligand addition and the start of the image collections. This scenario requires your cell culture incubator to be close to your microscope.

21. Do not collect background images with cells in them as the low-level and variable cellular autofluorescence will result in incorrect backgrounds. Autofluorescence introduces errors that are defined by other calculations (*see* **Note 28**) or by ensuring that your fluorescence levels collected are substantially above the autofluorescence background (**Section 3.5**, Step 2). Cellular autofluorescence is typically minimal in most cell nuclei and most intense in the Golgi apparatus. That generalization may vary with cell type.
22. In MetaMorph, the creation of the background image uses the "Add Plane" function to append the 15–20 field acceptor channel images into a single image "stack," which is repeated for the donor channel images and the FRET channel image. An average image is created for each of the three collection channels using the stack arithmetic function.
23. The acceptor channel measurements are not affected by FRET and can be used to quantify transport of the YFP-labeled AR in parallel with FRET determination. If using a red FP-tagged "segmentation" marker, initial FRET experiments must be conducted to ensure that neither the CFP-tagged nor the YFP-tagged factors transfer energy to the red FP attached to the marker. For the AR, automated routines in MetaMorph are used to define the RFP-tagged nucleus with the margins of the same cell defined by the YFP-tagged AR. Each cell therefore has matched nuclear and cytoplasmic ROIs for FRET determination, total AR level, and the proportion of AR in the cell nucleus.
24. If the backgrounds are so variable that inconsistent FRET values are obtained because of a result of incorrect background subtraction, the user may elect to establish individual backgrounds for each cell by manually drawing in an ROI into an empty area of the image near the cell to be measured. In that scenario, the spreadsheet then contains fluorescence levels from the background ROIs for which the background subtraction is done within the spreadsheet. This is laborious, but accurate.
25. The data spreadsheet should be set up to do all calculations by simply pasting the mathematical steps (*see* FRET determination) into a line for each set of matched data.

The spreadsheet should be capable of sorting ROIs for values for quality control criteria discussed in later sections. In Excel, the user should be familiar with the "*d*" functions (e.g., d_{average}) that allow data sets to be averaged only from data meeting the specified criteria.

26. Since all calculations start from the assumption that the level of fluorescence measured in the acceptor represents acceptor only, it is crucial that the calibrations from the donor FP-only cells show there to be zero bleed-through of the donor FP into the acceptor channel.
27. Particularly for areas of an image in which there is no energy transfer, the FRET data will be more accurate from the intensities averaged from the ROI and exported to the database. This is because, when directly subtracting one image from another in most software, negative pixel values are recorded as 0. Thus, "zero" FRET, which has measurement noise in which pixels are both below and above 0, will appear in the image as "above 0" noise.
28. In practice, intercepts marginally greater than 0 (3–5) are observed owing to cellular autofluorescence not corrected. Adding this subtraction into your data analysis can help improve FRET determination.
29. It is also possible to obtain the FRET/donor, FRET/acceptor, acceptor/donor, and donor/acceptor calibration ratios by conducting the simple division of the channel intensities within each ROI and averaging. In practice, variations in background correction result in more study-to-study differences when using the individual ROIs. We find the slope method to provide the most consistent readout of bleed-through.
30. If calculating E based on the physical properties of the FPs and instrument, rather than calibrating the instruments with calibration standards, be aware that you assume that (a) the FP properties used for your calculations are the same in your cell and (b) the optics of your microscope transmits light identically to the instrument used to establish the FP excitation and emission curves used in your calculations.
31. Acceptor photobleaching FRET is absolutely dependent on there being no bleed-through of the acceptor FP into the donor window. On wide-field systems, photobleaching typically is conducted by exposing the field for up to 5 min (depending on lamp intensity) through the same filter used to selectively excite YFP. Conduct tests on AR–YFP-expressing cells using progressive 30-s bleaches, followed by YFP image collection, to establish the optimal bleach time. On a confocal system, bleaching is much faster

owing to the availability of the 514-nm laser standard on many confocal instruments. However, that laser has some ability to photobleach CFP which affects your FRET determination unless you carefully calibrate that bleaching on the CFP-only control cells. The more major problem with acceptor photobleaching FRET on the confocal system is the large *z* shift in the pre- and post-bleach optical sections. Thus, one is not measuring donor FP fluorescence intensities from the same optical slice which causes large variations in the measured *E* values from different images.

32. To obtain relatively constant expression of CFP-labeled protein with a variable amount of YFP-labeled interacting partner, use a transfection reagent that delivers a constant amount of DNA into a cell (*see* **Note 6**). Engineer the expression vector for CFP-labeled protein in a way that it is poorly expressed (e.g., poor promoter, poor Kozak sequence), whereas the YFP-labeled protein is highly expressed. One therefore needs to have, e.g., 900 ng of CFP–AR expression vector in a 1000 ng total amount of DNA for transfection, which results in low variability in the amount of CFP–AR expressed per cell. Create three transfection mixes, each with 900 ng of CFP–AR vector but with 33, 67, or 100 ng of YFP–target protein vector (and 67, 33, or 0 ng of an inert DNA vector). After reagent is formed, mix all three together and plate onto cells to obtain cells expressing a constant level of CFP–AR and a wide variation in YFP-tagged target. Collect the YFP images at two exposure times (e.g., 100 and 400 ms). If the YFP signal is saturated at the 400 ms exposure time, multiply the background-subtracted fluorescence level obtained at 100 ms by 4 to accurately measure high expression levels of YFP-tagged protein.

References

1. Day, R.N., Davidson, M.W. (2009) The fluorescent protein palette: tools for cellular imaging. *Chem. Soc. Rev.* **38**, 2887–2921.
2. Day, R.N., Schaufele, F. (2008) Fluorescent protein tools for studying protein dynamics in living cells. *J. Biomed. Optics* **13** (031202), 1–6.
3. Giepmans, B.N., Adams, S.R., Ellisman, M.H., Tsien, R.Y. (2006) The fluorescent toolbox for assessing protein location and function. *Science* **312**, 217–224.
4. Shaner, N.C., Steinbach, P.A., Tsien, R.Y. (2005) A guide to choosing fluorescent proteins. *Nat. Methods* **2**, 905–909.
5. Jones, J.O., An, W.F., Diamond, M.I. (2009) AR Inhibitors identified by high-throughput microscopy detection of conformational change and subcellular localization. *ACS Chem. Biol.* **4**, 199–208.
6. Schaufele, F., Carbonell, X., Guerbadot, M., Borngraeber, S., Chapman, M., Ma, A., Miner, J., Diamond, M. (2005) The structural basis of androgen receptor activation: intramolecular and intermolecular amino–carboxy interactions. *Proc. Natl. Acad. Sci. USA* **102**, 9802–9807.

7. Heilemann, M. (2010) Fluorescence microscopy beyond the diffraction limit. *J. Biotechnol.* **149**, 243–251.
8. Hell, S.W. (2007) Far-field optical nanoscopy. *Science* **316**, 1153–1158.
9. Day, R.N., Schaufele, F. (2005) Imaging molecular interactions in living cells. *Mol. Endocrinol.* **19**, 1675–1686.
10. Gordon, G.W., Berry, G., Liang, X.H., Levine, B., Herman, B. (1998) Quantitative fluorescence resonance energy transfer measurements using fluorescence microscopy. *Biophys. J.* **74**, 2702–2713.
11. Vogel, S., Thaler, C., Koushik, S.V. (2006) Fanciful FRET. *Sci STKE* **2006** (331), re2.
12. Förster, T. (1948) Zwischenmolekulare energiewanderung und fluoreszenz. *Ann. Phys.* **6**, 54–75.
13. Förster, T. (1959) Transfer mechanisms of electronic excitation. *Discuss. Faraday Soc.* **27**, 1–17.
14. Stryer, L., Haugland. R.P. (1967) Energy transfer: a spectroscopic ruler. *Proc. Natl. Acad. Sci. USA* **58**, 719–726.
15. Patterson, G.H., Piston, D.W., Barisas, B.G. (2000) Förster distances between green fluorescent protein pairs. *Anal. Biochem.* **284**, 438–440.
16. van Royen, M.E., Cunha, S.M., Brink, M.C., Mattern, K.A., Nigg, A.L., Dubbink, H.J., Verschure, P.J., Trapman, J., Houtsmuller, A.B. (2007) Compartmentalization of androgen receptor protein–protein interactions in living cells. *J. Cell. Biol.* **177**, 63–72.
17. Kofoed, E.M., Guerbadot, M., Schaufele, F. (2010) Structure, affinity, and availability of estrogen receptor complexes in the cellular environment. *J. Biol. Chem.* **285**, 2428–2437.
18. Voss, T.C., Demarco, I.A., Day, R.N. (2005) Quantitative imaging of protein interactions in the cell nucleus. *Biotechniques* **38**, 413–424.
19. Wallrabe, H., Periasamy, A. (2005) Imaging protein molecules using FRET and FLIM microscopy. *Curr. Opin. Biotechnol.* **16**, 19–27.
20. Levitt, J.A., Matthews, D.R., Ameer-Beg, S.M., Suhling, K. (2009) Fluorescence lifetime and polarization-resolved imaging in cell biology. *Curr. Opin. Biotechnol.* **20**, 28–36.
21. Piston, D.W., Rizzo, M.A. (2008) FRET by fluorescence polarization microscopy. *Methods Cell. Biol.* **85**, 415–430.
22. Rizzo, M.A., Springer, G.H., Granada, B., Piston, D.W. (2004) An improved cyan fluorescent protein variant useful for FRET. *Nat. Biotechnol.* **22**, 445–449.
23. Nagai, T., Ibata, K., Park, E.S., Kubota, M., Mikoshiba, K., Miyawaki, A. (2002) A variant of yellow fluorescent protein with fast and efficient maturation for cell-biological applications. *Nat. Biotechnol.* **20**, 87–90.
24. Kofoed, E.M., Guerbadot, M., Schaufele, F. (2008) Dimerization between aequorea fluorescent proteins does not affect interaction between tagged estrogen receptors in living cells. *J. Biomed. Opt.* **13** (031207), 1–15.
25. Koushik, S.V., Chen, H., Thaler, C., Puhl, H.L., Vogel, S.S. (2006) Cerulean, Venus, and VenusY67C FRET reference standards. *Biophys. J.* **91**, L99–L101.
26. Elangovan, M., Wallrabe, H., Chen, Y., Day, R.N., Barroso, M., Periasamy, A. (2003) Characterization of one- and two-photon excitation fluorescence resonance energy transfer microscopy. *Methods* **29**, 58–73.
27. Chen, H., Puhl, H.L., Ikeda, S.R. (2007) Estimating protein-protein interaction affinity in living cells using quantitative Förster resonance energy transfer measurements. *J. Biomed. Opt.* **12** (054011), 1–9.
28. Hoppe, A., Christensen, K., Swanson, J.A. (2002) Fluorescence resonance energy transfer-based stoichiometry in living cells. *Biophys. J.* **83**, 3652–3664.
29. Brewer, J.A., Sleckman, B.P., Swat, W., Muglia, L.J. (2002) Green fluorescent protein–glucocorticoid receptor knockin mice reveal dynamic receptor modulation during thymocyte development. *J. Immunol.* **169**, 1309–1318.

Section IV

Analysis of Androgen Receptor Modifications and Interactions

Chapter 11

Analysis of Nuclear Receptor Acetylation

Chenguang Wang, Michael Powell, Lifeng Tian, and Richard G. Pestell

Abstract

Acetylation is an essential post-translational modification in which an acetyl group is covalently conjugated to a protein substrate. Histone acetylation was first proposed nearly half a century ago by Dr. Vincent Allfrey. Subsequent studies have shown that acetylated core histones are often associated with transcriptionally active chromatin. Acetylation at lysine residues of histone tails neutralizes the positive charge, which decreases the binding ability to DNA and increases the accessibility of transcription factors and co-activators to the chromatin template. In addition to histones, a number of non-histone substrates are acetylated. Acetylation of non-histone proteins governs biological processes, including cellular proliferation and survival, transcriptional activity, and intracellular trafficking. We demonstrated that acetylation of transcription factors can regulate cellular growth. Further, we have shown that nuclear receptors are acetylated at a phylogenetically conserved motif. Since our initial observations with the estrogen and androgen receptors, more than a dozen nuclear receptors have been shown to function as substrates for acetyltransferases with a variety of new methods (**Fig. 11.1**). This chapter focuses on the protocol used in the studies of NR acetylation and de-acetylation. We will discuss the potential pitfalls of each method.

Key words: Androgen receptor, acetylation, post-translational modification, histone, acetyltransferase, nuclear receptor.

1. Introduction

Acetylation was first proposed nearly half a century ago by Dr. Vincent Allfrey. Subsequent studies have shown that acetylation of core histones is often associated with transcriptionally active chromatin. Acetylation is an essential post-translational modification in which an acetyl is covalently conjugated to a protein substrate. Histone acetylation at lysine residues of histone tails neutralizes the positive charge, which decreases the binding

F. Saatcioglu (ed.), *Androgen Action*, Methods in Molecular Biology 776,
DOI 10.1007/978-1-61779-243-4_11, © Springer Science+Business Media, LLC 2011

ability to DNA and increases the accessibility of transcription factors and co-activators to the chromatin template.

In addition to histones, a number of non-histone substrates are acetylated. Acetylation of non-histone proteins governs biological processes, including cellular proliferation and survival, transcriptional activity, and intracellular trafficking. We have shown that nuclear receptors are acetylated at a phylogenetically conserved motif (1). We further demonstrated that acetylation of transcription factors (the androgen receptor) can regulate cellular growth (2, 3). Since our initial observations with the estrogen and androgen receptors, more than a dozen nuclear receptors have been shown to function as substrates for acetyltransferases with diverse functional consequences (**Fig. 11.1**). The androgen receptor plays an important role in normal development as well as in prostate cancer onset and progression. The AR is directly acetylated at lysine residues within the hinge domain, located in proximity to the second zinc finger of the DNA binding domain (DBD) (4). The point mutations of lysine residues on the AR (K630, 632, 633A) abrogate the p300-mediated regulation and reduce ligand-induced activity, revealing receptor acetylation as a key step in ligand-dependent receptor activation (4). The lysine to glutamine substitution (K630Q), which mimics acetylated lysines, has increased ligand-induced transcriptional activity compared to wild-type AR. This mutant also exhibits enhanced activity at lower concentrations of ligand when compared to wild-type AR and a relative resistance of the AR mutant to the antagonist flutamide (2). The significance of AR acetylation was investigated in great detail initially by our group and later by others. Mutation of the AR acetylation site abrogated MEKK1-induced apoptosis, DHT response, and regulation by co-activators including SRC1, p300, Ubc9, and TIP60 (5). Moreover, AR acetylation mutants, K630Q and K630T, which mimic constitutively acetylated receptor, enhanced p300 binding, reduced association with N-CoR/HDAC/Smad3 co-repressor complex, and increased AR transcriptional activity. Importantly, the AR acetylation mimics promoted prostate cancer cell growth and survival, and increased transcription of AR target genes (3). These studies were the first direct evidence that a single acetylation residue directly promotes contact-independent growth in vivo. Mutation of the AR acetylation site abrogated the regulation of the receptor by HDAC inhibitor TSA and by cAMP and AKT signaling (6). Moreover, mutation of the AR phosphorylation site showed a direct relationship to HDAC responsiveness and recruitment of the AR to an ARE in ChIP assays, defining acetylation and phosphorylation of the AR as functionally convergent events and demonstrating the importance of transcription factor acetylation sites in the recruitment to DNA binding sites in the context of local chromatin (6).

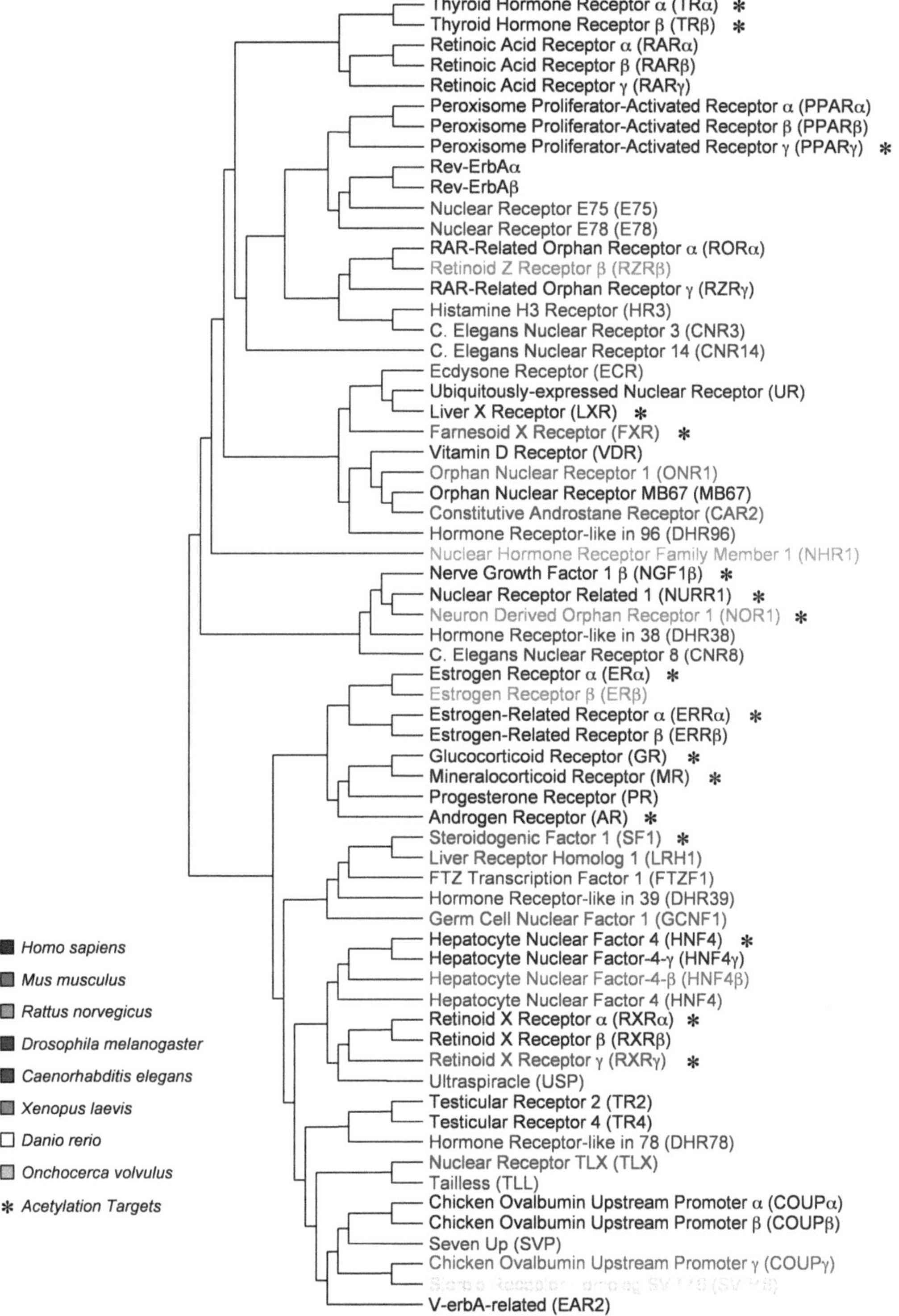

Fig. 11.1. Nuclear receptor acetylation. Phylogenetic tree indicating evolutionary relationships between nuclear receptors across eight different species. Receptors known to be targets of acetylation events are indicated with an *asterisk* (2–4, 12–26). Notably, acetylated receptors including the AR, ERα, TR, LXR, GR, HNF4, and SF1 all contain a conserved acetylation motif "R(K)XKK" (27) (Tree is adapted from Wang et al. (1)).

Recent studies have identified multiple distinct substrates for HATs, including transcription factors, nuclear transport proteins, and cytoskeletal proteins (7). Non-histone proteins that are directly acetylated include p53, the Kruppel-like factor (EKLF), HMGI (Y), GATA-1, E2F-1, nuclear receptors, and transcription co-activators including p300 and ACTR (7). Acetylation of transcription factors either enhances or represses transcriptional activity. Acetylation by p300/CBP enhanced the activity of the tumor suppressor p53, EKLF, and the erythroid cell differentiation factor, GATA-1 (8–10). In contrast, acetylation of the co-activator ACTR or ERα contributed to an inhibition of hormone-induced nuclear receptor signaling (1). Although the mechanisms by which acetylation regulates transcription factor function remain unclear, the possibilities include alterations in DNA binding, recruitment to chromatin, co-activator binding, dissociation with co-repressors including HDAC/N-CoR, as well as subcellular localization. In addition to changes in chromatin organization that occur during cell cycle transition, specific physical interactions occur between components of the cell cycle regulatory apparatus and proteins regulating histone acetylation (7). Transition through the G_1 phase of the cell cycle is regulated by cyclin-dependent kinase holoenzymes consisting of a regulatory subunit (cyclin D or cyclin E) and a catalytic subunit (cdk4/6). The p300/CBP associated factor (P/CAF) physically interacts with cyclin D1 to regulate activity of both the estrogen and the androgen receptor (11). In vivo HAT assays are used to understand the signaling pathways regulating HAT activity for specific substrates. They are also critical in understanding the biological significance of acetylation and its role in human diseases including cancer and metabolic syndrome.

2. Materials

2.1. Cell Culture

The Human Embryonic Kidney 293 cells (referred to as HEK 293) are maintained in Dulbecco's modified Eagle's medium (DMEM) supplemented with 10% (v/v) fetal bovine serum (FBS), 1% penicillin, and streptomycin. Cells are transfected with calcium phosphate-precipitated DNA.

2.2. Reagents

1. Core histones (Millipore, Catalogue Number: 13-107) are dissolved in ddH_2O and stored as 10 μL aliquots at –80°C.
2. Antibody against p300 (Santa Cruz Biotechnology).
3. Purified p300 proteins can be either purchased from commercial sources or prepared by baculovirus-mediated recombinant expression (4).
4. [1–^{14}C] acetyl-CoA, [^{3}H] acetyl-CoA, [^{3}H] acetic acid, sodium salt (GE Life Science).

2.3. Equipment

1. Sodium dodecyl sulfate-polyacrylamide gel electrophoresis (SDS-PAGE) gel set.
2. Liquid scintillation counter.
3. Orbital shaker.
4. PhosphorImager system.
5. Centrifuge.
6. Cell culture incubator.
7. Water bath.

2.4. Buffered Solutions

1. Phosphate-buffered saline (PBS): 0.01 M phosphate buffer, 0.0027 M potassium chloride, and 0.137 M sodium chloride, pH 7.4.
2. RIPA buffer (modified): 50 mM HEPES [*N*-(2-hydroxyethyl)piperazine-*N*′-(2-ethanesulfonic acid), pH 7.2], 150 mM NaCl, 1 mM ethylene glycol bis (2-aminoethyl ether)-*N*,*N*,*N*′*N*′-tetraacetic acid (EGTA), 1 mM ethylenediaminetetraacetic acid (EDTA). Buffer containing the above compounds can be made as a large volume, filtered through a 0.22 μm filter, and stored at room temperature (RT). Prior to use, the appropriate volume is aliquoted and chilled on ice. Tween-20 is added to a final concentration of 0.1% (v/v), 1 mM dithiothreitol (DTT), 1 mM phenylmethylsulfonyl fluoride (PMSF), 0.1 mM sodium orthovanadate, and 2.5 mM leupeptin just prior to use.
3. IB buffer: 10 mM Tris–HCl, pH 7.4, 2 mM $MgCl_2$, 3 mM $CaCl_2$, 10 mM sodium butyrate, and 1 mM PMSF.
4. NIB buffer: IB buffer supplemented with 1% Nonidet P-40.
5. In vitro HAT assay buffer (for 2X stock buffers): 100 mM Tris–HCl, pH 8.0, 20% glycerol, 100 mM NaCl, 20 mM butyric acid, 0.2 mM EDTA, 2 mM DTT, and 2 mM PMSF.
6. HDAC buffer: 50 mM Tris–HCl, pH 8.0, 150 mM NaCl, and 10% glycerol.

3. Methods

3.1. In Vitro Acetylation/Deacetylation Assays

3.1.1. Purification of Core Histone Substrate (See **Note 1**)

1. Six liters of HeLa cells are grown to a density of 5×10^5 cells/mL in DMEM supplemented with 10% newborn calf serum, 1% fetal calf serum (FCS), 0.1 g/L streptomycin, and 0.06 g/L penicillin.
2. The cells are centrifuged at $500 \times g$ and resuspended in 120 mL of cold PBS.

3. The solution is resuspended to a final concentration of 100 μg/mL cycloheximide, 10 mM sodium butyrate, and 0.2 mCi/mL [^{3}H] acetic acid and incubated for 1 h at 37°C with gentle stirring.
4. Cells are chilled on ice and centrifuged at 500×*g* for 5 min.
5. The cells are washed three times in 50 mL of PBS supplemented with 10 mM sodium butyrate.
6. Cells are then lysed in 40 mL of NIB buffer.
7. Nuclei are collected (500×*g*) and washed twice in 40 mL of NIB buffer followed by one wash with NIB buffer supplemented with 100 mM NaCl.
8. An additional wash is performed in 40 mL of 100 mM NaCl and IB buffer.
9. The nuclei are then extracted twice in high salt, 40 mL of 400 mM NaCl and IB buffer, followed by centrifugation.
10. The nuclear pellet is extracted twice in 10 pellet volumes of 0.2 M H_2SO_4 for 90 min on ice and centrifuged at 30,000×*g* for 25 min.
11. The supernatants are pooled and dialyzed extensively at 4°C against 100 mM acetic acid.
12. The extracted histones are lyophilized and resuspended in H_2O to a concentration of 4 mg/mL (*see* **Note 2**).

3.1.2. Preparation of Non-histone Substrates (See **Note 3)**

1. The bacteria are grown in 2X YT medium supplemented with 100 μg/mL ampicillin. The cells are grown to an optical density of $OD_{600} = 0.5$.
2. 1 μL/mL of 100 mM isopropyl-beta-D-thiogalactopyra noside (IPTG) is added to induce glutathione-S transferase (GST) fusion protein expression. For 500 mL medium, 500 μL of IPTG is added.
3. The cells are grown at either 37°C for 2 h or 30°C for 4 h.
4. The bacteria are pelleted at 3,000×*g* using a Beckman 10,500 rotor and washed once with PBS.
5. Resuspend in 15 mL of ice-cold buffer A (PBS supplemented with 2 mM DTT, 1 mM PMSF, 1 mg/mL pepstatin, 1 mg/mL leupeptin, and 1 mg/mL aprotinin).
6. The sample is kept on ice with 2 mg of lysozyme for 10 min.
7. Sonicate on ice until the solution appears clear.
8. The sample is mixed with 1 mL of 15% Triton X-100 in PBS and centrifuged at 40,000 rpm for 15 min at 4°C.
9. At the same time, 500 μL of Sepharose 4B beads are washed three times with buffer B (PBS supplemented with

0.5% Tween-20, 0.5% Triton X-100, 1 mM PMSF, and 1 mg/mL leupeptin).

10. The washing buffer is removed after the last wash, and the supernatant from step 8 is placed in the beads.
11. Place the sample on a rotator at 4°C for 6 h to overnight.
12. The supernatant is removed and the beads are washed with buffer C (PBS supplemented with 2 mM DTT and 0.5% Tween-20) three times. After the last wash, transfer the beads into an Eppendorf tube and remove supernatant.
13. The beads are washed with buffer D (20 mM Tris–HCl, pH 8.0, 2 mM DTT) three times. After the last wash, add back 400 μL buffer D.
14. The beads are incubated with 50 μL of glutathione (GSH) (100 mM in stock, Tris–HCl, pH 7.8) at RT for 30 min. The sample is centrifuged at 12,000×*g* in a bench-top centrifuge, and the supernatant is collected.
15. The GSH is dialyzed from the sample with dialysis buffer (20 mM Tris–HCl, pH 8.0, 25 mM NaCl, and 2 mM EDTA) using three changes of buffer.

3.1.3. In Vitro HAT Assay

1. A standard HAT assay was performed containing 5 μg of substrate and enzyme (200 ng of purified histone acetyl transferase (purified baculovirus p300 or P/CAF)).
2. The mixture was incubated at 30°C for 1 h in the presence of 90 pmol of [^{14}C]acetyl-CoA.
3. The reaction mix was resolved on a SDS-polyacrylamide gel and viewed following autoradiography of the gel.
4. Alternatively, [^{14}C]acetyl incorporation into the substrates can be also determined by liquid scintillation counting by following next three steps.
5. The reaction mixture is spotted onto phosphocellulose filter paper, such as Whatman P-81.
6. The filter paper is air-dried for 2–5 min and washed with 0.2 M sodium carbonate buffer (pH 9.2) at RT with five changes of the buffer for a total of 30 min.
7. The dried filter paper is counted in a liquid scintillation counter.

3.1.4. Immunoprecipitation (IP) HAT Assay

3.1.4.1. Cell Extract Preparation

1. Cells are grown in 150 mm culture dishes.
2. Approximately 24–48 h after transfection, cells are collected by scraping them into 1 mL of ice-cold PBS and pelleted by centrifugation.
3. The PBS is aspirated and the cells are resuspended in 300 μL of New RIPA buffer.

4. The sample of lysis mixture is placed in a microfuge tube on dry ice to freeze the sample and then thawed to release the cellular-soluble protein.
5. Samples are centrifuged at 12,000×*g* at 4°C for 10 min and the supernatant is transferred to a new tube.
6. Check the concentration using Bradford assay reagent (Bio-Rad).

3.1.4.2. Immunoprecipitation

1. Adjust the protein concentration to 1 mg/mL in 500 μL. Add relevant antibodies (2 μg per 500 μg extract) and incubate at 4°C for 2 h.
2. Add protein A-Sepharose or G-Sepharose beads (1:1 mix, 30 μL). Rotate the mixture at 4°C overnight. Pellet the beads–Ab complexes and wash with New RIPA buffer three times.

3.1.4.3. IP-HAT Assay

1. Wash the beads with HAT assay buffer.
2. Add back 30 μL HAT buffer, 1 μL of 5 mg/mL histones or non-histone substrates, enzyme, and 1 μL (6 pmol) of ^{3}H-acetyl-CoA.
3. The mixture is incubated at 30°C for 30 min. Flick the tube several times during the incubation.
4. ^{14}C-acetyl incorporation into the substrates is determined by liquid scintillation counting (*see* **Note 4**).

3.1.5. Identification of Acetylated Lysine Residues In Vitro by Edman Degradation Assay

1. In vitro acetylation assays were performed as described previously (1, 4).
2. A synthetic peptide corresponding to the androgen receptor (AR1 residues 623–640), NH2-GMTLGARKL KKLGNLKLQ-OH, and a control polypeptide, NH2-ELVHMINWAKRVPGFVDL-OH, are synthesized.
3. The peptide is acetylated in vitro by incubating with 5 mM acetyl-CoA and baculovirus-purified FLAG-p300 at 30°C for 1 h. After incubation, acetylated peptides were separated from contaminating p300 by passage through a 10 μm filter (Amicon) and further purified by analytical reversed-phase high pressure liquid chromatography.
4. The reaction products are analyzed with a PE Biosystems DE-STR matrix-assisted laser desorption ionization time-of-flight mass spectrometer.
5. Further analysis by Edman degradation is performed on a PE Biosystems Procise sequencer. The amount of phenylthiohydantoin-acetyl-lysine is measured by absorbance at 259 nm.

3.1.6. Identification of Protein Acetylation and Deacetylation by Proteomics

300 μL of acetylated AR peptide Ac-GMTLGAR(KAc)L(KAc)-(KAc)LGNLKLQ, identical to the androgen receptor sequence from 623 to 640 is incubated with 600 μM NAD+ and 1 μM mSIR2a or 1 μM hSIRT1 in potassium phosphate (50 mM, pH 7.6).

After 30 min of incubation at 35°C, the reaction solution is loaded onto a Vydac 1.0 by 50 mm C8 column (The Separations Group, Hesperia, CA). An HP 1100 high-performance liquid chromatography (HPLC) equipped with a degasser and a binary pump is used to generate acetonitrile gradients at a flow rate of 50 μL/min. Solvent A containing 0.1% trifluoroacetic acid (TFA) and solvent B contained 90% (vol/vol) acetonitrile – 0.1% TFA. The total solvent composition for all chromatography is given by the equation A + B = 100% (where A and B refer to the respective solvents).

1. The sample is de-salted at a 5% concentration of solvent B for 20 min, and the peptides are separated by a 3-min gradient in which the solvent B concentration increases from 5 to 15%, followed by a 35 min gradient in which the solvent B concentration increases from 15 to 50%, and a 5-min gradient in which the solvent B concentration increases from 50 to 95%.
2. The column effluent is delivered directly to an LCQ quadruple ion trap mass spectrometer (Thermo Finnigan, Riviera Beach, FL) equipped with an electrospray ionization source.
3. For liquid chromatography/mass spectrometry (LC/MS) analysis, the mass spectrometer is operated in normal MS scan mode to detect ions in the *m/z* range of 400–2,000. For LC/MS/MS analysis, the mass spectrometry is performed in the data-dependent mode. The mass spectrometer detects the intensity of the ions in the *m/z* range 700–1,500 and is switched to the collision-induced dissociation mode to acquire an MS/MS spectrum. The mass isolation window for the collision-induced dissociation mode is set at 3 mass units, and the relative collision energy is set at 30%. Reactions and analyses for yeast Sir2p and *Archaeoglobus fulgidus* Sir2af2 are performed similarly (2).
4. To determine the deacetylation selectivity of human SIRT1 on triacetylated AR peptide, the AR peptide (5 μL, 100 μM) and SIRT1 enzyme (2 μL, 3.85 μM) are added to 45 μL of NAD^+ (200 μM) dissolved in 50 mM phosphate buffer (pH 7.5).
5. The reaction is incubated at 25°C for 15 min and quenched by the addition of 8 μL of TFA (760 μM); 10 μL each of the reaction mixture and control are injected onto a C18 column and analyzed. To determine

the Michaelis–Menten curves of SIRT1-catalyzed deacetylation for p53 (KKQSTSRHKK[AC]LMFKTEG), p300 (ERSTELK[Ac]TEIK(Ac)EEEDQPSTS), and AR peptide, p53, p300, and AR peptide (from 5 μM to 525 μM) are added to NAD^+ (500 μM) in 50 mM phosphate buffer (pH 7.5) (2).

6. Deacetylation is initiated by the addition of 3 μL of SIRT1 enzyme (3.85 μM). The reaction mixture is incubated at 25°C for 30 min and then quenched with 8 μL of 760 μM TFA. 10 μL of each reaction mixture is injected onto the HPLC for analysis. The chromatograms are obtained by separation with 0.1% TFA in water using an analytical C18 column and 260 nm detection. The areas of resolved peaks for 2′- and 3′-*O*-acetyl-ADP-ribose are used to quantitate total deacetylation.

3.2. Identification of In Vivo Acetylation

1. The experiment can be conducted using either endogenous "target proteins" or proteins expressed in cells, using a mammalian expression vector.
2. Cell transfection: 293 T cells are transfected by Superfect Transfection reagent (Qiagen, Valencia, CA) on a 150 mm plate with the expression vector encoding the protein of interest.
3. Twenty-four hours after transfection, the cells are transferred to the fresh DMEM medium containing 1 mCi/mL of [^{3}H]-sodium acetate (Amersham) for 1 h before lysis.
4. Cells are washed twice with cold PBS and lysed in New RIPA buffer supplemented with protease inhibitors (*see* **Note 5**).
5. The lysates are centrifuge at 40,000 rpm for 30 min at 4°C.
6. Supernatants collected from five plates are immunoprecipitated, with the antibody to the relevant target protein that has been conjugated to protein A agarose beads, for 6 h to overnight at 4°C.
7. The beads are washed five times with 1 mL of New RIPA buffer supplemented with 0.5% Tween-20.
8. Immunoprecipitates are solubilized with SDS-PAGE sample buffer and resolved on a 8% SDS-PAGE gel.
9. Gels containing [^{3}H]-acetate-labeled "target" protein are fixed with 10% glacial acetic acid and 40% methanol for 1 h and enhanced by fluorography enhancing solution (Amplify, Amersham) for 30 min. Gels are then dried and subjected to autoradiography at –70°C for 15 days.

4. Notes

1. Core histones are made as substrates for acetyltransferase assay using HeLa cells labeled with [^{3}H] acetic acid.
2. Check radioactivity of labeled core histones by liquid scintillation counting. Typically, a specific activity of 2.5×10^6 cpm/mg of protein is obtained.
3. The fusion protein can be made either as a GST fusion in bacterial cells or from mammalian or insect cell-expression systems using other tags (FLAG, Myc, HIS, HA).
4. Alternatively, reaction mixtures can also be resolved by SDS-PAGE gel. Gels are dried, enhanced, and subjected to autoradiography.
5. Several protease inhibitors are also potent inhibitors of HATs. Samples assessed for HAT activity should not be exposed to Hg-containing compounds, *N*-ethylmaleimide or iodoacetamide. Other inhibitors including PMSF, leupeptin, aprotinin, bestatin, pepstatin, and benzamidine are reported to not affect HAT activity and can be used to protect against proteolysis.

Acknowledgments

This work was supported by grants from National Institutes of Health [R01CA70896, R01CA75503, and R01CA86072 to R.G.P.]. Work conducted at the Kimmel Cancer Center was supported by the NIH Cancer Center Core grant [P30CA56036 to R.G.P.]. This project is funded in part from the Dr. Ralph and Marian C. Falk Medical Research Trust and a grant from Pennsylvania Department of Health (to R.G.P. and C.W.). The Department specifically disclaims responsibility for an analysis, interpretations, or conclusions.

References

1. Wang, C., Fu, M., Angeletti, R.H., Siconolfi-Baez, L., Reutens, A.T., Albanese, C., Lisanti, M.P., Katzenellenbogen, B.S., Kato, S., Hopp, T., Fuqua, S.A., Lopez, G.N., Kushner, P.J., Pestell, R.G. (2001) Direct acetylation of the estrogen receptor alpha hinge region by p300 regulates transactivation and hormone sensitivity. *J Biol Chem* **276**, 18375–18383.
2. Fu, M., Liu, M., Sauve, A.A. Jiao, X., Zhang, X., Wu, X., Powell, M.J., Yang, T., Gu, W., Avantaggiati, M.L., Pattabiraman, N., Pestell, T.G., Wang, F., Quong, A.A., Wang, C., Pestell, R.G. (2006) Hormonal control of

androgen receptor function through SIRT1. *Mol Cell Biol* **26**, 8122–8135.
3. Fu, M., Rao, M., Wang, C., Sakamaki, T., Wang, J., Di Vizio, D., Zhang, X., Albanese, C., Balk, S., Chang, C., Fan, S., Rosen, E., Palvimo, J.J., Janne, O.A., Muratoglu, S., Avantaggiati, M.L., Pestell, R.G. (2003) Acetylation of androgen receptor enhances coactivator binding and promotes prostate cancer cell growth. *Mol Cell Biol* **23**, 8563–8575.
4. Fu, M., Wang, C., Reutens, A.T., Wang, J., Angeletti, R.H., Siconolfi-Baez, L., Ogryzko, V., Avantaggiati, M.L., Pestell, R.G. (2000) p300 and p300/cAMP-response element-binding protein-associated factor acetylate the androgen receptor at sites governing hormone-dependent transactivation. *J Biol Chem* **275**, 20853–20860.
5. Fu, M., Wang, C., Wang, J., Zhang, X., Sakamaki, T., Yeung, Y.G., Chang, C., Hopp, T., Fuqua, S.A., Jaffray, E., Hay, R.T., Palvimo, J.J., Janne, O.A., Pestell, R.G. (2002) Androgen receptor acetylation governs trans activation and MEKK1-induced apoptosis without affecting in vitro sumoylation and trans-repression function. *Mol Cell Biol* **22**, 3373–3388.
6. Fu, M., Rao, M., Wu, K., Wang, C., Zhang, X., Hessien, M., Yeung, Y.G., Gioeli, D., Weber, M.J., Pestell, R.G. (2004) The androgen receptor acetylation site regulates cAMP and AKT but not ERK-induced activity. *J Biol Chem* **279**, 29436–29449.
7. Fu, M., Wang, C., Wang, J., Zafonte, B.T., Lisanti, M.P., Pestell, R.G. (2002) Acetylation in hormone signaling and the cell cycle. *Cytokine Growth Factor Rev* **13**, 259–276.
8. Gu, W., Roeder, R.G. (1997) Activation of p53 sequence-specific DNA binding by acetylation of the p53 C-terminal domain. *Cell* **90**, 595–606.
9. Zhang, W., Bieker, J.J. (1998) Acetylation and modulation of erythroid Kruppel-like factor (EKLF) activity by interaction with histone acetyltransferases. *Proc Natl Acad Sci USA* **95**, 9855–9860.
10. Boyes, J., Byfield, P., Nakatani, Y., Ogryzko, V. (1998) Regulation of activity of the transcription factor GATA-1 by acetylation. *Nature* **396**, 594–598.
11. Reutens, A.T., Fu, M., Wang, C., Albanese, C., McPhaul, M.J., Sun, Z., Balk, S.P., Janne O.A., Palvimo, J.J., Pestell, R.G. (2001) Cyclin D1 binds the androgen receptor and regulates hormone-dependent signaling in a p300/CBP-associated factor (P/CAF)-dependent manner. *Mol Endocrinol* **15**, 797–811.
12. Fang, S., Tsang, S., Jones, R., Ponugoti, B., Yoon, H., Wu, S.Y., Chiang, C.M., Willson, T.M., Kemper, J.K. (2008) The p300 acetylase is critical for ligand-activated farnesoid X receptor (FXR) induction of SHP. *J Biol Chem* **283**, 35086–35095.
13. Faus, H., Haendler, B. (2006) Post-translational modifications of steroid receptors. *Biomed Pharmacother* **60**, 520–528.
14. Feige, J.N., Auwerx, J. (2007) DisSIRTing on LXR and cholesterol metabolism. *Cell Metab* **6**, 343–345.
15. Han, L., Zhou, R., Niu, J., McNutt, M.A., Wang, P., Tong, T. (2010) SIRT1 is regulated by a PPAR{gamma}-SIRT1 negative feedback loop associated with senescence. *Nucleic Acids Res* **38**, 7458–7471.
16. Kang, S.A., Na, H., Kang, H.J., Kim, S.H., Lee, M.H., Lee, M.O. (2010) Regulation of Nur77 protein turnover through acetylation and deacetylation induced by p300 and HDAC1. *Biochem Pharmacol* **80**, 867–873.
17. Kemper, J.K., Xiao, Z., Ponugoti, B., Miao, J., Fang, S., Kanamaluru, D., Tsang, S., Wu, S.Y., Chiang, C.M., Veenstra, T.D. (2009) FXR acetylation is normally dynamically regulated by p300 and SIRT1 but constitutively elevated in metabolic disease states. *Cell Metab* **10**, 392–404.
18. Leader, J.E., Wang, C., Fu, M., Pestell, R.G. (2006) Epigenetic regulation of nuclear steroid receptors. *Biochem Pharmacol* **72**, 1589–1596.
19. Li, X., Zhang, S., Blander, G., Tse J.G., Krieger, M., Guarente, L. (2007) SIRT1 deacetylates and positively regulates the nuclear receptor LXR. *Mol Cell* **28**, 91–106.
20. Lin, H.Y., Hopkins, R., Cao, H.J., Tang, H.Y., Alexander, C., Davis, F.B., Davis, P.J. (2005) Acetylation of nuclear hormone receptor superfamily members: thyroid hormone causes acetylation of its own receptor by a mitogen-activated protein kinase-dependent mechanism. *Steroids* **70**, 444–449.
21. Sanchez-Pacheco, A., Martinez-Iglesias, O., Mendez-Pertuz, M., Aranda, A. (2009) Residues K128, 132, and 134 in the thyroid hormone receptor-alpha are essential for receptor acetylation and activity. *Endocrinology* **150**, 5143–5152.
22. Soutoglou, E., Katrakili, N., Talianidis, I. (2000) Acetylation regulates transcription factor activity at multiple levels. *Mol Cell* **5**, 745–751.
23. Viengchareun, S., Le Menuet, D., Martinerie, L., Munier, M., Pascual-Le Tallec, L., Lombes, M. (2007) The mineralocorticoid receptor: insights into its molecular and

(patho)physiological biology. *Nucl Recept Signal* **5**, e012.

24. Wang, C., Powell, MJ., Popov, V.M., Pestell, R.G. (2008) Acetylation in nuclear receptor signaling and the role of sirtuins. *Mol Endocrinol* **22**, 539–545.
25. Wilson, B.J., Tremblay, A.M., Deblois, G., Sylvain-Drolet, G., Giguere, V. (2010) An acetylation switch modulates the transcriptional activity of estrogen-related receptor alpha. *Mol Endocrinol* **24**, 1349–1358.
26. Zhao, W.X. Tian, M., Zhao, B.X., Li, G.D., Liu, B., Zhan, Y.Y., Chen, H.Z., Wu, Q. (2007) Orphan receptor TR3 attenuates the p300-induced acetylation of retinoid X receptor-alpha. *Mol Endocrinol* **21**, 2877–2889.
27. Fu, M., Wang, C., Zhang, X., Pestell, R.G. (2004) Acetylation of nuclear receptors in cellular growth and apoptosis. *Biochem Pharmacol* **68**, 1199–1208.

Chapter 12

Analysis of Androgen Receptor SUMOylation

Miia M. Rytinki, Sanna Kaikkonen, Päivi Sutinen,
and Jorma J. Palvimo

Abstract

Androgen receptor (AR) is a ligand-controlled transcription factor that is deregulated and therefore targeted in prostate cancer. In addition to androgens, AR is regulated by post-translational modifications (PTMs). SUMOylation, conjugation of small ubiquitin-related modifier (SUMO) protein 1, 2, or 3, is a bulky PTM regulating several important physiological processes. We have shown that AR is modified by SUMO-1 at two conserved lysine residues in its N-terminal domain. This agonist-enhanced modification represses the transcriptional activity of the receptor in a reversible and target gene-selective fashion. Acceptor sites for SUMOs are also found in several other nuclear receptors. Since the cellular steady-state level of SUMO modifications of most substrates, including AR, is very low, transfection- and SUMO overexpression-based protocols are often needed to render the modifications clearly detectable. This chapter describes protocols for analyzing AR SUMOylation in cultured cells by immunoblotting, gel mobility shift assays, and immunoprecipitation. These methodologies are generally applicable for determining whether a particular protein is SUMOylated and for identifying the lysine residue(s) modified.

Key words: Androgen receptor, SUMO, post-translational modification, lysine residue, steroid hormone, immunoblotting, immunoprecipitation, transfection, cell culture, prostate cancer cell.

1. Introduction

Androgens, testosterone and 5α-dihydrotestosterone, control the development and maintenance of male reproductive organs, including prostate, through the androgen receptor (AR) (1, 2). Androgens also play a critical role in controlling the malignant growth of prostate and androgen ablation is a standard therapy for prostate cancer (3). In addition to the receptor and the receptor's cognate DNA-binding sites, transcriptional regulation by the androgens requires several AR-interacting coregulator proteins,

F. Saatcioglu (ed.), *Androgen Action*, Methods in Molecular Biology 776,
DOI 10.1007/978-1-61779-243-4_12, © Springer Science+Business Media, LLC 2011

coactivators, and corepressors. The functions of many coregulator proteins are linked to post-translational modifications (PTMs) that often target nucleosomal histones. PTMs are also used to directly modify the structure and activity of AR.

Covalent conjugation of proteins by small ubiquitin-related modifier (SUMO), SUMOylation, has emerged as a significant regulatory mechanism in cell physiology (4, 5). The modification can regulate subcellular localization, protein–protein interactions, and transcriptional activity, often resulting in transcriptional repression. Mammals contain three ~100-amino acid long SUMO proteins, SUMO-1, -2, and -3, that form isopeptide linkages with lysine ε-amino groups of their target proteins. SUMO-1 is ~50% identical with SUMO-2/3, whereas SUMO-2 and -3 are nearly (~97%) identical. The SUMO-1 is not thought to form polymers, whereas the SUMO-2 and -3 can form chains. The SUMO conjugation pathway is analogous with ubiquitin conjugation (ubiquitylation), but it requires E1, E2, and E3 activities that are distinct from the enzymes in ubiquitylation. SUMOs are activated by the SAE1 and -2 dimer (E1 activity) and subsequently conjugated by the Ubc9 (E2 activity). For example PIAS1, -2, -3 and -4 proteins can stimulate SUMOylation in a manner that resembles the action of E3 ligases (6). Interestingly, both Ubc9 and PIAS proteins can function as transcriptional coregulators for AR. SUMOylation is dynamically adjusted by forward and reverse reactions. The reverse reaction is potentially catalyzed by six SUMO-specific proteases (SENP1, -2, -3, -5, -6, and -7) (7).

Like AR, several other nuclear receptors, such as progesterone receptor, mineralocorticoid receptor, glucocorticoid receptor, liver X receptor β, and peroxisome proliferator-activated receptor α and γ, are shown to be SUMOylated (8, *for other references, see* (9)). Their SUMO acceptor lysines typically conform to a minimal consensus motif Ψ*KxE* (where Ψ is a large hydrophobic residue, *x* is any residue). As in the case of most of the other SUMO targets, these proteins are modified only to a small percentage at steady state, probably due to their rapid de-modification. Yet, disruption of the SUMOylation sites in each of these nuclear receptors leads to clearly enhanced activity, implying that SUMOylation has a general role in the regulation of nuclear receptors.

The low level of SUMOylated protein hampers the detection of SUMOylation in vivo. Coexpression of the protein of interest-encoding construct with and without a SUMO expression vector can be used to circumvent this obstacle. Even though the physiological significance of such SUMO overexpression-based experiments can be criticized, the common opinion in the field is that these types of approaches have yielded significant and biologically relevant data which are generally in agreement with SILAC (stable isotope labeling by amino acids in cell culture) and mass spectrometry-based identifications of endogenous SUMO targets. The usage of SUMO protease inhibitor *N*-ethylmaleimide (NEM) in protein extraction buffers, availability of both epitope-tagged and untagged expression vectors for SUMOs, and good commercially available antibodies form the basis for straightforward protocols for assaying AR SUMOylation (or SUMOylation of a given protein) in mammalian cells by immunoblotting and immunoprecipitation.

2. Materials

2.1. Cell Culture and Sample Collection

1. Cell lines used: African Green monkey kidney (COS-1) cells and Vertebral-Cancer of the Prostate (VCaP) cells from ATCC (Manassas, VA, USA).
2. For maintenance of the COS-1 cells: Dulbecco's Modified Eagle's Medium (DMEM) (Gibco®, Invitrogen, part of Life Technologies, Carlsbad, CA, USA) supplemented with 10% (v/v) fetal bovine serum (FBS) and 25 U/ml penicillin and 25 μg/ml streptomycin. For maintenance of the VCaP cells: DMEM containing 10% (v/v) US defined FBS (HyClone, Logan, Utah, USA), 25 U/ml penicillin, and 25 μg/ml streptomycin in a 5% CO_2 atmosphere at 37°C.
3. For hormone/anti-hormone treatment of the cell lines: DMEM supplemented with 2.5% (v/v) charcoal-stripped FBS (*see* **Note 1**).
4. Trypsin (0.25%) with 1 mM ethylenediaminetetraacetic acid (EDTA).
5. Transfection reagent: TransIT®-LT1 (Mirus Bio LLC, Madison, WI, USA).
6. Expression vectors for mammalian cells: pcDNA3.1-FLAG, pcDNA3.1-FLAG-AR, pCMV-MYC, pCMV-MYC-SUMO-1, and pCMV-MYC-SUMO-2 (10). (Comparable SUMO expression plasmids can be obtained from Addgene (www.addgene.org), a non-profit plasmid repository.) The plasmids are purified from *Escherichia coli* JM109 (Promega Corporation, Madison, WI, USA) strain cultures using QIAGEN Plasmid Maxi Kit according to the manufacturer's instructions (Qiagen, Hilden, Germany). The purified plasmids are dissolved in TE buffer (10 mM Tris–HCl, pH 8, and 1 mM EDTA, pH 8) at 1 μg/μl.
7. AR ligands: Testosterone (Steraloids Inc., Newport, RI, USA) is used (*see* **Note 2**). Testosterone is stored in glass tube as 0.1 mM stock in absolute ethanol (EtOH) at

–20°C and bicalutamide (Bidragon Pharmaservice LLC, Burlingame, CA, USA) as 10 mM stock in EtOH at –20°C.

8. Cell-harvesting solution: Phosphate-buffered saline (PBS, 137 mM NaCl, 2.7 mM KCl, 100 mM Na_2HPO_4, and 2 mM KH_2PO_4, pH 7.4) containing 10 mM *N*-ethylmaleimide (NEM, added just before use from 2 M stock in EtOH stored at –20°C) (*see* **Note 3**).

2.2. SDS-Polyacrylamide Gel Electrophoresis (SDS-PAGE) and Immunoblotting

1. Separating buffer (4X): 1.5 M Tris–HCl, pH 8.8. Stored at room temperature.
2. Stacking buffer (4X): 0.5 M Tris–HCl, pH 6.8. Stored at room temperature.
3. 30% (w/v) acrylamide/bis solution (37.5:1 with 2.6% C) (Bio-Rad Laboratories, Inc., Hercules, CA, USA).
4. 10% (w/v) sodium dodecyl sulfate (SDS), stored at room temperature.
5. Ammonium persulfate: 10% solution in water, aliquots stored at –20°C (Bio-Rad Laboratories, Inc., Hercules, CA, USA).
6. *N*,*N*,*N*,*N*′-tetramethyl-ethylenediamine (TEMED).
7. SDS running buffer (10X): 250 mM Tris Base, 1920 mM glycine, 1% (w/v) SDS, stored at room temperature.
8. Prestained molecular weight marker: PageRuler^{TM} Prestained Protein Ladder (Fermentas International Inc., Burlington, Ontario, Canada).
9. Transfer buffer: 25 mM Tris, 192 mM glycine, and 20% (v/v) methanol stored at 4°C. A 10X stock solution can be prepared: 250 mM Tris and 1920 mM glycine, stored at room temperature.
10. Nitrocellulose membrane, 0.45 μm (Pierce, part of Thermo Fisher Scientific Inc., Waltham, MA, USA).
11. Three mm thick chromatography paper.
12. Tris-buffered saline with 0.1% (v/v) Tween-20 (TBST). 10X: 1.4 M NaCl, 0.1 M Tris, pH 7.4; Tween-20 added to the diluted solution. Also TBS without Tween-20 is needed for a final washing step in the immunoblotting.
13. Blocking buffer: 5% (w/v) nonfat dry milk in TBST.
14. Primary antibodies: mouse monoclonal AR (441) (sc-7305; Santa Cruz Biotechnology, Inc., Santa Cruz, CA, USA) (*see* **Note 4**).
15. Secondary antibody: Horseradish peroxidase-conjugated goat anti-mouse IgG (Molecular Probes Inc., Eugene, OR, USA) (*see* **Note 5**).

16. Thermo Scientific Pierce ECL Western Blotting Substrate (Pierce).
17. Kodak BioMax Light Film (Carestream Health, Inc., Rochester, NY, USA).

2.3. Stripping and Re-blotting for α-Tubulin

1. Stripping buffer: 62.5 mM Tris–HCl, pH 6.7, 2% (w/v) SDS, 100 mM β-mercaptoethanol. Store at room temperature.
2. Primary antibody: mouse monoclonal anti-α-tubulin antibody (sc-5286; Santa Cruz Biotechnology, Inc., Santa Cruz, CA, USA).

2.4. Sample Preparation for Immunoblotting

1. 5X SDS-SB: 312 mM Tris–HCl, pH 6.8, 62.5% glycerol, 10% (w/v) SDS, 0.025% (w/v) bromophenol blue, 1425 mM β-mercaptoethanol (added before use) (*see* **Note 6**).
2. Cell lysis buffer: 1X SDS-SB containing 1:200-diluted Protease Inhibitor Cocktail P8340 (PIC; Sigma-Aldrich, St. Louis, MO, USA) and 10 mM NEM.

2.5. Immunoprecipitation

1. Radioimmunoprecipitation assay (RIPA) buffer: 50 mM Tris–HCl, pH 7.8, 150 mM NaCl, 15 mM $MgCl_2$, 5 mM EDTA, and 0.5% (v/v) Triton X-100. Before use, add 1:200 volume of PIC and NEM to 10 mM.
2. Protein G Agarose.
3. Mouse monoclonal anti-cMYC (9E10) antibody (sc-40; Santa Cruz Biotechnology, Inc.).

3. Methods

The SUMO modifications of ectopically expressed AR can be analyzed by using different approaches. The direct immunoblotting of cell lysates- and immunoprecipitation-based procedures described in detail below can be used for analyzing the AR SUMOylation in any laboratory possessing basic cell culture and protein electrophoresis and transfer equipment. An alternative protocol, histidine pull-down (Ni^{2+}-nitroloacetic acid (NTA) chromatography) assay, for isolating His_6-tagged SUMO or ubiquitin proteins under denaturing buffer conditions has been described in detail elsewhere (11). Due to the usage of denaturing conditions that do not tolerate noncovalent protein–protein interactions, only proteins covalently bound to SUMO are detected in the latter assay, whereas the immunoprecipitation assay does not necessary differentiate between noncovalent and covalent SUMO–target protein interactions. To better prove the SUMO modification(s) of a particular protein, the target lysines should be identified by site-directed mutagenesis approaches. Web-based tools can be used for prediction of consensus SUMOylation sites in the target protein (e.g., SUMOplot algorithm at www.abgent.com.cn/doc/sumoplot/login.asp), with keeping in mind that the SUMOylation sites are usually conserved in vertebrates. The putative acceptor lysines are thereafter mutated to a non-conjugated amino acid residue, usually to arginines (preserving the basic nature of the residue) by site-directed mutagenesis of expression vectors. However, recent proteomics studies suggest that a significant portion of SUMOylation occurs also at non-consensus sites, but the consensus and the nonconsensus site SUMOylation do not seem to occur in the same protein.

The SUMO modifications are generally labile, being effectively de-conjugated by SUMO protease activity in cell lysates. Therefore, the cell lysates should be either quick denatured by heating in the presence of 2% SDS at 95°C or they should be prepared in the presence of a sufficient concentration of cysteine protease inhibitor, such as *N*-ethylmaleimide (NEM), that blocks the SUMO protease activity. In our protocols, already the harvest of cells is carried out in the presence of the NEM. The NEM-dependent upshifted bands (i.e., bands migrating slower than unmodified protein and seen only in the presence of the NEM in extraction buffers) in the immunoblots are indicative of SUMO or ubiquitin modifications. If a protein is SUMOylated, the bulky modification results in a clear upshift in the mobility of the protein on SDS-PAGE. The upshift is generally much more than could be anticipated on the basis of SUMOs' relative molecular mass (~11 kDa), and it is markedly influenced by the localization of the SUMO acceptor site in the target protein, with the modifications around the proteins' center regions upshifting more than the ones near the proteins' ends.

3.1. Preparation of Samples for Immunoblotting

3.1.1. Ectopically Expressed Proteins

1. COS-1 cells are passaged (1:8) by trypsinization at near confluency (twice a week) to new maintenance cultures on 10 cm^2 dishes. Experimental cultures are seeded into 6-well plates (300,000 cells and 3 ml medium/well). The seeded experimental cultures are left to settle in the maintenance culture media with 10% FBS for 20 h. The culture media is replaced with 2 ml/well of transfection culture media containing 2.5% charcoal-stripped FBS. The cells are transfected 4 h after the medium replacement.
2. Transfection of AR (pcDNA3.1-FLAG-AR) and SUMO-1 or SUMO-2 (pCMV-MYC-SUMO-1, pCMV-MYC-SUMO-2) is performed according to the manufacturer's

guidelines for TransIT®-LT1 transfection reagent. The corresponding empty vectors should be used in samples without the respective protein-expressing vectors to maintain equal amount of DNA. Dilute the stock DNAs (1 μg/μl) 1:10 with H_2O to 1.5 ml microcentrifuge tubes so that they contain 300 ng of each of the appropriate plasmid (*see* **Note 7**). Prepare one DMEM–TransIT®-LT1 mixture that is sufficient for all the wells to be transfected (1.8 μl of TransIT®-LT1 with 98 μl of DMEM is required for each well of the 6-well plate), mix well by pipetting gently, and incubate for 5 min. Add 100 μl of the TransIT®-LT1–DMEM mixture to each of the tubes of the plasmid mixtures, mix well by pipetting gently, and incubate for 20 min. Distribute the transfection mixture evenly by dropwise pipetting to the cells and incubate the cells for 24 h.

3. Ligand treatment: Dilute 0.1 mM testosterone stock 1:10 in DMEM and add 21.2 μl (1:100) of the dilution (or the same volume of DMEM containing the same concentration of vehicle, EtOH) to 6-well plate (final concentration 100 nM in the well). After 24 h incubation, the cells are ready to be harvested.
4. Transfer cells to ice, wash once with 1 ml of ice-cold 1× PBS, and add 400 μl of 1× PBS containing 10 mM NEM. Scrape the cells with the larger end of a 200 μl pipette tip and collect the cells into 1.5 ml tubes (keep tubes always on ice) and spin the cells down in a microcentrifuge (~16,000×*g* for 10 s). Remove the supernatant carefully.
5. Suspend the pellet in 150 μl of 1X SDS-SB containing 1:200 PIC, 10 mM NEM, and 285 mM β-mercaptoethanol. The samples are sonicated for 15 s with a small probe sonicator (Ultrasonic processor UP50H, Hielscher Ultrasonics GmbH, Teltow, Germany), placed directly back on ice, and subsequently heated for 5 min at 95°C. Centrifuge the samples at 16,000×*g* for 5 min. The samples are now ready for SDS-PAGE.

3.1.2. Endogenous Proteins

1. VCaP cells are passaged (1:2) by trypsinization at near confluency (once a week) to new maintenance cultures on 10 cm^2 dishes. Remove trypsin by centrifugation at 69×*g* for 4 min and resuspend in fresh medium before seeding. Fresh maintenance medium is changed once a week to the maintenance cultures. Experimental cultures are seeded into 6-well plates (800,000 cells/well). The seeded experimental cultures are left to settle in maintenance culture media containing 10% FBS for 48 h. The culture media is replaced with

2 ml of transfection culture media containing 2.5% charcoal-stripped FBS. The cells are kept in the transfection culture media for 72 h before treatment with ligands.

2. Ligand treatment: Testosterone and vehicle are added as in **Section 3.1.1**, Step 3. Bicalutamide is diluted similarly as testosterone, but yielding a final concentration of 10 μM in the well. After for 4 h, the cells are harvested as in previous section starting from Step 4.

3.2. SDS-PAGE

1. The following instructions are for the use of Bio-Rad Mini-PROTEAN Electrophoresis System.
2. Prepare a 7.5% gel (5 ml is enough for one 1.5 mm thick gel): mix 2.42 ml H_2O, 1.25 ml of 1.5 M Tris, pH 8.8, 50 μl 10% SDS, 1.25 ml 30% acrylamide/bis solution, 25 μl 10% ammonium persulfate, and 2.5 μl TEMED. Pour the separation gel, but leave ~2 cm space (the short glass plate) for the stacking gel. Fill the gel cassette up with H_2O and let the gel polymerize for 30 min.
3. Pour off the water and dry briefly the remaining water with a piece of 3 mm paper.
4. Prepare the stacking gel (2 ml is enough for one gel) by mixing 1.21 ml H_2O, 0.5 ml of 0.5 M Tris, pH 6.8, 20 μl 10% SDS, 260 μl 30% acrylamide/bis solution, 25 μl 10% ammonium persulfate, and 1 μl TEMED. Pour the stacking gel, insert the comb, and let it polymerize for 30 min (*see* **Note 8**).
5. Once the gel has polymerized it can be inserted to the electrode assembly in the tank. Fill the inner and outer chamber with the SDS running buffer and remove the comb from the gel.
6. Load 3 μl of PageRuler™ Prestained Protein Ladder and 12 μl of each sample into the wells (the amount is adequate for visualization of both the exogenous and the endogenous AR).
7. Place the lid to the tank and connect the electrodes to a power supply. Let the samples run through the stacking gel at 160 V (~10 min), thereafter increase to 200 V (~45 min), and the dye front can be run off the gel.

3.3. Immunoblotting of SUMO-Modified AR

1. The following instructions are for the use of Bio-Rad Mini Trans-Blot® Electrophoretic Transfer Cell.
2. The SDS-PAGE resolved samples are electrophoretically transferred to a nitrocellulose membrane. Cut two pieces of 3 mm paper at the size of the fiber pads; cut also a piece of the nitrocellulose membrane just a bit larger than the

separating gel. Pre-wet the 3 mm papers and the membrane in a small tray of transfer buffer. In a larger tray with transfer buffer, assemble the gel sandwich starting with placing the black side of the gel cassette down. On top of the black side, place a pre-wetted fiber pad, and a 3 mm paper is placed on the fiber pad.

3. Disassemble the gel from the electrode assembly. Separate the glass plates and remove the stacking gel. To be able to follow the orientation of the gel, cut one corner of the separating gel.
4. The separating gel is put on the 3 mm paper. The pre-wetted membrane is carefully placed over the separating gel, add the last 3 mm paper and remove air bubbles between the gel and membrane by rolling, e.g., a pen gently over the 3 mm paper. The last fiber pad is added and the cassette containing the gel sandwich is closed.
5. Be sure that the cassette is placed in the electrode module so that the black side of the cassette is toward the black part of the module, so that the nitrocellulose membrane will be between the gel and the anode.
6. Place a Bio-Ice Cooling Unit (kept frozen at −20°C) and fill the transfer tank with pre-cooled transfer buffer. Start a magnetic stir bar, place the lid to the tank, and connect the electrodes to a power supply. Set the current to 250 mA, and let the proteins transfer for 1 h.
7. After the transfer, the cassette is opened so that the black side is down. Remove gently the top fiber pad and 3 mm paper, and the membrane should be visible. The contours of the gel under the membrane should be visible, cut with a scalpel the overflow of the membrane, and mark the orientation by cutting one of the corners. Control that the transfer has been successful by checking that the protein ladder is clearly visible on the membrane.
8. Block unspecific binding sites by incubating the membrane (the protein side up) in 25 ml of blocking buffer at room temperature on a rocking platform for 1 h.
9. Shortly rinse off the blocking buffer and add a 1:1000 dilution of anti-AR antibody in TBST (*see* **Note 9**). Routinely, the membrane is incubated with the primary antibody for overnight at 4°C, but 1 h at room temperature should be sufficient for visualizing the AR.
10. The membrane is briefly rinsed with TBST and washed with ~20 ml of TBST three times for 5 min on a rocking platform (*see* **Note 10**).

11. The appropriate secondary antibody is diluted 1:10,000 in TBST and incubated for 45–60 min at room temperature on a rocking platform.
12. The membrane is briefly rinsed with TBST and washed two times with ~20 ml of TBST for 10 min each and once with TBS without Tween-20 for 10 min.
13. During the final wash, prepare the ECL reagent. Mix equal amount of the solutions 1 and 2; 2 ml of the mixed solution is enough for one membrane. Prepare also a plastic pocket, in which the ECL-treated membrane is put before placing it into an X-ray film cassette. Put a piece of a plastic film on the table, lift the membrane from the TBS buffer, and dry the buffer droplets in a tissue paper before placing the membrane on the plastic film. Spread the mixture of the ECL reagent equally over the membrane and incubate for 1 min.
14. Lift the membrane, dry the droplets with a tissue paper, and place the membrane in a plastic pocket, tape the pocket on one side of the cassette, and close the cassette.
15. In a dark room under safe light conditions, place an X-ray film in the cassette and take a few different exposure times of each blot (*see* **Note 11**). Usually from 30 s to 2.5 min exposure yields bands of sufficient intensity. After the films have been developed, mark the positions of protein ladder markers to the films. An example of the results produced is shown in **Fig. 12.1**.

3.4. Stripping and Re-blotting of α-Tubulin

1. To control the loading of samples, the membrane is re-blotted with an anti-α-tubulin antibody. Before re-blotting, the previous antibody should be removed by stripping.
2. Transfer the membrane to a box with a cover that can be tightly closed (*see* **Note 12**). Add 20 ml of stripping buffer to the membrane, cover the box tightly, and put the membrane to a 50°C incubator with gentle shaking for 30 min.
3. Rinse the membrane twice in a fume hood with TBST. Wash the membrane further two times with 20 ml of TBST for 5 min by shaking on a rocking platform.
4. Now, the membrane is ready for the blocking buffer and the following steps are the same as in the **Section 3.3**, Steps 8–15 except for the fact that the anti-α-tubulin is used as the primary antibody in Step 9.

3.5. Immunoprecipitation of AR

1. The starting material for immunoprecipitation is prepared as in **Section 3.1.1**. However, to obtain enough starting material for detection of minor SUMO modifications, pool cells from 2 wells of a 6-well plate (for endogenous AR, *see*

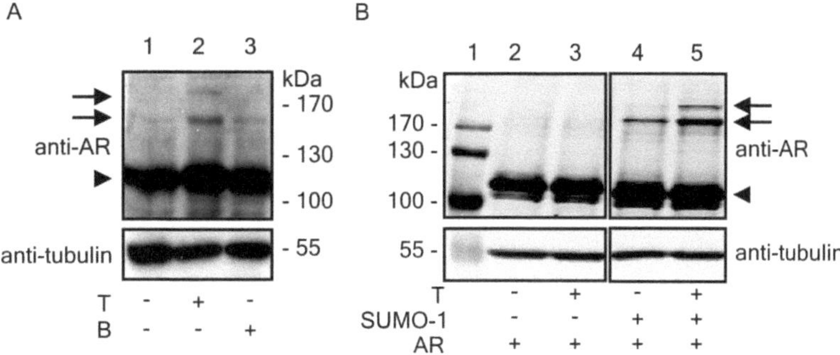

Fig. 12.1. SUMOylation shifts the gel mobility of AR as assessed by immunoblotting of whole cell lysates with anti-AR antibody. (**a**) VCaP cells were treated with vehicle (–), 100 nM of testosterone (T) or 10 μM bicalutamide (B, an anti-androgen) for 4 h as indicated, and whole cell lysates were prepared and analyzed by immunoblotting with the ECL detection system. Evidence for the SUMOylation of the endogenous AR is visible as upshifted AR immunoreactive bands in the testosterone-treated samples (depicted by *arrows*; the position of un-conjugated AR marked by an *arrowhead*; *cf. lane* 2 to *lanes* 1 and 3). (**b**) Ectopic coexpression of SUMO with AR results in upshifted AR bands in COS-1 cells. Cells were transfected with AR in the presence and absence of SUMO-1 and the cells were treated with 100 nM T or vehicle as indicated. AR immunoreactivity was visualized by Odyssey Infrared Imaging system. The prestained protein ladder is visible in the infrared imaging system (lane 1). *Panels* below the anti-AR immunoblots showing immunoblots with anti-α-tubulin antibody verify equal protein loading.

Note 13). Otherwise, follow the **Section 3.1.1.**, until Step 4. All the following steps are done on ice.

2. After removal of the supernatant, resuspend the cell pellet in 0.5 ml of RIPA (with 1:200 PIC and 10 mM NEM) and incubate for 5 min on ice.
3. Centrifuge the samples at ~16,000×*g* for 20 min at 4°C. Prepare the 50% Protein G agarose slurry during the centrifugation step. Wash the ethanol storage buffer off the agarose beads and resuspend the agarose beads in RIPA buffer to yield ~50% (v/v) agarose slurry.
4. Transfer the supernatants into fresh tubes; take also 50 μl into a separate tube as an input sample. Add 12.5 μl of 5X SDS-SB (with 1425 mM β-mercaptoethanol) to the input samples with 50 μl of lysates, heat the samples for 5 min at 95°C, and store them until analysis at –20°C (for long term, more than a week, storage at –70°C is recommended). Preclear the samples by adding 15 μl of 50% Protein G agarose beads to each sample and rotate at 4°C for 30 min.
5. Centrifuge at ~3500×*g* for 1 min. Transfer the supernatant into fresh tubes. Add 1 μg of anti-MYC antibody to each sample (*see* **Note 14**). Incubate by rotation, preferentially overnight, but at least 1 h at 4°C.

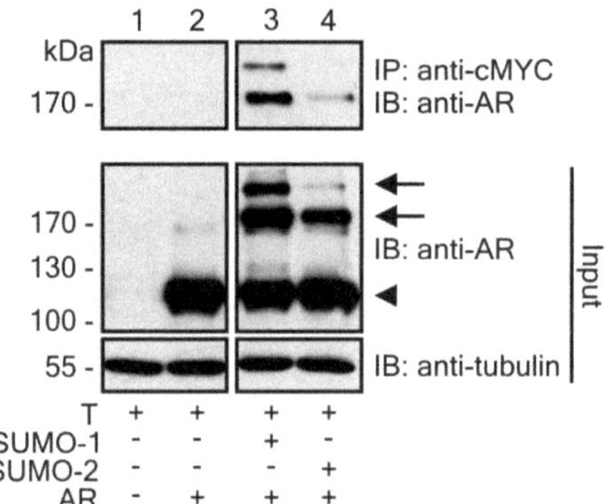

Fig. 12.2. Analysis of AR SUMOylation by immunoprecipitating AR-SUMO-conjugates. COS-1 cells were cotransfected with and without MYC-SUMO-1 or MYC-SUMO-2, and the cells were grown in the presence of 100 nM T for 24 h. The anti-MYC-immunoprecipitated (IP) proteins were immunoblotted (IB) with anti-AR (*upper panels*). Immunoblots of input AR samples and those of α-tubulin levels (loading control) are shown in *middle* and *lower panels*, respectively.

6. Add 25 μl pre-washed 50% Protein G agarose slurry to each sample and rotate at 4°C for 1 h.
7. Collect the beads by centrifugation at ~3500×*g* for 1 min. Wash the beads three times with 0.5 ml of RIPA (with 1:200 PIC and 10 mM NEM). After the last wash step, remove the buffer carefully.
8. Release the bound proteins by adding 50 μl of 1X SDS-SB (with 1:200 PIC, 10 mM NEM, and 285 mM β-mercaptoethanol) and heating the samples for 5 min at 95°C. Continue with the protocol for SDS-PAGE and immunoblotting. Loading of 10 μl of the immunoprecipitates and input samples to the gels is adequate. An example of the result is shown in **Fig. 12.2**.

4. Notes

1. Charcoal stripping of FBS: Add 10 g of activated charcoal (Charcoal activated GR for analysis; Merck, Darmstadt, Germany) and 1 g of dextran to 100 ml of DMEM in 1000 ml flask. Mix the DMEM–charcoal–dextran slurry well before adding 500 ml of FBS to the flask. Magnetic stir at room temperature for 1 h. Centrifuge in two 500 ml bottles at ~17,000×*g* for 20 min. Pour supernatants to clean centrifuge bottles and repeat the centrifugation step. Sterile filter, aliquot, and store at –20°C.

2. A more stable non-aromatizable synthetic androgen, R1881 (Perkin-Elmer, Waltham, MA, USA) at 10 nM concentration or 5α-dihydrotestosterone (Steraloids Inc.) can be used instead.
3. Handle the NEM with care, as it is a highly toxic substance. Note also that the NEM is unstable at $pH > 7$. Therefore, it should be added just before use. Also iodoacetamide can be used to inhibit cysteine proteases.
4. It is good to keep in mind that different commercially available anti-AR antibodies have been raised against different AR regions, and the bulky SUMO modification may differentially influence their ability to bind and therefore detect or immunoprecipitate the AR.
5. For quantitative analysis, an Odyssey Infrared Imaging System (LI-COR Inc., Lincoln, NE, USA) can be used. For this approach, a different secondary antibody is needed: DyLight™ 680 (or 800) conjugated goat anti-mouse IgG (Pierce). To avoid background in this detection method, a blocking buffer without Tween-20 is recommended. An example of an immunoblot from Odyssey Infrared Imaging system is shown in **Fig. 12.1b**.
6. Prepare the 5X stock solution, however, without the addition of β-mercaptoethanol. Aliquot the solution, e.g., in 900 μl (add 100 μl of β-mercaptoethanol (14.2 M) before use); store the aliquots at –20°C. 5X SDS-SB stock solution is especially useful to minimize the dilution of the input samples in immunoprecipitations, but otherwise a 2X SDS-SB stock solution is more user friendly.
7. The plasmids (stored in TE buffer) are diluted in H_2O just before the transfection.
8. Several manufacturers provide ready-made gels that are convenient to use. Make sure to choose gels that are compatible with your SDS-PAGE running apparatus.
9. The primary antibody for AR gives a clean result. Therefore, an antibody recognizing tubulin (sc-5286; Santa Cruz Biotechnology, Inc) can be added to the same primary antibody solution controlling the protein loading control of the blot. In this way, the stripping of the membrane can be avoided.
10. The primary antibody solution can be re-used a couple of times. To avoid bacterial growth, 0.05% (w/v) sodium azide should be added if stored for longer than a few days. A stock solution of 2% can be made, to ease the addition of sodium azide (remember that the sodium azide is highly toxic!).

11. To assure the right orientation of the film, a good practice is, e.g., to always bend the left upper corner of the film. In that way, you always keep track of the orientation of your exposed films. Luminescence marker stickers from, e.g., Stratagene (Agilent Technologies, Santa Clara, CA, USA) provide a convenient means for exact positioning of the samples in relation to the marker protein ladder.
12. Since the β-mercaptoethanol is a very smelly compound, it is good to work in a fume. Secure the covers tightly before transferring the box into the incubator. The stripping buffer is re-used once to decrease the consumption of β-mercaptoethanol. Alternatively (more expensive) dithiothreitol can be used instead of β-mercaptoethanol.
13. For immunoprecipitation of the endogenous AR, at least one 10 cm^2 dish of cells is recommended for each precipitation. The SUMO-modified AR is relatively insoluble in RIPA buffer, resulting in a poor recovery.
14. The immunoprecipitation can also be performed by precipitating AR (with anti-FLAG or anti-AR antibodies) and immunoblotting the precipitates with anti-cMYC.

Acknowledgments

This work was supported by Academy of Finland, Finnish Cancer Organisations, Sigrid Jusélius Foundation. We thank Dr. D. Owerbach for SUMO constructs.

References

1. Gao W, Bohl CE, Dalton JT (2005) Chemistry and structural biology of androgen receptor. *Chem Rev* **105**: 3352–3370
2. Heemers HV, Tindall DJ (2007) Androgen receptor (AR) coregulators: a diversity of functions converging on and regulating the AR transcriptional complex. *Endocr Rev* **28**:778–808
3. Knudsen KE, Scher HI (2009) Starving the addiction: new opportunities for durable suppression of AR signaling in prostate cancer. *Clin Cancer Res* **15**:4792–4798
4. Hay RT (2005) SUMO: a history of modification. *Mol Cell* **18**:1–12
5. Geiss-Friedlander R, Melchior F (2007) Concepts in sumoylation: a decade on. *Nat Rev Mol Cell Biol* **8**:947–956
6. Rytinki M, Kaikkonen S, Pehkonen P, Jääskeläinen T, Palvimo JJ (2009) PIAS proteins: pleiotropic interactors associated with SUMO. *Cell Mol Life Sci* **66**: 3029–3041
7. Hay RT (2007) SUMO-specific proteases: a twist in the tail. *Trends Cell Biol* **17**:370–376
8. Poukka H, Karvonen U, Jänne OA, Palvimo JJ (2000) Covalent modification of the androgen receptor by small ubiquitin-like modifier 1 (SUMO-1). *Proc Natl Acad Sci USA* **97**:14145–14150

9. Kaikkonen S, Jääskeläinen T, Karvonen U, Rytinki M, Makkonen H, Gioeli D, Paschal BM, Palvimo JJ (2009) SUMO-specific protease 1 (SENP1) reverses the hormone-augmented SUMOylation of androgen receptor and modulates gene responses in prostate cancer cells. *Mol Endocrinol* **23**: 292–307
10. Owerbach D, McKay EM, Yeh ET, Gabbay KH, Bohren KM (2005) A proline-90 residue unique to SUMO-4 prevents maturation and sumoylation. *Biochem Biophys Res Commun* **337**:517–520
11. Rytinki MM, Palvimo JJ (2009) SUMOylation attenuates the function of PGC-1α. *J Biol Chem* **284**:26184–26193

Chapter 13

Analysis of Ligand-Specific Co-repressor Binding to the Androgen Receptor

Claudia Gerlach*, Daniela Roell*, and Aria Baniahmad

Abstract

The recruitment of co-repressors to the androgen receptor is an important mechanism for reducing androgen-mediated gene activation. Importantly, co-repressors play a major role in the treatment of hormone-dependent growing tissue, such as prostate cancer and breast cancer. In line with this, co-repressor dysfunction seems to be a major player for development of castration-resistant prostate cancer or therapy-resistant breast cancer. The molecular basis of hormone therapy by particular antihormones (antagonists) for the androgen receptor (AR) is mediated by enhanced recruitment and activity of co-repressors that cause repression of AR target genes that regulate proliferation and alteration of cancer cells. Therefore co-repressor recruitment is a crucial molecular mechanism of gene repression as well as inhibition of cancer growth. Here we describe different strategies to investigate co-repressor recruitment to the AR. First, we developed a modified mammalian two-hybrid system to investigate the recruitment of co-repressors to the AR within mammalian cells. This assay is very useful for the identification of the molecular mechanism of new AR antagonists and for molecular analysis of castration-resistant prostate cancer expressing the AR. Second, we describe a technique to analyze the interaction of AR isolated from human prostate cancer cells with a newly generated AR-specific co-repressor peptide, which is bacterially expressed and affinity purified by glutathione-S-transferase affinity precipitation assays in vitro. In summary, these methods can greatly facilitate the study of AR–co-repressor interactions.

Key words: Co-repressor, hormone-dependent prostate cancer, antagonist, androgen receptor, mammalian one-hybrid system, GST affinity precipitation.

1. Introduction

Prostate cancer (PCa) is one of the major health problems in men and one of the most often diagnosed malignancies. In western countries it has now emerged as the second leading cause of

*Both the authors contributed equally to this work.

F. Saatcioglu (ed.), *Androgen Action*, Methods in Molecular Biology 776,
DOI 10.1007/978-1-61779-243-4_13,

cancer death for men (1). As the androgen receptor (AR) is the main effector of PCa cell proliferation (2, 3) inhibition of the AR plays an important goal in PCa therapy. A common treatment of PCa is the combination of hormone ablation and treatment with AR antagonists. One molecular mechanism for androgen antagonism was shown to involve co-repressors (CoR) (4–8) that bind to and transcriptionally inactivate the AR. However, antihormone treatment is effective only for a limited period of about 16–24 months after which PCa becomes androgen independent and castration resistant (9, 10). Different mechanisms are involved in this process (11). One important underlying mechanism is a shift in co-activator/CoR ratio, a reduced level of CoR activity or CoR binding to AR (5, 8, 12, 13).

The AR CoR mainly interact with the receptor amino-terminus (5, 7, 14) in a ligand-dependent manner. Binding of natural AR agonists leads to dissociation of CoR, such as that observed for the CoR Alien (7). AR antagonists, such as cyproterone acetate (CPA), induce the interaction of Alien with AR. Furthermore, synthetic antihormones that are used in PCa therapy induce the binding and activity of CoR. To analyze the molecular mechanism of novel AR antagonists, the interaction of CoR with AR is an important approach.

To detect the recruitment of CoR to the AR in the presence of AR antagonists within mammalian cells, we used a modified mammalian two-hybrid system. Originally, the two-hybrid screen (or yeast two-hybrid system) was evaluated by Fields and Song in yeast (1989), as a technique to detect protein–protein interactions. Domains from two proteins are used to generate a fusion protein: the DNA-binding domain (DBD) of GAL4, a strong transcriptional activator of the yeast *Saccharomyces cerevisiae,* as the DNA-binding moiety, and a potent transactivation domain such as that from herpes viral transcriptional activator VP16, forming the transactivation moiety. The system has later been adapted to investigate protein–DNA and DNA–DNA interactions in bacteria as well as in eukaryotic cells.

To avoid interference of a heterologous DBD on the AR in an AR-fusion protein, we used the full-length, unfused AR. The strong viral activation domain VP16 was used fused to several AR CoR in a mammalian expression vector containing the CMV promoter/enhancer. This expression vector is transiently co-transfected into CV-1 cells (African green monkey kidney epithelial cells) with the AR-responsive reporter expression vector pMMTV-luc or other AR-responsive reporters (14). Interestingly CoR exhibit both a ligand and a response element-specific interaction with the AR (15). CV-1 cells are additionally co-transfected with the human unmodified AR gene because they lack expression of endogenous AR as well as expression of other functional steroid receptors that might interfere with the assay system. Thus, CV-1 cells are a preferable test system excluding possible side effects by

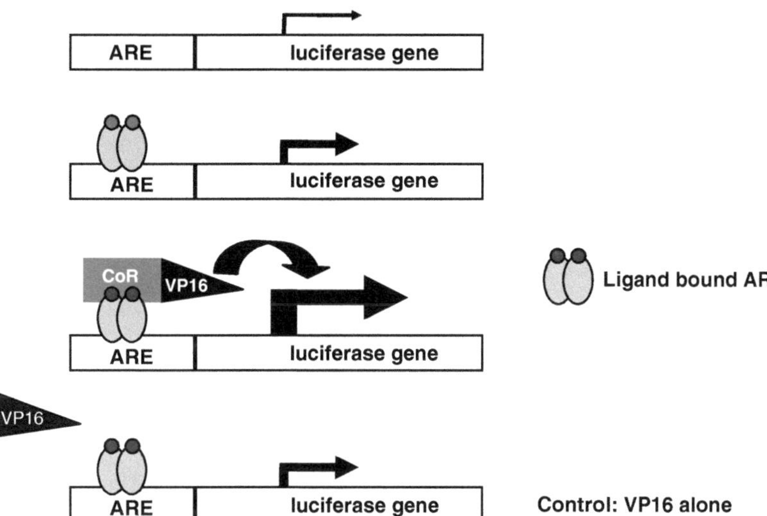

Fig. 13.1. Mechanism of the modified mammalian two-hybrid system to investigate ligand-induced CoR recruitment to the AR. The promoters themselves exhibit a slight basal activity (indicated by the *thin arrow*). In case of androgen (e.g., R1881) treatment, the luciferase gene expression is induced. If recruitment of a CoR that is fused to the VP16 transactivation domain by this ligand occurs, the luciferase gene expression is further enhanced. As control VP16 alone or VP16 fusions not binding to the AR should not enhance androgen-mediated luciferase gene activation.

other steroid receptor signaling pathways. Upon ligand binding, the AR is translocated into the nucleus and binds to the androgen response element of the reporter, such as one containing the pMMTV promoter, to induce the expression of the luciferase gene (**Fig. 13.1**). If AR antagonists are added to the cell medium and taken up by the cells, the co-transfected VP16-CoR fusion protein might be recruited to the AR and the luciferase expression may thereby be enhanced. This is due to the fact that the potent VP16 activation domain is tethered in the proximity of the promoter (**Fig. 13.1**). For optimal results, the silencing domains of CoR are removed, so that only the receptor interaction domains of CoR are fused to VP16 to avoid silencing of the reporter expression (e.g., cSMRT **Fig. 13.2**). The ligand-induced VP16-CoR recruitment and AR-dependent activation is measured as relative luciferase units (RLU, 10 s measurement). As internal normalization for transfection efficiency, pCMV-lacZ expression vector is co-transfected into the cells and the measured RLU are normalized using the β-galactosidase activity assay (**Fig. 13.2**).

Another possibility to compensate for the loss of CoR activity in advanced PCa is to increase the expression of highly AR-specific CoR. These CoR can potentially be ectopically expressed in PCa with the goal to permanently inhibit the AR. To that end, we identified small peptides from a peptide aptamer library that bind specifically to the AR as a basis for the generation of novel AR-specific CoR, called AB–CoR (*a*ptamer-*b*ased *CoR*). Here,

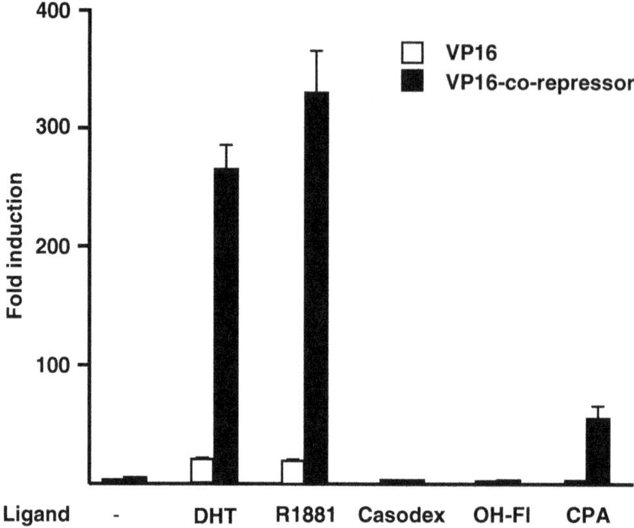

Fig. 13.2. Ligand-specific recruitment of the AR CoR cSMRT (only C-terminus of SMRT, devoid of the silencing domains) to the AR. CV1 cells were transfected with the pMMTV-luc, the pSG5-hAR, and the pCMV-lacZ (internal control) plasmids. Additionally the empty pCMV-VP16 was transfected as negative control or the pCMV-VP16-cSMRT plasmid (exchangeable by any other CoR plasmid). Cells were treated with the indicated ligands (end concentration 10^{-7} M). Luciferase and β-galactosidase units were measured, normalized, and obtained values are indicated as fold inhibition. The androgens DHT, R1881, and the partial androgen receptor antagonist CPA show a clear recruitment of cSMRT, whereas the antagonists casodex and hydroxyflutamide (OHFl) induce no recruitment of cSMRT (figure from Dotzlaw et al. 2002 (5)). The increased units in the presence of VP16-cSMRT, and ligand such as CPA or agonist indicates interaction of cSMRT with the AR.

we analyze the interaction of the AR protein using human PCa cell (LNCaP) extracts with the expressed GST–AB-CoR fusion proteins in vitro via GST affinity precipitation experiments. The method described here can also potentially be used for any other AR-interacting protein or a CoR protein domain.

2. Materials

2.1. Materials for the Modified Mammalian Two-Hybrid System

2.1.1. Cell Culture for CV-1 Cells

1. Dulbecco's modified Eagles Medium (DMEM) supplemented with 10% fetal calf serum (FCS) that is heat inactivated, 2.5% HEPES (1 M, pH 7.0), and 1% penicillin/streptomycin
2. Tissue culture incubator (37°C, 5% CO_2, and 78% humidity)

3. Heat-inactivated FCS is prepared by thawing a 500 ml bottle of FCS (overnight at 4°C in the refrigerator or at 37°C in the water bath only until thawing) and following heat inactivation for 30 min at 56°C in a water bath. FCS is then aliquoted into sterile 50 ml plastic tubes under sterile conditions in a laminar flow hood and stored at –20°C.
4. 1× PBS: Prepare 10× PBS (1 l) by dissolving 80 g NaCl, 2 g KCl, 26.8 g $Na_2HPO_4{\cdot}7H_2O$, and 2.4 g KH_2PO_4 in 800 ml H_2O (unless otherwise noted, all H_2O is Millipore quality 18.2 MΩ). Adjust pH to 7.4 with HCl and then adjust volume to 1 l with H_2O. After autoclaving, the 10× PBS is stored at room temperature (RT). Dilute 10× PBS buffer 1:10 with H_2O and autoclave before use.
5. Trypsin: 10× trypsin is diluted to 1× trypsin with the indicated dilution buffer (*see* below) under a laminar flow, aliquoted in 50 ml plastic tubes, and stored at –20°C.
6. Trypsin dilution buffer: 0.4 mg/ml KCl, 0.06 mg/ml KH_2PO_4, 8 mg/ml NaCl, 0.35 mg/ml $NaHCO_3$, 0.048 mg/ml Na_2HPO_4, 1 mg/ml D-glucose; sterilize with a 0.2 μm filter, do not autoclave.

2.1.2. Seeding Cells

1. For the modified two-hybrid assay, cells are seeded in DMEM with 1% penicillin/streptomycin, 2.5% HEPES (1 M, pH 7.0), and 10% heat-inactivated and charcoal-stripped FCS.
2. To prepare the heat-inactivated and charcoal-stripped FCS, follow first the instructions in **Section 2.1.1**, Step 3. The 500 ml heat-inactivated FCS is then mixed in a 1 l glass beaker with 30 g charcoal (Fluka) and stirred with a paddle agitator for 30 min at RT to remove the steroids from the serum. Centrifuge the mixture at 11,800×*g* for 15 min at RT. Carefully pipette off supernatant into new plastic flasks (e.g., 50 ml plastic tubes) and spin again at 11,800×*g* for 15 min at RT. Pipette supernatant into a glass bottle and filter with a two step filter (0.8 and 0.2 μm) (PALL) in sterile conditions under laminar flow. Aliquot heat-inactivated and charcoal-stripped FCS in 50 ml plastic tubes and store at –20°C.
3. 6-well plates.

2.1.3. Transient Transfection with $CaCl_2$

1. DNA expression vectors (cloned in our laboratory or kindly obtained from other groups (13)): pMMTV-luc, pCMV-lacZ, AR expression vector (e.g., pSG5-hAR), VP16-CoR fusion expression vectors (e.g., pCMV-VP16-cSMRT), and pCMV-VP16 empty vector as control.

2. 10× calf thymus (CT) DNA: Dissolve CT DNA (Rockland) directly in its original flask in TE buffer (10 mM Tris, 1 mM EDTA, pH 7.6) to 10 mg/ml; after adding TE, let it dissolve by stirring overnight at 4°C. Transfer the solution into a 15 ml plastic tube. Sonicate the DNA solution with a narrow tip for the sonicator for 30 s on ice. Vortex the solution. Test 10 μg DNA for fragment length on a 1% agarose gel. Sonicated DNA will run as a smear. The main amount should have 1 kb in length. Repeat sonication and mixing until this is achieved. Aliquot and store at –20°C.
3. 2 M $CaCl_2$: Aliquot and store at –20°C.
4. 10× HEBS: Prepare a solution of 800 ml containing 200 mM HEPES, 1.37 M NaCl, 60 mM glucose, 50 ml KCl, and 7 mM Na_2HPO_4 in H_2O. Adjust pH to 7.2 with NaOH (5 M) and then adjust volume to 1 l with H_2O. Sterilize the solution by filtering with a syringe through a 0.2 μM filter (Sartorius).

2.1.4. Medium Change

1. AR ligands: R1881 (methyltrienolone) (Perkin Elmer), dihydrotestosterone (DHT) (Dr. Ehrenstorfer GmbH), CPA (Sigma), hydroxyflutamide (OHFl) (Chemos) dissolved in dimethyl sulfoxide (DMSO).
2. DMEM medium containing charcoal-stripped and heat-inactivated FCS (prepared in **Section 2.1.2**, Step 2).

2.1.5. Cell Lysis

1. Acetate (Ac)-lysis buffer: mix 50 ml 1 M Tris/Ac (pH 7.8), 10 ml 1 M MgAc, 200 μl 0.5 M EDTA (pH 8.0), 10 ml Triton X-100, and 150 ml glycerol, and adjust volume to 1 l with H_2O. Mix and store at 4°C.
2. Dithiothreitol (DTT) (Sigma): Prepare a 1 M DTT solution in H_2O and aliquot in 1.5 ml plastic tubes and store at –20°C.
3. Phenyl methyl sulfone acid fluoride (PMSF) (Sigma): Prepare a 0.05 M PMSF solution in isopropanol, aliquot in 1 ml tubes and store at –20°C.

2.1.6. Luciferase Assay

1. Luciferin-adenosine triphosphate (ATP) solution: Dissolve 5 mg D-luciferin in 500 μl of 100 mM ATP and add to 10 ml 1 M Tris/Ac (pH 7.8). Then adjust volume to 50 ml with H_2O and store aliquots at –20°C.
2. Luciferin-coenzyme A (CoA) solution: Add 15 mg CoA to 50 ml luciferin-ATP solution from Step 1, dissolve it, and aliquot in 15 ml plastic tubes. Store at –20°C.
3. Luminometer

2.1.7. β-Galactosidase Assay

1. Z-buffer: Dissolve 10.68 g $Na_2HPO_4 \cdot 2H_2O$, 5.5 g $NaH_2PO_4 \cdot H_2O$, 0.75 g KCl, 0.246 g $MgSO_4 \cdot 7H_2O$ in 800 ml H_2O, and adjust pH to 7.0 with sodium hydroxide (5 M). Add 2.7 ml β-mercaptoethanol under a hood, mix it, and adjust volume to 1 l with H_2O. Store buffer at 4°C.
2. K_xPO_4 solution (*x* indicates different valencies of the K ion in the solution): To obtain a 1 M K_xPO_4 solution prepare two different solutions by dissolving 27.2 g KH_2PO_4 (Merck) in 200 ml H_2O (final concentration 1 M) and 41.6 g $K_2HPO_4 \cdot 3\ H_2O$ (Merck) in another 200 ml H_2O (final concentration 1 M). Mix 38.5 ml 1 M KH_2PO_4 with 61.5 ml of 1 M K_2HPO_4 to obtain the 1 M K_xPO_4 solution.
3. ONPG (ortho-nitrophenyl-β-galactoside): ONPG is 0.1 M in K_xPO_4. Dissolve 800 mg ONPG in 20 ml 1 M K_xPO_4 and 180 ml H_2O. Aliquot in 50 ml plastic tubes and store at –20°C.
4. Spectrophotometer and cuvettes.

2.2. Materials for GST Affinity Purification of AR

2.2.1. Bacterial Protein Expression

1. Bacterial expression vector for GST-fusion protein of interest (insertion of cDNA of interest in frame with the GST coding sequence). Bacterial GST expression vectors harbor a multiple cloning site 3′of the GST-cDNA and are commercially available. As negative control, use expression plasmid for GST alone.
2. Chemically competent *Escherichia coli* strain HB101 or BL21.
3. LB (Luria Bertani) Medium (liquid and agar plates): 10 g Bactotrypton, 5 g yeast extract, 5 g NaCl, fill up to 1 l with H_2O, adjust to pH 7.0 with sodium hydroxide (5 M).
4. Ampicillin is dissolved in H_2O at 100 mg/ml and aliquots (~100 μl) are stored at –20°C.
5. Isopropyl-β-D-thiogalactopyranoside (IPTG) is dissolved in H_2O to 0.2 M and stored at –20°C.
6. NETN buffer for bacterial lysis: 100 mM NaCl, 20 mM Tris–HCl, pH 8.0, 1 mM EDTA, 0.5% NP-40, and 4 mg/ml lysozyme (add fresh).
7. Lysozyme.
8. Liquid nitrogen.

2.2.2. Affinity Purification and GST Pull-down to Analyze the CoR–AR Interactions

1. Bacterial protein extract (from **Section 2.2.1**)
2. Glutathione-Sepharose 4B-beads (GE Healthcare) is equilibrated as 50% slurry with 1× PBS. After centrifugation (500×*g*, 5 min) the supernatant is carefully discarded and the beads are washed with 1× PBS in a ratio of 10 ml PBS

to 1.33 ml of the used original beads. After inversion and again centrifugation the beads are dissolved in NETN buffer in a ratio of 1 ml NETN buffer to 1.33 ml of the used original beads. This step results in a 50% slurry.

3. 1× PBS (*see* **Section 2.1.1**).
4. NETN buffer (*see* **Section 2.2.1**) without lysozyme.
5. STE buffer: 150 mM NaCl, 10 mM Tris, pH 8.0, 1 mM EDTA.
6. Wash buffer: 100 mM Tris–HCl, pH 8.0, 150 mM NaCl.
7. LNCaP whole cell extract (*see* **Section 2.2.3**) as a source for human AR protein.
8. 5% non-fat milk for blocking unspecific binding.
9. AR agonist: R1881 (methyltrienolone) (Perkin Elmer); AR antagonists: hydroxyflutamide (LKT Laboratories Inc.)

2.2.3. Cell Culture, Whole Cell Extraction, and Protein Quantification for GST Affinity Precipitation

1. LNCaP cells are human PCa cells from a lymph node metastasis and are androgen-sensitive, obtained from ATCC.
2. RPMI 1640 supplemented with 10% FCS and 1% sodium pyruvate.
3. PBS (*see* **Section 2.1.1**).
4. 10× trypsin solution (0.5% trypsin, 5.3 mM EDTA) diluted 1:10 with trypsin dilution buffer (*see* **Section 2.1.1**)
5. Protease inhibitor cocktail tablets (Roche), one tablet is dissolved in 10 ml of 1× PBS that is used for cell harvesting.
6. NETN buffer (*see* **Section 2.2.1**) without lysozyme, supplemented with protease inhibitor cocktail.
7. Rotiquant solution (Roth), or equivalent, for protein concentration measurement

2.2.4. SDS-PAGE and Coomassie Staining

1. 10× Sodium dodecylsulfate polyacrylamide gel electrophoresis (SDS) running buffer: 250 mM Tris, 1910 mM glycine, and 1% SDS.
2. 30% acrylamide/bis-acrylamide solution.
3. Ammonium persulfate (APS).
4. Tetramethylethylenediamine (TEMED).
5. 4× SDS Loading Buffer (Roth)
6. Fixing solution: 10% acetic acid, 10% methanol, and 40% ethanol.
7. Sensitization solution: 1% acetic acid and 10% ammonium sulfate.
8. Sensitive coomassie staining solution: 0.125% Coomassie Brilliant Blue (CBB) R-250 (Roth), 45% ethanol, and 5% acetic acid.

9. Coomassie destaining solution I: 40% ethanol and 5% acetic acid.
10. Coomassie destaining solution II: 30% ethanol and 3% acetic acid.

2.2.5. Ponceau Staining and Western Blotting for Detection of Human AR

1. Ponceau solution: 0.1% Ponceau S, 5% acetic acid
2. For semidry blot:
   anode buffer I: 300 mM Tris–HCl, pH 10.4, 20% (v/v) methanol

   anode buffer II: 25 mM Tris–HCl, pH 10.4

   cathode buffer: 25 mM Tris–HCl, pH 9.4, 40 mM capronic acid, 20% methanol, 0.01% SDS
3. TBS-Tween: 50 mM Tris, pH 7.5, 150 mM NaCl, 0.1% Tween 20 (antibody diluent)
4. Blocking buffer: 10% non-fat milk in TBS-Tween
5. HRP-coupled secondary antibodies
6. Enhanced chemiluminescence (ECL) reagents (GE Healthcare)

3. Methods

3.1. The Modified Mammalian Two-Hybrid System

In the experimental setup to analyze CoR binding in mammalian cells using a modified mammalian two-hybrid system, several controls have to be included. In general the empty vector expressing only the VP16 activation domain is the negative control. This VP16 expression vector (pVP16) has to be treated in similar experimental setups with all ligands that are tested. As further controls, cells transfected with pVP16 as well as cells transfected with the CoR expression vector are treated with those ligands known to recruit the particular CoR (e.g., CPA for Alien). As control, an enhancement of the AR-mediated transactivation should be measured under these conditions. Each experiment is performed in triplicate. For AR expression vector, we use the natural, unmodified human AR (pSG5-hAR), or the human AR point mutant T877A (pSG5-T877A). The experiment can also be performed with other AR mutants, e.g., deletion mutants.

3.1.1. Cell Culture for CV-1 Cells

1. CV-1 cells are cultured in DMEM medium containing 10% heat-inactivated FCS, 2.5% 1 M HEPES, and 1% penicillin/streptomycin on 15 cm dishes. Store the prepared medium at 4°C.
2. CV-1 cells grow adherent and should be split three times a week (Mondays, Wednesdays, and Fridays) in a proportion of 1:4, 1:5, or 1:6 according to their growth.

3. To split the cells remove medium and wash cells with 2 ml 1× PBS. Wash cells with 2 ml trypsin, remove, and add again 2.5 ml trypsin. Let the cells detach in the CO_2 incubator (37°C, 5% CO_2, 78% humidity). Check after 5 min under microscope for cells getting round shape and detach (*see* **Note 1**).
4. Add 9.5 ml prepared DMEM medium (RT) and singularize the cells by pipetting the cell suspension up and down with a 10 ml pipette.
5. Take the appropriate amount of the cell solution (*see* Step 2) to seed new 15 cm dish/dishes and adjust total volume to 15 ml with the prepared DMEM medium in **Section 2.1.1**, Step 1.

3.1.2. Seeding Cells

1. Prepare DMEM medium containing 10% FCS heat inactivated and charcoal stripped, 2.5% 1 M HEPES, and 1% penicillin/streptomycin. Store at 4°C.
2. To detach cells, remove medium and wash cells with 2 ml 1× PBS. Add 2 ml trypsin to cells for washing, remove, and add again 2.5 ml trypsin. Let the cells detach in the CO_2 incubator (*see* **Notes 1** and **2**).
3. Add 9.5 ml of DMEM medium prepared in Step 1 and disperse the cells by pipetting cell suspension up and down with a 10 ml pipette (*see* **Note 3**). Transfer cells to a 10 ml or 50 ml plastic tube depending on the total volume when trypsinizing several 15 cm dishes (*see* **Note 4**).
4. Invert the tube with the cell solution to mix, make a Neubauer chamber ready and count two 16-square areas.
5. For each well in a 6-well plate, 2 ml of medium is required (*see* **Note 5**). Calculate the number of 6-wells required for the experiment and thus the total volume necessary (**Fig. 13.3**).
6. Calculate number of cells per ml (cells/ml) in the cell suspension with the corresponding formula for the Neubauer cell chamber.
7. Seed 100,000 cells (aspired cell number) per each well.
   Calculation:
   (aspired cell number):(cells/ml (from Step 6)) = (cells for each well) (*see* **Note 6**)
   (cells for each well) × (number of 6–wells to be seeded) = Volume of cell solution
   The number of wells to be seeded has to be the same number as used for calculation in Step 5.
8. Subtract the volume of cell suspension from the calculated total volume in Step 5. The result is the volume of

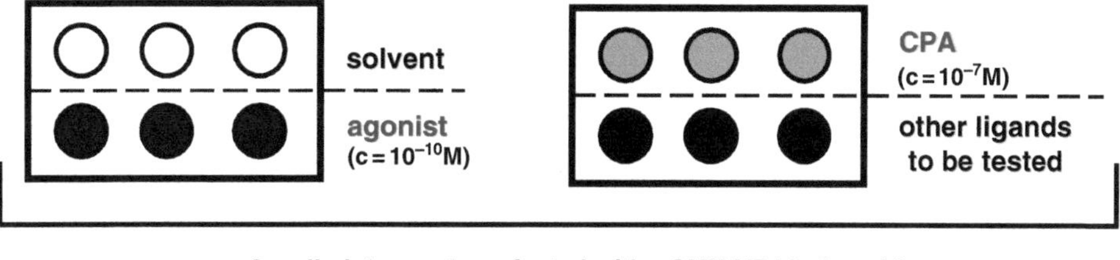

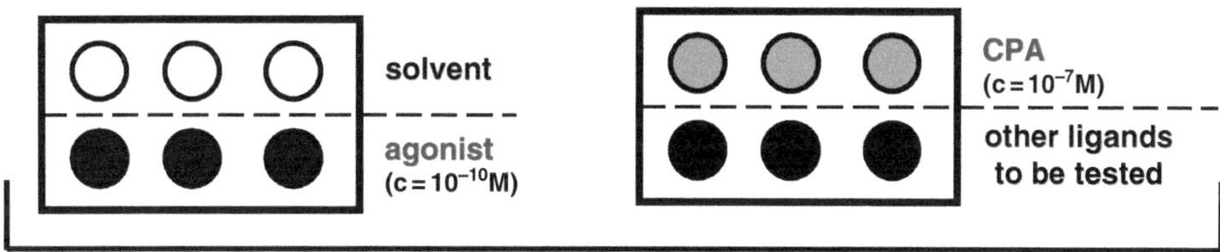

Fig. 13.3. Experimental setup for the modified mammalian two-hybrid assay. The scheme depicts the minimal setup for an experiment. Cells are treated with the indicated substances (on the right side of each triplicate well). All single setups are performed as triplicates and each six-well plate is transfected with one transfection mix. The transfected pCMV-VP16 plasmid serves as control and has to be treated with all used AR ligand either hormones or antihormones. Another control is the implication of an androgen receptor ligand known to recruit the VP16-CoR fusion protein (e.g., for Alien with CPA).

DMEM medium containing charcoal-stripped FCS (prepared in Step 1) needed.

9. Mix the volume of DMEM medium containing heat-inactivated and charcoal-stripped FCS and the volume of cell suspension from Step 6 in a sterile glass bottle (*see* **Note 7**).
10. Label 6-well cover plates with name, date, cell type, and the later used expression vectors for transfection plus later applied substances.
11. Mix total cell suspension from Step 9 by swirling the glass bottle in a circle and seed two 6-well plates with a 10 ml pipette by releasing 2 ml of cell suspension in every 6-well.
12. Repeat Step 11 for all 6-well plates.
13. Distribute cells by gently inclining 6-well plates slowly and for at least three s to the front, the back, left, and right (*see* **Note 8**).
14. Let the cells attach for at least 4 h in a CO_2 incubator.

3.1.3. Transient Transfection Using the $CaPO_4$ Co-precipitation Method

1. For one well in a 6-well plate, 5.4 μg DNA is required in total. Use 1 μg pMMTV-luc, 2 μg VP16-CoR expression vector, or vector pVP16 as control, 0.2 μg AR expression vector and 0.2 μg pCMV-lacZ and adjust DNA amount to 5.4 with 2 μg CT DNA (*see* **Note 9**).
2. Prepare 1 μg/μl DNA solutions with H_2O of the required expression vector. CT DNA can be used in a 10 μg/μl concentration.
3. The transfection mix for one 6-well contains:
   - 1 μl pMMTV-luc (1 μg/μl)
   - 2 μl VP16 expression vector/VP16-CoR expression vector (1 μg/μl)
   - 0.2 μl AR expression vector (1 μg/μl)
   - 0.2 μl pCMV-lacZ (1 μg/μl)
   - 0.2 μl CT DNA (10 μg/μl)
   - 180 μl H_2O (Roth)
   - 21.6 μl 10× HEBS (*see* **Section 2.1.3.**, Step 4)
   - 10.8 μl $CaCl_2$ (2 M) (*see* **Section 3.1.2.**, Step 3)
4. The transfection mix for one 6-well plate (calculate for 7 wells to ensure equal volume for each well, as shown below) contains:
   - 7 μl pMMTV-luc (1 μg/μl)
   - 14 μl VP16 expression vector/VP16-CoR expression vector (1 μg/μl)
   - 1.4 μl AR expression vector
   - 1.4 μl pCMV-lacZ (1 μg/μl)
   - 1.4 μl CT DNA (10 μg/μl)
   - 1260 μl H_2O (Roth)
   - 151.2 μl 10× HEBS
   - = 1436 μl total volume
   - 75.6 μl $CaCl_2$ (2 M)
5. Super transfection mix: Calculate the volume of the super transfection mix (implying all wells) by multiplying the volumes in Step 4 as many times as 6-well plates seeded, excluding the $CaCl_2$ volume.
6. Mix the calculated volume of all components in a plastic tube by starting with pipetting the H_2O, excluding the $CaCl_2$. Vortex all components before pipetting. Vortex the super transfection mix thoroughly.
7. Pipette 1436 μl (*see* Step 4) of the transfection super mix in 2 ml plastic tubes (first 1 ml and then add the remaining

436 μl). Vortex the transfection super mix each time before pipetting (*see* **Note 10**).

8. Label all 2 ml tubes with consecutive numbers and add to tube number one 75.6 μl $CaCl_2$ (2 M) and start a stop watch directly afterward. Vortex the tube continuously for exactly 10 s (*see* **Note 11**). Incubate exactly for 20 min at RT (*see* **Note 12**).
9. Wait at least 1 min, and then add 75.6 μl $CaCl_2$ (2 M) to tube number 2 and follow Step 8 again.
10. Repeat this procedure with each transfection mix.
11. When all transfection mixes contain $CaCl_2$ take the 2 ml tubes with the rack to the laminar flow and make the laminar flow ready (*see* **Note 13**).
12. After 20 min incubation time of transfection mix number one, vortex tube for about 7 s, and then add 220 ml of the transfection mix to every well of the first 6-well plate (*see* **Note 14**).
13. Wait until 20 min are runned down after adding $CaCl_2$ to tube number two and follow Step 12 again.
14. Repeat Steps 12 and 13 until DNA–$CaPO_4$ solution has been added to all wells (*see* **Note 15**).
15. Incubate in the CO_2 incubator for about 20 h.

3.1.4. Medium Change and Hormone Treatment

1. Every well will be treated with 2 ml heat-inactivated and charcoal-stripped DMEM medium containing the appropriate hormone. As negative control, use only DMEM medium heat-inactivated and charcoal-stripped and/or medium containing solvent of the used compounds. Prepare one hormone solution for each substance by taking into account all wells treated with the same substance and add 1 ml to the total volume. Prepare hormone solutions under the laminar flow by mixing DMEM medium containing heat-inactivated and charcoal-stripped FCS with the appropriate hormone or antiandrogen. Vortex hormone solution (*see* **Note 16**).
2. Remove medium with transfection mix from the wells and wash cells with 2 ml 1× PBS (*see* **Note 17**).
3. Remove PBS and add 2 ml of the appropriate hormone mix to each well.
4. Repeat Steps 2 and 3 for all 6-well plates.
5. Return the cells to the incubator for about 3 days.

3.1.5. Cell Lysis

1. Prepare 1.5 ml tubes in a rack on ice and label them.
2. Calculate the required volume of Ac-lysis buffer (400 μl per well) and add at least 4 extra wells in the calculation.

3. Calculate the needed volume of DTT (final concentration 4 mM) and PMSF (final concentration 2.2 mM) (*see* **Note 18**).
4. Pipette the calculated Ac-lysis buffer volume from Step 1 in a beaker and add the calculated volumes of DTT and PMSF solutions from Step 3. Mix and put on ice.
5. Take the 6-well plates out of the CO_2 incubator and take a look at each well to make sure that none of the added compounds had a toxic effect.
6. Remove the medium of two plates and add 400 μl Ac-lysis buffer prepared in Step 4 to each well with a multipipette. Let the buffer incubate for 20 min (*see* **Note 19**).
7. Repeat Step 6 with the remaining 6-well plates.
8. After 20 min, pipette the cell extracts (about 400 μl) from each well in a 1.5 ml tube on ice (*see* Step 1).

3.1.6. Measurement of Luciferase Activity

1. Switch on luminometer 10 min before using.
2. Thaw the luciferin–CoA solution in a 37°C water bath.
3. Parameters for measurement are 100 μl luciferin–CoA and 10 s light. Set this parameters for the measurement.
4. Prime luminometer with luciferin–CoA solution.
5. Measure luciferin–CoA solution alone to obtain basal RLU values which must not exceed 120 RLU. If they do, wash the luminometer lines with water and subsequently with 70% ethanol and start again from Step 4.
6. Pipette 50 μl cell lysis suspension into a luminometer test tube and measure (*see* **Note 20**).
7. Repeat Step 6 with all cell lysates (*see* **Note 21**).
8. Clean the luminometer lines with water and/or ethanol and put an empty tube in the slot.
9. Freeze the remaining luciferin–CoA solution at –20°C.

3.1.7. Measuring the β-Galactosidase Activity

1. Switch on spectrophotometer 10 min before using.
2. Thaw up the ONPG solution in a 37°C water bath.
3. Place 1.5 ml tubes in a rack for as many probes as needed plus two additional ones for negative controls.
4. Pipette 700 μl Z-buffer into the tubes for negative controls or 650 μl Z-buffer in the remaining tubes for the probes using a multipipette.
5. Pipette 50 μl of a cell extract in a tube containing 650 μl Z-buffer.
6. Pipette 200 μl ONPG solution in every tube including the two negative controls with a multipipette.

7. Mix the tubes in the rack by turning the rack several times around.
8. Put the rack with the tubes in a 37°C incubator and turn on a stopwatch (*see* **Note 22**).
9. Let the solutions incubate until a yellow staining is observed (*see* **Note 23**).
10. Note the time period when taking out the probes out of the incubator.
11. Set the spectrophotometer at 420 nm.
12. Prime the spectrophotometer with the negative controls (not containing any cell lysate) (*see* **Note 24**).
13. Measure the OD value at 420 nm for every probe and protocol them (*see* **Note 25**).
14. Clean the spectrophotometer and switch it off.

3.1.8. Computational Analysis

1. Prepare an excel table with one column for the setup, one for the measured RLU, one for the basal RLU, one for netto RLU values, one for mean value, and one for the standard deviation (STABWN).
2. Enter in all obtained RLU values in the appropriate column.
3. Calculate the mean value of the measured basal RLU and write it in the column for the basal RLU value.
4. Set the program to subtract the mean basal RLU value from every measured RLU value and use the results in the column for the netto RLU values.
5. Set the program to calculate the mean of the netto RLU value of every triplicate and use the appropriate column for the results.
6. Set the program to calculate the standard deviation from the netto RLU value of every triplicate.
7. Design bar diagram with the mean values and include the standard deviation. This is the not normalized diagram.
8. Prepare an excel table with one column for the setup, one for the measured OD values, one for the calculated lacZ values (*see* Step 10), one for the netto RLU values, one for the normalized netto RLU values, one for mean values, and one for the standard deviation (STABWN).
9. Enter in all obtained OD values in the appropriate column.
10. Set the program to calculate the lacZ value for every OD value with the formula: $(OD_{420\ nm} \times 1000)/(\text{incubation time in min})$.

11. Copy the netto RLU values from Step 4 to the appropriate column.
12. Set the program to divide the netto RLU values by the calculated lacZ value from Step 10 to obtain the normalized RLU values.
13. Calculate the mean of the normalized RLU values of every triplicate and use the appropriate column for the results.
14. Calculate the standard deviation from the normalized RLU value of every triplicate.
15. Design bar diagram with the mean values and include the standard deviation. This is the normalized diagram and the final result.
16. Evaluate the normalized diagram (*see* **Note 26**).

3.2. Methods for GST Affinity Precipitation

Here, the interaction of the endogenous human AR of LNCaP cells with the bacterially expressed and affinity-purified GST–AB-CoR protein is assayed in vitro. In parallel, GST alone and another adequate negative control are used to indicate specificity of binding. Furthermore, GST affinity precipitation is performed in the presence of different AR ligands using LNCaP cell extracts. A scheme for the procedure is shown in **Fig. 13.4**.

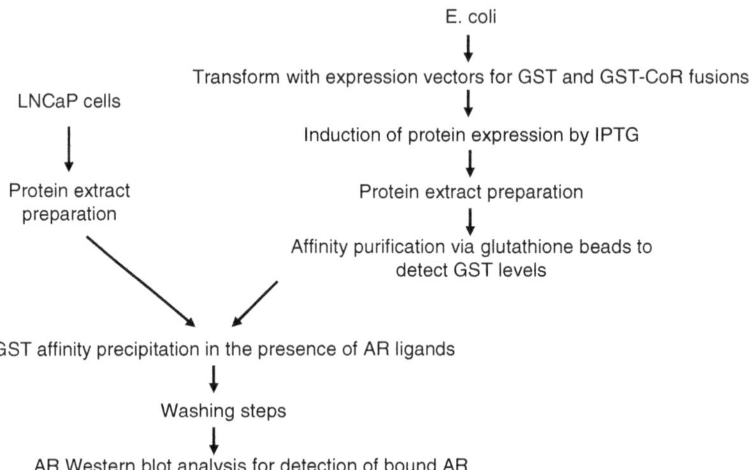

Fig. 13.4. Scheme of GST affinity precipitation to detect an interaction between GST–CoR fusion protein and the human AR. First, LNCaP extract is prepared. *E. coli* are transformed with the GST–CoR or GST coding plasmids and expression is induced with IPTG. Then, bacterial extracts are prepared and the recombinant proteins are affinity purified by glutathione beads. Subsequently, the isolated GST-fusion proteins or GST alone are incubated with LNCaP cell extracts in the presence of AR agonists or antagonists. After removing unbound proteins, Western blot with an AR-specific antibody is performed. Detection of AR indicates AR interaction with CoR.

3.2.1. Bacterial GST–AB-CoR 524 Expression

1. After transformation of pGST-linker (16), pGST-AB-CoR or GST-control fusion plasmids (ampicillin resistant) in chemically competent *E. coli* strain HB101 (*see* **Note 27**), a single cell clone of each plasmid is grown in 3 ml LB medium at 37°C overnight and ampicillin is added at a concentration of 100 μg/ml two times (after 3 and 6 h) to ensure that plasmid is not lost.
2. The overnight culture is transferred to 400 ml pre-warmed LB medium and grown at 37°C until an $OD_{600\ nm}$ of 0.6–0.8 (takes 3–5 h). In this exponential growth phase, GST and GST–AB-CoR or GST-control protein expression is induced via induction of the tac-promoter by adding 0.1 mM (or 0.2 mM) isopropyl-D-thiogalactopyranoside (IPTG) (*see* **Note 28**) to the culture medium and incubated overnight at 16°C (*see* **Notes 29–31**).
3. Bacteria are pelleted via centrifugation (15 min, 7500×*g*, 4°C). From this step, all ensuing steps should be carried out at 4°C to inhibit protein degradation. Bacteria from the 400 ml culture are resuspended in 8 ml (*see* **Note 32**) of NETN buffer with freshly added 4 mg/ml lysozyme (if required, supplemented with 1 tablet of protease inhibitor cocktail per 10 ml buffer) (*see* **Notes 33** and **34**).
4. Cells are lysed by three cycles of freezing in liquid nitrogen and thawing at 37°C. Be sure that the suspension is not warmed up. Incubate at 37°C under gentle shaking for just enough time so that the frozen suspension will thaw, and then quickly freeze samples again in liquid nitrogen.
5. Insoluble components are removed by centrifugation (40 min, 35,000×*g*, 4°C). As the pellet is not very compact, the supernatant can not be taken, rather the pellet is removed by pipetting. A second centrifugation step may be useful, if pellet cannot be well dislodged.
6. The supernatant (protein extract) is supplemented with glycerol (10% end concentration) and stored in aliquots at –80°C.

3.2.2. Affinity Purification of the GST–AB-CoR Fusion Protein and GST Affinity Precipitation of AR

This method is based on the specific interaction between the enzyme (GST, glutathione-S-transferase) that is fused to the protein of interest and its substrate glutathione, which is immobilized to sepharose beads.

1. Before starting, the glutathione-sepharose beads have to be washed to remove the ethanol and to be equilibrated with the buffer. Therefore, the original sepharose beads (75% concentrated) are taken with a wide-bored tip in a new tube and washed by gently inverting with PBS in the following ratio (10 ml 1× PBS per 1.33 ml sepharose beads)

and centrifuged (500×g, 5 min). After discarding the supernatant very carefully (leave some millimetres of fluid over the pellet) the sepharose beads are resuspended in NETN buffer (1 ml buffer per 1.33 ml sepharose). This step results in a 50% sepharose slurry (*see* **Note 35**).

2. It is important as controls to use similar amounts of GST protein, GST–AB-CoR, and GST–control fusion proteins in the interaction assays. Therefore, pre-tests must be performed first to determine the amount of GST-fusion proteins loaded on the beads, which is determined by SDS-PAGE and Coomassie staining. To that end, 40 μl of the equilibrated 50% sepharose beads slurry is incubated with 400 μl of the bacterial protein extract (gained in **Section 3.2.1**) for 2 h on a rotator at 4°C (*see* **Note 36**).
3. The sepharose beads are then washed (by gently inverting) three times with 1 ml NETN buffer (500 g) and then applied on a SDS-PAGE following by Coomassie staining (*see* **Section 3.2.4, Fig. 13.5**)
4. With the knowledge of the protein isolation efficiency, the affinity purification prior to the GST precipitation of AR can be performed with similar protein amounts of GST and GST–CoR by taking 10 μl 50% sepharose beads slurry (also in 1.5 ml tubes).
5. Rotating incubation is again performed for 2 h at 4°C following washing twice (each 500 g, 5 min) with 1 ml NETN buffer, twice with 1 ml STE buffer, and finally twice with 1 ml wash buffer (*see* **Note 37**).

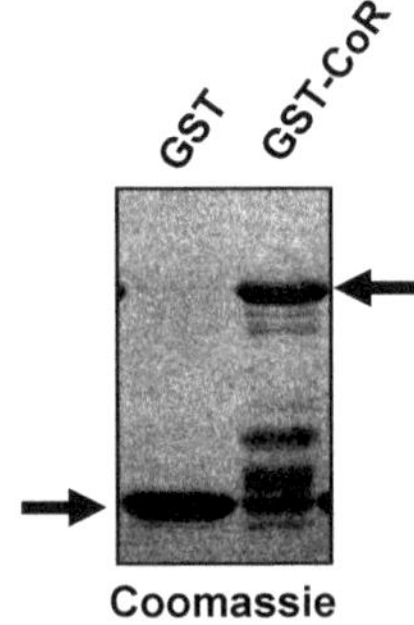

Fig. 13.5 Affinity purification of a GST–CoR fusion protein. GST or GST–CoR was bacterially expressed in the strain BL21 via induction of the lac promoter by adding 0.2 mM IPTG to the culture medium. Proteins were purified on a glutathione-sepharose resin which was washed stringently. Bound proteins were separated by SDS-PAGE and stained with Coomassie brilliant blue.

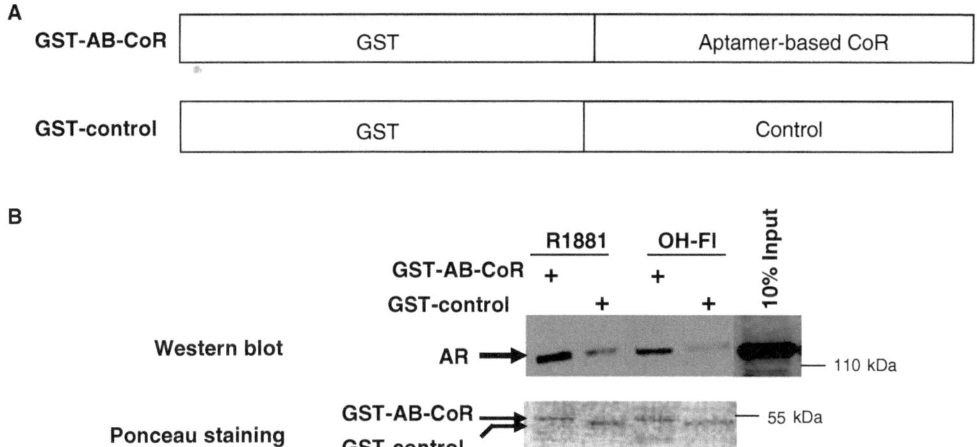

Fig. 13.6 Interaction of AB–CoR with the human AR in vitro (GST affinity precipitation). Bacterially expressed and affinity-purified GST–AB-CoR and GST-control fusion protein were incubated with 0.5 mg LNCaP cell extracts in the absence or presence of 10^{-8} M R1881 or 10^{-7} M OH-F. (**a**) Scheme of the isolated GST-fusion proteins GST–AB-CoR and the appropriate GST-control, which lacks the AR-specific aptamer. (**b**) Western blot analysis was performed to detect the human AR using α-AR (F39.4.1, Biogenex) in the presence of the AR ligands, R1881 or hydroxyflutamide (OHFl), that both activate the endogenous LNCaP AR protein. GST–AB-CoR interacts with the human AR, whereas AR binds to the GST-control only weakly. The membrane was stained with Ponceau S solution to show equal loading.

6. The washed GST protein and GST-fusion protein bound to sepharose is now incubated with 0.5 mg of LNCaP whole cell extract (*see* **Section 3.2.3**) as these have significant levels of the human AR as a natural source.
7. The rotating incubation is performed in a final volume of 200 μl, filled up with PBS, in the absence or presence of AR agonist (R1881 10^{-8} M) or AR antagonist (OH-F, 10^{-7} M) for 1 h at 4°C.
8. After this, the treated sepharose beads are washed five times with 1 ml NETN buffer, which, as in the case of previous hormone treatments, is also supplemented with the appropriate ligand to maintain the interaction during the wash steps.
9. SDS loading buffer (Roth) is then added, denatured at 95°C for 5 min, centrifuged at 26,000×*g* for 1 min, and the bound proteins are separated by SDS-PAGE (*see* **Section 3.2.4**). The AR protein is detected by Western blotting. For visualization of protein loading and efficiency of blotting, the membrane is stained with Ponceau red solution (*see* **Section 3.2.5, Fig. 13.6**).

3.2.3. Cell Culture, Whole Cell Extraction, and Protein Quantification for GST Affinity Precipitation

1. The adherent PCa cell line LNCaP is passaged on 15 cm dishes. In case reaching confluence, cells are detached with trypsin. The RPMI medium was supplemented with 10% untreated FCS, which consequently contains androgens. For whole cell extraction, confluent plates without additional treatment are used.
2. Whole cell extraction is performed with NETN buffer. For this purpose, the cells are first washed with cold 1× PBS and then are scraped off the plate with 1 ml PBS, supplemented with protease inhibitors (PI, 1 tablet per 10 ml).
3. After gentle centrifugation (5 min, 3000×*g*, 4°C), the cells are resuspended and lysed in the 5× packed cell volume with NETN-PI buffer and incubated on ice for 10 min.
4. For further lysis, the cells are frozen at –20°C (in the case of unstable proteins) or alternatively three times freezing in liquid nitrogen and thawing at 37°C. Avoid warming up the solution.
5. Finally, the insoluble components are pelleted (15 min, 35,000×*g*, 4°C). The supernatant contains the whole cell extracts. Supplemented with glycerol (10% final concentration) the extract can be stored in aliquots at –80°C.
6. For quantifying the cellular proteins Rotiquant is used that is based on the biuret reaction. The whole cell extracts are diluted 1:10 and 1 ml of the working solution (generated after the manufacturer's protocol) is added. The reaction mixes are incubated for 30 min at 37°C and are spectrophotometrically measured at 492 nm. In parallel, a BSA (bovine serum albumin) standard series is measured and the resulting calibration curve is used for quantification (*see* **Note 38**).

3.2.4. SDS-PAGE and Coomassie Staining

For detecting the separated proteins in the SDS gel, the sensitive Coomassie staining is used as follows:

First the proteins in the gel are fixed for 1 h with the fixing solution. The gel is then treated with the sensitization solution for 2 h and then stained with the CBB R-250 (at least 4 h) and finally destained one after another with the Coomassie destaining solutions I and II, each for 1–2 h (*see* **Note 39**).

3.2.5. Western Blotting for Human AR and Ponceau Staining

For detection of the separated proteins on the membrane, Ponceau staining is performed to assess the relative loading for GST and GST-fusion proteins. Ponceau staining, in contrast to Coomassie, is reversible and therefore does not interfere with subsequent Western blotting analyses.

The membrane is incubated with Ponceau solution by slightly shaking for 5 min. Protein staining is visible after washing the membrane with H_2O for a few minutes. Further washing leads to disappearance of the staining.

4. Notes

Notes for the modified mammalian two-hybrid system

1. In case CV-1 cells do not detach properly, gently knock the 15 cm dish against a solid surface. Avoid medium to be spilt inside the lid.
2. It is very important to calculate the number of 6-wells needed to be seeded before cells are trypsinized. One 15 cm dish with 90% confluency will be sufficient for about 60 × 6 wells.
3. This step is crucial for seeding cells. If the cells are not dispersed properly, they will grow in clusters which may affect the transfection efficiency.
4. If more than one 15 cm dish is trypsinized, all cell suspensions can be combined in one 50 ml plastic tube or a glass bottle.
5. Calculate cell number with at least two more additional wells than needed.
6. Divide 100,000 through cells/ml because 100,000 cells are seeded per well.
7. Use big glass bottles for a comparatively small volume (e.g., use a 500 ml bottle for a mixture of 70 ml cell suspension). If the radius of the flask is too small, cells do not mix well.
8. Do not shake plates in a circle. In this case cells would not distribute equally.
9. The amounts of CoR and AR expression vector can be modified. In our experience, 2 μg for CoR and 0.2 μg for the AR expression vector worked best. If the CoR DNA amount is too high, there might be AR activation without ligand addition.
10. The transfection of one 6-well plate with one transfection mix is the maximum number of 6-wells transfectable with one mix, because a major volume is needed for the $CaPO_4$ crystals to form properly.
11. Make sure that the whole content in the 2 ml tubes is mixed by holding the tube only on the top and not at the side. This is important for all vortexing steps.
12. After adding the $CaCl_2$, the $CaPO_4$ crystals begin to precipitate. Do not vortex during this formation process before the 20 min incubation time is over.
13. The formation size of the $CaPO_4$ crystals is mainly dependent on the pH value of the HEBS, which is very critical. Small differences on second decimal place for the pH result in the formation of smaller or bigger crystals.

The $CaCl_2$ also contributes to the crystal size. CV-1 cells are best transfected with medium size crystals. Check the crystal size of the used HEBS and $CaCl_2$ combination by applying the indicated volume for transfection mix of the two components to three wells containing 2 ml heat-inactivated and charcoal-stripped serum containing medium before the experiment is done the first time. Wait for 3 h and observe the crystal size under the microscope. This is only to be done once and not before every $CaCl_2$ transfection.

14. When the time between the pipetting steps is quite long, vortex the transfection mix again and then continue.
15. When the time between the transfection mixes was chosen too short and there is a great hurry, discard the experiment. Next time choose longer breaks between adding the $CaCl_2$ to transfection mixes.
16. The final concentrations of the hormones, and if required antihormones, depend on the CoR and/or the antihormone. For all experimental setups we use a final concentration range of the synthetic androgen R1881 of 10^{-8} M to10^{-10} M and of the partial AR agonist CPA of 10^{-7} M.
17. Do not process more than two 6-well plates at one time. The cells could otherwise dry out.
18. DTT and PMSF are required to ensure protein stability.
19. Do not process more than two 6-well plates at one time to make sure that every plate has nearly the same lysis time.
20. Some luminometers indicate sometimes within the measurement of a single probe low and high values. Pay attention to the measuring process during the 10 s and if a high deviation can be seen repeat the measuring of the probe.
21. In case there are high variations (above 50%) among the obtained RLU in a triplicate setup, measure all three probes again to exclude a measuring error by the luminometer.
22. In case there is no 37°C incubator available, a 37°C water bath can also be used.
23. Pay attention to the staining. If the staining is to strong, the measuring with the spectrophotometer is not in the linear range and the prepared lacZ-solutions have to be diluted with Z-buffer for re-measurement.
24. The negative controls are used as reference and must not have any staining. If so there was a pipetting mistake or the used solutions are contaminated.
25. Look in the manual of the spectrophotometer for the linear area. If the measured OD values are not in this area the

lacZ-assay probes must be diluted (if values are too high) or put back in the 37°C incubator (if values are too low).

26. Negative controls (added solvent or medium alone) with the pVP16 expression vector should only show a slight basal activity. An important positive control includes the use of a known VP16-CoR fusion that binds to the AR in the presence of a suitable AR ligand (e.g., VP16-cSMRT plus CPA, **Fig. 13.2**).

Notes for GST pull-down

27. The *E. coli* strain BL21 is also appropriate for protein expression, as it is deficient for some proteases.
28. IPTG concentration can be used in a range of 0.1–1 mM. Optimal concentration is 0.1 mM for GST–CoR fusion proteins.
29. For some GST–CoR expression the induction by IPTG is not optimal at 37°C for 3 h indicated as standard protocol. Therefore, overnight incubation at 16°C should be tested, which was shown to be more efficient. An even better expression level can be achieved with 3 days culturing at 4°C. These conditions can be pre-tested for each protein.
30. With the aim to compensate for an eventual lack of substrates for bacterial growth or efficient protein translation such as seldom loaded tRNAs, a more eutrophic medium that is enriched in amino acids can also be used: YTA medium (16 g Bactotrypton, 10 g yeast extract, and 5 g NaCl, pH 7.0). This medium optimizes the metabolism and thereby the amount of amino acid loaded tRNAs is increased.
31. Heterologously expressed proteins tend to build insoluble aggregates and inclusion bodies, which can be avoided by adding osmolytes (660 mM sorbitol and 2.5 mM betaine) to the bacteria medium (17).
32. Final protein concentration can be optimized by resuspending the cell pellet in an appropriate volume of cell lysis buffer. The pellet from a 400 ml overnight culture (at 16°C) is dissolved in 8 ml of lysis buffer, whereas less (ca. 4 ml) is needed for a 4°C culturing over 3 days.
33. In order to increase the protein yield the cell lysis step could be performed with the addition of protease inhibitors. Therefore 10 ml of NETN buffer (containing lysozyme) can be supplemented with one tablet of protease inhibitor (Roche), which in this case leads to about a 2.5-fold higher amount of affinity-purified protein.
34. For cell lysis another buffer can alternatively be used, the lysis buffer (60 mM KCl, 20 mM Hepes, pH 7.8, 1 mM

EDTA, 2 mM DTT, freshly added 4 mg/ml lysozyme), which was, however, in this case less efficient for protein isolation.

35. In general, to avoid removing sepharose beads, it is very important after centrifugation to remove the supernatant very carefully and to leave some millimetres of solution above the beads.

36. For affinity purification of the GST-fusion proteins using the glutathione sepharose it is not advised to scale up the preparation sizes. Normally, 40 μl of 50% sepharose and 400 μl of protein extract are sufficient for protein binding to beads; however, incubating 2 ml of 50% sepharose with 20 ml protein extract resulted only in moderate GST–AB-CoR yield.

37. To enhance AR binding specificity in the GST affinity precipitation, the GST and GST–CoR fusion protein-bound sepharose beads can be blocked after 2× washing with 1 ml NETN buffer supplemented with 5% denatured milk (400 μl milk for 40 μl sepharose beads). 5% non-fat milk is prepared by mixing 5 g non-fat milk powder per 100 ml TBS-Tween (*see* **Section 2.2.5**) in a closed bottle overnight, denaturing at 95°C for 10 min, centrifugation at 26,000×*g* for 2 min, and then the aliquots are stored at –20°C.
This step is also performed on a rotator for 30 min at 4°C following the same washing procedure as before (2× NETN, 2× STE, 2× wash buffer).

38. For quantification of the whole cell extract it is necessary to take care that a detergent-resistant method for protein detection is used, as the cell lysis buffer NETN contains 0.5% NP-40.

39. For visualization of the SDS-PAGE-separated proteins the gel is stained first with CBB G-250, but for enhanced protein detection the CBB R-250 was employed, which is five times more sensitive and visualizes about 0.1 μg of protein.

References

1. Jemal, A., Siegel, R., Ward, E., Hao, Y., Xu, J., Murray, T., Thun, M.J. (2008) Cancer statistics. *CA Cancer J. Clin.* **58**:71–96.
2. Brinkmann, A.O., Trapman, J. (2000) Prostate cancer schemes for androgen escape. *Nat. Med.* **6**:628–629
3. Ntais, C., Polycarpou, A., Tsatsoulis, A. (2003) Molecular epidemiology of prostate cancer: androgens and polymorphisms in androgen-related genes. *Eur. J. Endocrinol.* **149**:469–477.
4. Burke, L.J., Baniahmad, A. (2000) Co-repressors 2000. *FASEB J.* **14**: 1876–1888
5. Dotzlaw, H., Moehren, U., Mink, S., Cato, A.C., Iniguez Lluhi, J.A., Baniahmad, A. (2002) The amino terminus of the human AR is target for corepressor action and antihormone agonism. *Mol. Endocrinol.* **16**: 661–673.
6. Burd, C.J., Morey, L.M., Knudsen, K.E. (2006) Androgen receptor corepressors and

prostate cancer. *Endocrine-related-cancer* **13**:979–994.

7. Moehren, U., Papaioannou, M., Reeb, C.A., Hong, W., Baniahmad, A. (2007) Alien interacts with the human androgen receptor and inhibits prostate cancer cell growth. *Mol. Endocrinol.* **21**:1039–1048.
8. Eisold, M., Asim, M., Eskelinen, H., Linke, T., Baniahmad, A. (2009) Inhibition of MAPK-signaling pathway promotes the interaction of the corepressor SMRT with the human androgen receptor and mediates repression of prostate cancer cells growth in the presence of antiandrogens. *J. Mol. Endocrinol.* **42**:429–435.
9. Feldman, B.J., Feldman, D. (2001) The development of androgen-independent prostate cancer. *Nat. Rev. Cancer* **1**:34–45.
10. Denmeade, S.R., Isaacs, J.T. (2001) A history of prostate cancer treatment. *Nat. Rev. Cancer* **2**:389–396.
11. Dehm, S.M., Tindall, D.J. (2007) Androgen receptor structural and functional elements: role and regulation in prostate cancer. *Mol. Endocrinol.* **21**:2855–2863.
12. Baek, S.H., Ohgi, K.A., Nelson, C.A., Welsbie, D., Chen, C., Swayers, C.L., Rose, D.W., Rosenfeld, M.G. (2006) Ligand-specific allosteric regulation of coactivator functions of androgen receptor in prostate cancer cells. *Proc. Natl. Acad. Sci.* **103**: 3100–3105.
13. Asim, M., Siddiqui, I.A., Hafeez, B.B., Baniahmad, A., Mukhtar, H. (2008) Src kinase potentiates androgen receptor transactivation function and invasion of androgen-independent prostate cancer C4-2 cells. *Oncogene* **27**:3596–3604.
14. Baniahmad, A. (2005) Nuclear hormone receptor co-repressors. *J. Steroid Biochem. Mol.* **93**:89–97.
15. Papaioannou, M., Reeb, C., Asim, M., Dotzlaw, H., Baniahmad, A. (2005) Co-activator and co-repressor interplay on the human androgen receptor. *Andrologia* **37**(6): 211–212.
16. Dressel, U., Thormeyer, D., Altincicek, B., Paululat, A., Eggert, M., Schneider, S., Tenbaum, S.P., Renkawitz, R., Baniahmad, A. (1999) Alien, a highly conserved protein with characteristics of a corepressor for members of the nuclear hormone receptor superfamily. *Mol. Cell. Biol.* **37**(6):1881–1884.
17. Blackwell, J.R., Horgan, R. (1991) A novel strategy for production of a highly expressed recombinant protein in an active form. *FEBS Lett.* **295**(1–3):10–12.

Chapter 14

Detection of Ligand-Selective Interactions of the Human Androgen Receptor by SELDI-MS-TOF

Thomas Linke, Martin Scholten, and Aria Baniahmad

Abstract

The human androgen receptor (AR) is expressed in nearly all prostate cancers (PCa) and is known to participate in tumor progression through the expression of genes involved in the proliferation and differentiation of PCa. It is suggested that different types of ligands induce a distinct AR conformation that would lead to a specific set of interacting partners for the AR, such as coactivators (CoA) and corepressors (CoR), heat shock proteins (HSP), remodeling factors, kinases, phosphatases, and transcription factors resulting in various degrees of AR activity and stability. The natural ligand of the AR, dihydrotestosterone (DHT), induces a transcriptionally active conformation of the AR while the steroidal antiandrogen cyproterone acetate (CPA) and the nonsteroidal compounds hydroxyflutamide (OHF), bicalutamide (Cas), and atraric acid (AA) prevent acquisition of a transcriptionally active conformation. The AR has, in addition to transactivation, other functional properties. However, the current known interaction partners of AR cannot explain the multitude of AR-mediated functions. Thus, many of the ligand-specific AR-interacting proteins still remain unidentified. Here we provide an assay system to assess AR interactions in LNCaP PCa cells. LNCaP cells were treated with the AR-agonist R1881 or AR-antagonists Cas or AA to induce ligand-specific cofactor (CoF) binding to the AR in vivo. Here we describe a method for the identification of ligand-selective interaction partners of AR combining immunological methods with surface-enhanced laser desorption/ionization (SELDI)–time of flight (TOF)–mass spectrometry (MS). Exemplified here is the interaction of a novel AR-CoF, the cell-cycle regulating protein *cell division cycle-associated protein 2* (CDCA2) with AR in the presence of antagonist which is verified by a protein–protein interaction assay in vivo. This scheme can provide further insights into the molecular mechanisms of AR ligand selectivity.

Key words: Prostate cancer, androgen receptor, androgens, antiandrogens, cofactor, coactivator, corepressor, SELDI, mass spectrometry.

F. Saatcioglu (ed.), *Androgen Action*, Methods in Molecular Biology 776,
DOI 10.1007/978-1-61779-243-4_14, © Springer Science+Business Media, LLC 2011

1. Introduction

Prostate cancer is the most commonly diagnosed malignancy and the second leading cause of male cancer death in most western countries (1). The growth and development of the prostate is regulated by androgens and the androgen receptor (AR) (2, 3).

The AR is a member of the nuclear hormone receptor superfamily (4), a large group of ligand-dependent transcription factors, and is activated by androgens, such as dihydrotestosterone (DHT) (5). Antiandrogens (AR-antagonists), such as Cas and AA, reduce AR-mediated transactivation of gene expression (6).

The binding of androgens induces a conformational change in the AR, allowing nuclear translocation, increased phosphorylation, homodimer formation, interaction with DNA, and the recruitment of a pool of coactivators (CoA) resulting in regulation of gene expression (7–10). Rapid signaling of MAPK by liganded AR has also been described (11). A number of cofactors (CoF) as well as CoF complexes that support androgen-dependent transcriptional control have been identified (12).

In PCa treatment various ligands for AR are utilized. The application of AR-antagonists (antihormones) induces corepressor (CoR) recruitment and results in reduced expression of AR target genes and reduced tumor growth. AR-antagonists can act as complete antagonists, such as Cas or OHF, or as mixed agonists/antagonists, such as CPA (13–15).

Thus, the regulation of AR activity includes complex and alternating mechanisms that are not yet fully understood. Therefore, the identification of ligand-specific AR interaction partners and proteins, or protein complexes, that mediate AR activity is essential for understanding AR function in the normal prostate, benign prostate hyperplasia (BPH), and PCa in vivo. Here we provide a method to discover and study such interactions.

2. Materials

2.1. Cell Culture

2.1.1. For Detection of Ligand-Selective AR-Interacting Proteins and for Quantification of AR Protein Levels

1. LNCaP cells (androgen-responsive PCa cell line) (available from American Type Tissue Collection; kindly provided by A. Protopopov).
2. CO_2 incubator (37°C, 5% CO_2, 78% humidity) (Thermo Scientific Heraeus®).

3. Purified water ($R \cdot 18$ M cm at 25°C) (Synergy Ultrapure Water Systems, Millipore™).
4. Phosphate-buffered saline (PBS) (10×): 123 mM NaCl, 17 mM Na_2HPO_4, and 2.5 mM KH_2PO_4, adjust pH to 7.4 with HCl, sterilize by autoclaving, and store at room temperature (RT).
5. 1× PBS: 10× PBS diluted with purified water, stored at RT.
6. Trypsin buffer for trypsin solution (10×): 4 g KCl, 0.6 g KH_2PO_4, 80 g NaCl, 3.5 g $NaHCO_3$, 0.48 g Na_2HPO_4, and 10 g glucose, sterilized with VacuCap® 90 PF Filter Unit (Pall Corporation).
7. Trypsin solution (1×): 10× trypsin (Gibco/Invitrogen) is diluted to 1× trypsin with the indicated dilution buffer recipe from the manufacturer under a laminar flow and sterilized with VacuCap® 90 PF Filter Unit (Pall Corporation). It is then aliquoted in 50 ml plastic tubes and stored at –20°C.

2.1.2. For Detection of Ligand-Selective AR-Interacting Proteins

1. Cell culture dish, 145×20 mm, sterile.
2. osewell Park Memorial Institute (RPMI) 1640 medium (Gibco®) supplemented with 10% fetal calf serum (FCS) (Invitrogen) and 25 mM 4-(2-hydroxyethyl)-1-piperazineethanesulfonic acid (HEPES) buffer (pH 7.8), 1% penicillin/streptomycin (Gibco/Invitrogen), and 1% sodium pyruvate.

2.1.3. For Quantification of AR Protein Level

1. Multiwell cell culture plates.
2. Rosewell Park Memorial Institute (RPMI) 1640 medium (Gibco®) supplemented with 5% fetal calf serum (FCS) (Invitrogen) and 25 mM HEPES buffer (pH 7.8), 1% penicillin/streptomycin (Gibco/Invitrogen), and 1% sodium pyruvate.
3. Neubauer chamber (Roth).

2.2. Cell Treatment

2.2.1. For Detection of Ligand-Selective AR-Interacting Proteins

1. Methyltrienolone (R1881) (PerkinElmer®, Inc.): stock solution (10^{-5} M) dissolved in dimethyl sulfoxide (DMSO), stored at –20°C, and final concentration: 10^{-8} M.
2. Cas (Interpharma Praha): stock solution (10^{-4} M) dissolved in ethanol, stored at –20°C, and applied final concentration: 10^{-7} M.
3. AA (Merck): stock solution (10^{-2} M) dissolved in EtOH/DMSO (ethanol/dimethyl sulfoxide) (1:1, v/v), stored at –20°C, and applied final concentration: 10^{-5} M (6).
4. Control (EtOH or DMSO).

2.2.2. For Quantification of AR Protein Level

1. Flufenamic acid (FA) as positive control (Sigma), stock solution (10^{-4} M) dissolved in DMSO and stored at –20°C and applied final concentration: 10^{-7} M.
2. AR-antagonist.
3. Control (DMSO).

2.3. Cell Harvest

2.3.1. For Detection of Ligand-Selective AR-Interacting Proteins and Quantification of AR Protein Level

1. Purified water: *see* **Section 2.1**, Step 3.
2. 10× PBS: *see* **Section 2.1**, Step 4.
3. 1× PBS: *see* **Section 2.1**, Step 5.
4. *NanoDrop® ND-1000* spectrophotometer (Thermo Scientific).
5. Centrifuge *Universal 320R* (Hettich) (or equivalent).
6. PBS-PI (PBS-Protease Inhibitor): 1 pellet of complete Protease Inhibitor Cocktail Tablets (Roche) added into 10 ml 1× PBS prepared immediately before use and stored on ice.
7. Cell scraper, 28 cm, sterile.

2.3.2. For Detection of Ligand-Selective AR-Interacting Proteins

1. PBS-PI + ligands: according to cell treatments, the respective ligands are added to ice-cold 1× PBS and stored on ice (*see* **Note 1**).
   - 5 ml PBS-PI supplemented with 5 μl R1881 (10^{-5} M)
   - 5 ml PBS-PI supplemented with 5 μl Cas (10^{-4} M)
   - 5 ml PBS-PI supplemented with 5 μl AA (10^{-2} M)
   - 5 ml PBS-PI control

2.4. Cell Lysis

2.4.1. For Detection of Ligand-Selective AR-Interacting Proteins

1. Lysis buffer stock solution: 0.1 M Na_3PO_4, 5 mM EDTA, and 2 mM $MgCl_2$.
2. Lysis buffer: 97% lysis buffer stock solution (v/v), 500 μM leupeptin (Serva), 0.1 mM phenylmethylsulfonyl fluoride (PMSF), and 1% 3-[(3-cholamidopropyl)dimethylammonio]-1-propane sulfonate (CHAPS), prepared immediately before use and stored on ice. According to cell treatment, the respective ligands are added to the lysis buffer (*see* **Note 2**).
   - 1 ml lysis buffer supplemented with 1 μl R1881 (10^{-5} M)
   - 1 ml lysis buffer supplemented with 1 μl Cas (10^{-4} M
   - 1 ml lysis buffer supplemented with 1 μl AA (10^{-2} M)
   - 1 ml PBS-PI control
3. Centrifuge *Universal 320R* (Hettich) (or equivalent).

2.4.2. For Quantification of AR Protein Level

1. NETN lysis buffer: 200 mM NaCl, 1 mM EDTA, 20 mM Tris–HCl (pH 8.0), 0.5% Nonidet-P40 (NP-40) substitute

(Serva), and 1 pellet of complete Protease Inhibitor Cocktail Tablets (Roche) added into 10 ml lysis buffer, prepared immediately before use and stored on ice.

2. Centrifuge *Universal 320R* (Hettich) (or equivalent).

2.5. Co-immunoprecipitation (CoIP) with Interaction Discovery Mapping (IDM) Affinity Beads

1. Antibodies:
   - rabbit anti-AR (Upstate/Millipore™, PG-21)
   - rabbit IgG (Santa Cruz Biotechnology®, Inc.)
2. Protein-A Agarose (Sigma-Aldrich®).
3. Sodium acetate buffer (SAB) 50 mM, pH 5.0 adjusted with NaOH.
4. PBS, *see* **Section 2.1**.
5. Washing buffer: 0.5 M $NaCl_2$ + 0.05% Triton X-100 in 1× PBS (*see* **Note 3**).
6. Elution buffer: 50% acetonitrile (ACN) + 0.5% trifluoroacetic acid (TFA).
7. SDS loading buffer *Roti®-Load 1* (4×), diluted 1:1 in purified water.
8. Fixed Speed Rotator SB2 (Stuart®).
9. Centrifuge *Universal 320R* (Hettich) (or equivalent).

2.6. CoIP with Dynabeads® Protein-A (Invitrogen)

1. Antibodies: *see* **Section 2.5**, Step 1.
2. Sodium phosphate buffer 0.1 M, pH 8.0 (SPB): dilute 93.2 ml of 1 M Na_2HPO_4 and 6.8 ml of 1 M NaH_2PO_4 to 1 l with purified water.
3. Washing buffer (ice cold): 1× PBS, 0.5 M $NaCl_2$, and 0.1% Triton X-100 + ligands (*see* **Section 2.2**) (*see* **Notes 1** and **3**).
4. Elution buffer: 0.1 M citrate buffer, pH 2.0, adjusted with HCl, stored at RT.
5. Fixed Speed Rotator SB2 (Stuart®) (or equivalent).

2.7. CoIP with Protein-A Agarose Beads (Roche)

1. Antibodies: *see* **Section 2.5**, Step 1.
2. Protein-A Agarose, fast flow (Upstate, Millipore™).
3. Washing buffer: PBS-PI (ice cold) + ligands (*see* **Section 2.2** and **2.3**) (*see* **Note 1**).
4. Fixed Speed Rotator SB2 (Stuart®) (or equivalent).
5. Centrifuge *Universal 320R* (Hettich) (or equivalent).

2.8. Surface-Enhanced Laser Desorption/Ionization (SELDI)–Time of Flight (TOF)–Mass Spectrometry (MS)

1. *ProteinChip System Series 4000* mass spectrometer (Ciphergen Biosystems, Inc.).
2. ProteinChip arrays (Ciphergen Biosystems, Inc.):
   2a. *H50-ProteinChip®* array (reverse phase); the surfaces of these chips are coated with methylene chains developing hydrophobic interactions with proteins.
   2b. *NP20-ProteinChip®* array (normal phase); the surfaces of these chips are coated with silicon oxid mediating the binding of proteins via serine, threonine, and lysine due to hydrophilic interactions.
3. Energy absorbing matrix (EAM): 20% sinapinic acid (SPA) (Ciphergen Biosystems, Inc.), 50% acetonitrile (ACN) (Sigma-Aldrich®), and 0.5% trifluoroacetic acid (TFA) (Sigma-Aldrich®).
4. H50 washing buffer: 30% ACN + 0.1% TFA.
5. NP20 washing buffer: 1× PBS, 200 mM NaCl, and 0.02% Triton X-100.
6. Software: *Ciphergen Express Client 3.0* (series 4000) (Ciphergen Biosystems, Inc.) and *ProteinChip Software 3.2.1* (PBSIIc) (Bio-Rad).

2.9. Sodium Dodecyl Sulfate (SDS)–Polyacrylamide Gel Electrophoresis (PAGE)

1. *Mini-Protean Tetra Cell* system (Bio-Rad Laboratories™)
2. Purified water: *see* **Section 2.1**
3. 30% acrylamide mix 37.5:1
4. SDS (ultrapure)
5. Ammonium persulfate (APS) 99%, to be diluted in water and stored at –20°C (Serva Electrophoresis)
6. *N*,*N*,*N*,*N*′-tetramethyl-ethylenediamine (TEMED) (*see* **Note 4**)
7. Tris base
8. 2-Propanol (≥99.8%)
9. SDS loading buffer *Roti®-Load 1* (4×)
10. PeqGold prestained protein marker IV (Peqlab Biotechnologie GmbH™) (or equivalent)
11. Glycine
12. Centrifuge *Universal 320R* (Hettich) (or equivalent)
13. Power supply *Elite 300 Plus* (Schütt Labortechnik) (or equivalent)

2.9.1. Separation Gel

The concentration of acrylamide in the gel depends on the molecular weight of the protein to be analyzed; for the AR (about 110 kD), a final concentration of 8% is recommended.

Scheme to obtain 10 ml:

Reagent/final concentration	8%	12%	15%
1.5 M Tris–HCl, pH 8.8	2.5 ml	2.5 ml	2.5 ml
30% acrylamide mix	2.7 ml	4.0 ml	5.0 ml
10% SDS	0.1 ml	0.1 ml	0.1 ml
10% ammonium persulfate (APS)	0.1 ml	0.1 ml	0.1 ml
Purified water	4.6 ml	3.3 ml	2.3 ml
N,N,N,N′-tetramethyl-ethylenediamine (TEMED)	6 μl	4 μl	4 μl

Attention: Acrylamide is neurotoxic and mutagenic when unpolymerized – avoid contact and wear gloves

2.9.2. Stacking Gel

The loaded proteins are initially condensed in a first step in the stacking gel before they are separated in a second step in the separation gel that is located below the stacking gel.

Scheme to obtain 4 ml:

Reagent	Volume
1 M Tris–HCl, pH 6.8	0.5 ml
30% acrylamide mix	0.67 ml
10% SDS	0.04 ml
10% ammonium persulfate (APS)	0.04 ml
Purified water	2.7 ml
N,N,N,N′-tetramethyl-ethylenediamine (TEMED)	4 μl

2.10. Western Blotting

2.10.1. For Detection of Ligand-Selective AR-Interacting Proteins and Quantification of AR Protein Level

1. Tween 20.
2. Sodium chloride ≥99.5%.
3. HCl.
4. Nonfat dry milk (Valio™).
5. ECL™ Western blotting detection reagents (GE Healthcare™).
6. HRP-juice (PJK™ GmbH).
7. Whatman™ 3MM paper.
8. Polyvinylidene fluoride (PVDF) membrane (Millipore™).
9. Ponceau S (Sigma-Aldrich®).
10. Acetic acid.

11. β-mercaptoethanol ≥99%.
12. Ethanol.
13. Methanol.
14. Mini-Blot system.
15. Transfer buffer: glycine, 39 mM; Tris base, 48 mM; SDS, 0.037% (w/v); and methanol, 10% (v/v). If the concentration of the acrylamide in the gel is higher than 10%, use 20% (v/v) methanol instead. This buffer should be prepared freshly and used only once.
16. Tris-buffered saline (TBS) buffer (10×): Tris base, 500 mM; sodium chloride, 1,500 mM; adjusted to pH 7.5 with HCl and stored at 4°C.
17. TBS-T: TBS (10×) diluted with purified water to 1× and 0.1% (v/v) Tween 20 (*see* **Note 5**).
18. Blocking buffer: 10% (w/v) nonfat dry milk in 1× TBS-T.
19. Amersham ECL™ Western blotting detection reagents, or alternatively, for increased sensitivity, HRP-juice.
20. As primary antibody (AB) against AR we used mouse anti-AR (BioGenex™), diluted 1:1,000 in 5 ml TBS-T and stored at 4°C. As secondary AB we used goat anti-mouse AB conjugated to horseradish peroxidase (HRP) (Santa Cruz Biotechnologies™), diluted 1:5,000 in 5 ml TBS-T and stored at 4°C.
21. Orbital shaker *Polymax 2040* (or equivalent).
22. Shaking water bath.

2.10.2. Detection of Ligand-Selective AR-Interacting Proteins

To detect ligand-specific AR-interacting protein CDCA2 we used rabbit anti-CDCA2 AB (Abcam®), diluted 1:1,000 in 5 ml TBS-T and stored at 4°C short term (1–2 weeks). Aliquot and store at –20 or –80°C and avoid repeated freeze/thaw cycles. As secondary AB we used bovine anti-rabbit AB conjugated to HRP (Santa Cruz Biotechnologies™), diluted 1:2,000 in 5 ml TBS-T and stored at 4°C.

2.10.3. Quantification of AR Protein Level

To quantify AR protein amount we used anti-AR AB as mentioned in **Section 2.10**, Step 20. To perform loading control we used rabbit anti-β-actin AB (Cell Signaling), diluted 1:2,000 in 5 ml TBS-T and stored at 4°C. To detect this primary AB we used as a secondary AB bovine anti-rabbit AB (Santa Cruz Biotechnologies™) diluted 1:2,000 in 5 ml TBS-T and stored at 4°C.

3. Methods

3.1. Cell Culture

3.1.1. Detection of Ligand-Selective AR-Interacting Proteins and Quantification of AR Protein Level

1. Monolayer cultures of LNCaP cells were cultured at 37°C.
2. Medium was changed every third day after gently washing the cells with 1× PBS and discarding it. Fresh medium is then added.
3. When the cells reach 70% confluency, they are split 1:4 using trypsin. Therefore each dish was incubated with trypsin solution for about 2–5 min at 37°C in 5% CO_2 atmosphere (*see* **Note 6**).

3.1.2. Detection of Ligand-Selective AR-Interacting Proteins

1. Cells were cultured in 14.5 cm dishes in 15 ml medium containing 10% FCS.
2. Wash cells with 7 ml PBS.
3. Remove PBS and trypsinize cells with 2 ml trypsin per dish.

3.1.3. For Quantification of AR Protein Level Only

1. LNCaP cells were plated in 14.5 cm dishes in 15 ml of medium and cultured at 37°C (one dish for testing six substances, including a positive and a negative control).
2. After reaching 70% confluency (of about 2 days of growth) the medium is discarded and the cells are washed once with PBS.
3. After treatment with trypsin we determine the cell concentration in a Neubauer chamber.
4. Cells are split and a total of 64,000 cells are seeded in 2 ml medium in each well of a six-well plate.
5. Cells are cultured for 48 h at 37°C.

3.2. Cell Treatment

3.2.1. Detection of Ligand-Selective AR-Interacting Proteins

1. LNCaP cells are treated either with AR-agonist R1881 or AR-antagonists Cas or AA.
2. To test for ligand selectivity, we used four ligand-specific approaches (I–IV):
   - I: AR-agonist (R1881)
   - II: AR-antagonist I (Cas)
   - III: AR-antagonist II (AA)
   - IV: Control (EtOH or DMSO)
3. LNCaP cells are cultured as described in **Section 3.1** (4 dishes for each approach, in total 16 dishes). After reaching confluency of 60% (about 2 days of growth), the medium is discarded and the cells are washed once with PBS (7 ml for each dish).

4. For each of the four approaches (I–IV) 40 ml of medium (10 ml for each dish) is freshly prepared. Therefore, in total 160 ml of medium is partitioned into 4 × 40 ml and the ligands are added as follows: A volume of 40 μl of the R1881 stock solution was added into 40 ml medium (approach I); 40 μl of the Cas stock solution added to 40 ml medium (approach II); 40 μl of the AA stock solution added to 40 ml medium (approach III); and 40 μl of EtOH or DMSO is added to 40 ml medium as control (approach IV).
5. Finally, according to the four approaches (I–IV), 10 ml of the freshly prepared and ligand-supplemented medium is added to each dish. The cells are incubated with the ligands for 2 h.

3.2.2. Quantification of AR Protein Level

1. LNCaP cells were treated with either FA- or AR-antagonist or control for 48 h. FA, known to inhibit the androgen receptor expression at mRNA and protein levels (16), was used as a positive control.
2. To test influences of AR-antagonists to AR protein levels, three approaches were employed:

   I: AR-antagonist I (FA)

   II: AR-antagonist II

   III: Control (DMSO)
3. A volume of 4 μl of the stock solution to be tested was added to 4 ml medium. LNCaP cells were incubated with the freshly prepared and separately supplemented (approaches I–III) medium (2 ml for each well) for 48 h at 37°C.

3.3. Cell Harvest

3.3.1. Detection of Ligand-Selective AR-Interacting Proteins

1. Cells were harvested by discarding the medium of each dish and washing once with 1 ml ice-cold washing buffer (PBS-PI). Notice that in the preceding cell treatment the washing buffer was supplemented with the adequate ligand (*see* **Note 1**).
2. Using a scraper the cells were harvested in 1 ml ice-cold PBS-PI and transferred into a 2 ml microcentrifuge tube stored on ice (*see* **Note 7**). Cells were pelleted by centrifugation at 389×*g* for 4 min at 4°C and afterward the supernatant was discarded.
3. The maintained cell pellet can be frozen in liquid nitrogen and afterward stored at –80°C (*see* **Note 8**) ore used directly for further preparation (*see* **Section 3.4**, Step 2).

3.3.2. Quantification of AR Protein Level

1. Cells were harvested as described above (*see* **Section 3.3**, Steps 1–3). Notice that the washing buffer (400 μl ice-cold PBS-PI each well) does not contain any AR ligands.

3.4. Cell Lysis

3.4.1. For Detection of Ligand-Selective AR-Interacting Proteins Only

1. Cell pellets were thawed on ice (*see* **Note 9**).
2. A 5× packed cell volume of lysis buffer was added and the mix was resuspended very gently (*see* **Note 10**). Lysis buffer was prepared fresh before use. Notice that the respective ligands were added to the lysis buffer according to the ligand-specific approaches (I–IV) assuring stable protein–protein interactions (*see* **Note 2**).
3. Cell pellets were incubated in the lysis buffer for 30 min on ice, resuspending very gently every 10 min (*see* **Note 10**).
4. Insoluble components were separated from soluble proteins by centrifuging at 4°C with 21,382×*g* for 15 min and the supernatants were transferred into new tubes on ice.
5. Whole cell extracts were frozen once in liquid nitrogen and thawed very gently on ice to mechanically disrupt cell membrane and afterward stored at –80°C.

3.4.2. Quantification of AR Protein Level

1. Cell pellets were thawed gently on ice (*see* **Note 9**) and resuspended in 100 μl NETN lysis buffer (approximately fivefold pellet volume).
2. After incubation on ice for 10 min, do three cycles of shock freezing in liquid nitrogen and subsequent thawing at 37°C (*see* **Note 11**).
3. Insoluble components were separated from soluble proteins by centrifuging at 4°C with 21,382×*g* for 15 min; supernatants were transferred into new tubes on ice.
4. Whole cell extracts were frozen once in liquid nitrogen and thawed very gently on ice to mechanically disrupt cell membrane and afterward stored at –80°C.

3.5. Co-immunoprecipitation (CoIP)

Co-immunoprecipitation allows the specific isolation of proteins out of complex biological solutions. Here we exemplify this with the CoIP of AR and few interaction partners. Specific anti-AR AB was coupled to affinity matrices such as IDM Affinity Beads, Dynabeads® Protein-A, and Protein-A Agarose beads. IDM Affinity Beads and Dynabeads® Protein-A were used as a proteomic approach for AR protein complex isolation and CoF detection via SELDI-TOF-MS. For specific AR-CoF detection and identification such as Western blot (WB), we used Protein-A Agarose beads.

1. Each of the four ligand-specific approaches (I–IV) was divided into two samples. One sample was used for CoIP with anti-AR AB for precipitation of the AR with its bound CoF and the other sample was used with IgG AB as a negative control (*see* **Note 12**).

2. The total protein in the samples was measured using NanoDrop® ND-1000 spectrophotometer (Thermo Scientific). For the CoIP, similar protein amounts and concentrations for the samples of the four approaches (I–IV) should be used. Thus samples used in CoIP should contain an identical amount of at least 1 mg protein for each sample.

3.5.1. CoIP with IDM Affinity Beads

Precipitation of the AR-specific proteins and interacting factors using the matrix IDM Affinity Beads takes 3 days. Precipitates were used for measurement with SELDI-TOF-MS and protein identification via Western blotting.

3.5.1.1. Day 1

Preparation of the affinity matrix and coupling of Protein-A to the matrix followed by repeated wash steps:

1. In each sample a volume of 20 μl of the IDM Affinity Beads was transferred into 0.5 ml tube (*see* **Note 13**).
2. The beads were washed six times with 200 μl purified water. After discarding the supernatant, the beads were incubated with 4 μl Protein-A and 40 μl of 50 mM SAB pH 5.0 overnight at RT in a fixed speed rotator.

3.5.1.2. Day 2

AB coupling to the Protein-A-coupled beads and incubation with the cell lysate:

1. Discard the supernatant and wash the bead–Protein-A complex twice with 400 μl SAB, pH 5.0, followed by three washing steps using 400 μl PBS for removal of unbound Protein-A (*see* **Note 1**).
2. For binding of AB to Protein-A-coupled beads 10 μl anti-AR antibody or 3 μl IgG as control was incubated with beads for 3 h at 4°C in a fixed speed rotator. Antibodies were dissolved in 50 μl SAB, pH 5.0, for each sample. Note that pre-tests were performed to detect the amount of anti-AR AB and IgG for comparison.
3. After AB binding, the beads were washed with 400 μl PBS for removal of unbound antibodies. For more stringent conditions Triton X-100 was added to washing buffer (*see* **Note 3**).
4. Incubation with gently thawed cell lysates (*see* **Note 9**) was performed overnight in a fixed speed rotator at 4°C.

3.5.1.3. Day 3

Elution of the specifically bound proteins and protein complexes:

1. The beads were washed twice with ice-cold PBS-PI and afterward with 400 μl washing buffer containing different concentrations of the detergent Triton X-100 (*see* **Note 3**) depending on Protein-A–AB–antigen affinity (*see* **Note 1**). At this step we used a concentration of 0.5% Triton X-100.

2. Elution of proteins off the beads occurred by incubation in 40 μl elution buffer for 25 min in a fixed speed rotator at RT followed by rigorous vortexing for 1–2 min.
3. Centrifuge at 389 × *g* for 30 s and transfer supernatant into a new tube and mix with 2× SDS loading buffer.
4. The samples were frozen in liquid nitrogen and thereupon immediately stored at –80°C.

3.5.2. CoIP with Dynabeads® Protein-A

The CoIP of the AR and bound CoF with the Dynabeads® Protein-A protocol requires 2 days of work. The resulting AR–CoIP samples are used for AR-CoF detection via SELDI-TOF-MS and immunological techniques such as Western blotting.

3.5.2.1. Day 1

Preparation of the beads including the coupling of antibodies and afterward the binding of the antigen to the beads:

1. For each sample, 100 μl of the beads (*see* **Note 13**) was transferred into 0.5 ml tubes. Using the magnetic character of the beads the separation of matrix from the storage solution occurs by placing the tubes in a magnetic column.
2. After this step the beads were successively washed for three times in 400 μl sodium phosphate buffer 0.1 M, pH 8.0, and afterward resuspended in 90 μl SPB supplemented with antibodies; a volume of 10 μl anti-AR AB or a volume of 3 μl IgG AB is used as a negative control. The incubation time of the beads with the antibodies was 2 h at RT in a fixed speed rotator.
   Notice that in pre-tests the optimal AB concentration was detected in Western blot (*see* **Note 14**).
3. After washing the Dynabeads® Protein-A–AB complex in 400 μl 0.1 M sodium phosphate buffer, pH 8.0, the complex was incubated with the gently thawn-up cell lysates (*see* **Note 9**) overnight at 4°C on a fixed speed rotator.

3.5.2.2. Day 2

Different washing steps and the elution of the antigen off the beads:

1. The samples with Dynabeads® Protein-A–AB–antigen complex were washed twice with washing buffer for 10 min at 4°C followed by a wash step with 400 μl ice-cold PBS-PI (*see* **Notes 1** and **3**).
2. Elution of proteins was performed by incubation with 40 μl elution buffer for 25 min at RT followed by vortexing. Supernatants were transferred into new tubes and 2× SDS loading buffer was added. Because of low pH value of elution buffer the added 2× SDS loading buffer turns yellow. To neutralize, 0.1 M Tris–HCl buffer, pH 9.0, was carefully added until the color reversed back to blue.

3. The maintained eluate was frozen in liquid nitrogen and immediately stored at –80°C.

3.5.3. CoIP with Protein-A Agarose Beads

These beads were primarily used to obtain Co-immunoprecipitates for detection and identification of AR-bound CoF via immunological processes like Western blotting.

3.5.3.1. Day 1

Includes a preclearing step, the coupling of antibodies to the Protein-A Agarose beads, and finally the binding of the antigen to AB:

1. For the preclearing step 40 μl Protein-A Agarose beads were added in each gently thawn-up sample (*see* **Note 9**) and incubated for 1 h on a fixed speed rotator at 4°C. Precleared lysates were separated from beads by centrifugation (608 ×*g* for 5 min at 4°C).
2. The supernatants were transferred into new tubes and antibodies were added; for each of the four approaches (I–IV) two samples were prepared. Sample one of each approach was supplemented with anti-AR AB (4 μl) and to sample two of each approach 1 μl of IgG AB was added (*see* **Note 12**).
3. The incubation of the lysates with antibodies was performed for 3 h at 4°C on a fixed speed rotator.
4. After this a volume of 30 μl Protein-A Agarose beads into each sample was added and incubated overnight at 4°C C on a fixed speed rotator.

3.5.3.2. Day 2

Removal of unspecific bound proteins and the elution of the specific bound proteins off the beads:

1. The first wash step was performed with ligand-supplemented ice-cold PBS (*see* **Note 1**). This step was repeated twice to ensure the removal of unspecific bound proteins (*see* **Note 3**).
2. The supernatant was transferred into new tubes and its AR protein level was detected via Western blotting to check the IP efficiency (*see* **Note 15**).
3. To elute the proteins off the matrix the samples were incubated in 2× SDS loading buffer and afterward vortexed very gently. Tubes were incubated on ice for 5 min and afterward the samples were boiled up to 95°C for 5 min. The samples were centrifuged about 10–20 s in a microcentrifuge to pellet the beads.
4. The maintained supernatant was transferred into new tube and frozen in liquid nitrogen and immediately stored at –80°C.

3.6. SELDI-TOF-MS

Surface-enhanced laser desorption/ionization (SELDI) allows the retention of proteins on a solid phase chromatographic surface (ProteinChip Array) with direct detection of retained proteins by time of flight (TOF) mass spectrometry (MS) (16) (**Fig. 14.1**).

3.6.1. Chip Preparation

1. To activate the chip each spot on the chip was washed with H50 or NP20 washing buffer according to the used chip.

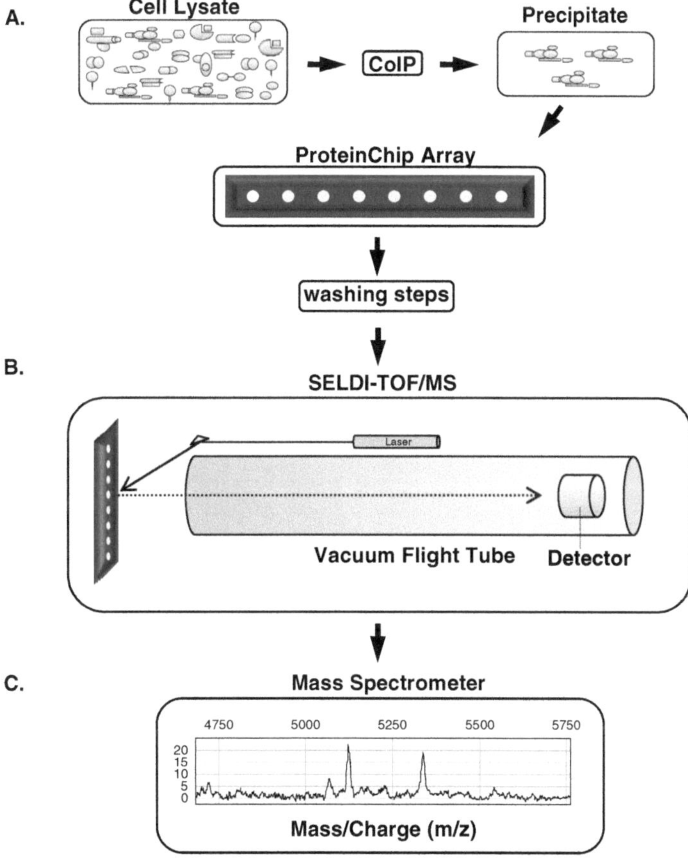

Fig. 14.1. Analysis of proteins using ProteinChip arrays and SELDI-TOF-MS. (**a**) The crude serum sample is processed via co-immunoprecipitation (CoIP). The precipitate is placed on a ProteinChip array which contains chemically (cationic, anionic, hydrophobic, hydrophilic, etc.) or biologically treated surfaces for specific interaction with proteins of interest. Proteins, thus, bind to chemical or biological "docking sites" on the ProteinChip surface. Non-binding proteins, salts, and other contaminants are washed away, eliminating sample "noise". (**b**) Retained proteins are "eluted" from the ProteinChip array by SELDI. Ionized proteins are detected and their mass accurately determined by TOF-MS (18). (**c**) Samples are spotted onto NP20 or H50 chips® (Ciphergen Biosystems). The ProteinChip is irradiated with a laser, resulting in ionization of the adherent molecules. The ions travel through a vacuum tube and their mass-to-charge ratios are calculated from their time of flight through the vacuum chamber (19).

Therefore, 5 μl washing buffer was applied onto each spot of the chip and removed immediately by carefully pipetting.

2. Chips were dried at 37°C for about 5 min using a thermoblock. It is visible if the chip is dry.
3. Next step was to add a volume of 5 μl eluate on the spots of the chip. Again chips were dried at 37°C using a thermoblock avoiding temperatures above 37°C.
4. Afterward, for removal of saline, the chips were washed by adding and immediately removing a volume of 5 μl purified water.
5. The last step of the chip preparation was to add 0.5 μl of EAM on each spot used on the chip. And finally the chip had to dry up again.

3.6.2. Measurement

1. The dried chip was placed in the right direction into the measuring instrument (*ProteinChip System Series 4000* mass spectrometer, Ciphergen Biosystems, Inc., Fremont).
2. Two different measuring protocols were used depending on the protein size. The first protocol afforded accurate measurement of proteins with size 1–20 kDa and is implemented in the software (*see* **Section 2.7**, Step 6) and named "cal2300fm10" (*2300* terms the laser intensity in nJ and *fm10* defines the *f*ocus*m*ass in kDa). The second protocol we used was related to accurate measuring proteins with a range of size from 20 up to 200 kDa and was named "cal3500fm50."

3.6.3. MS Data Interpretation and Database Research

1. Numerous signals (peaks) occurred in the spectrum. Significant peaks are defined by having no counterpart in control or a much lower intensity (**Figs. 14.2**, **14.3**, and **14.4**).
2. Note that specific peaks occurring at 70 and at 140 kDa are likely to be AB. The 70 kDa signal in the spectrum tagged one heavy and one light chain of AB; the peak at 140 kDa represents the entire AB.
3. Significant and AR-specific peaks (**Figs. 14.2**, **14.3**, and **14.4**), ligand-selective signals (**Figs. 14.2** and **14.3**), and also ligand-independent signals (**Fig. 14.4**) occurred.
4. For further investigation, the molecular weights of putative AR-interacting proteins identified via SELDI-TOF-MS were further examined via database search on the "ExPASy Proteomics Server" (http://ca.expasy.org/), with the tool TagIdent that identifies proteins with isoelectric point (pI),

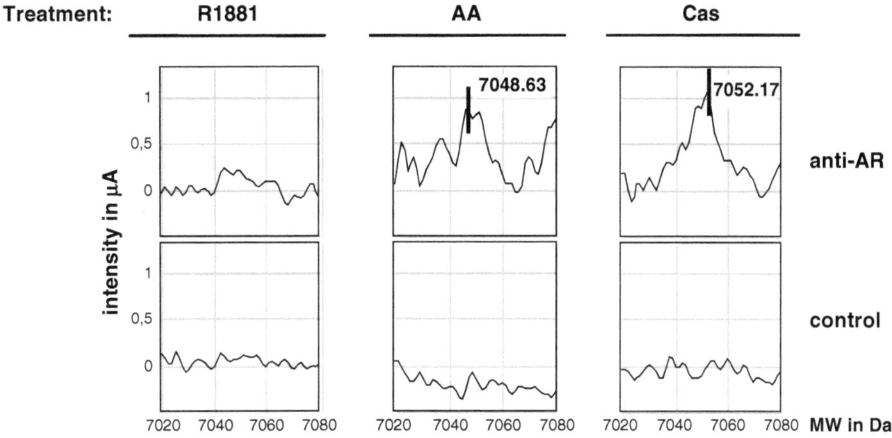

Fig. 14.2. LNCaP cells treated with the AR-antagonists AA or Cas yielded specific signal at about 7.0 kDa in the SELDI-TOF spectrum. LNCaP cells were treated with the agonist R1881 or AR-antagonists AA or Cas for 2 h, harvested and lysed. CoIP was performed with specific antibody against AR. As negative control IgG was used. Immunoprecipitates were measured with SELDI-TOF-MS. The molecular weight (MW) is shown on the *x*-axis in [Da], the intensity of the signal is shown on the *y*-axis in [μA]. Specific peaks are indicated with the correspondent MW.

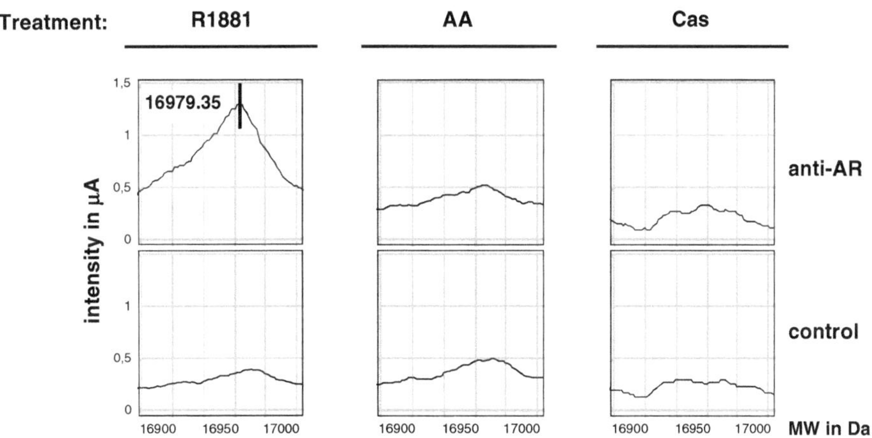

Fig. 14.3. A ligand-selective signal was detected at 17 kDa exclusively in the presence of the androgen R1881. LNCaP cells were treated with the AR-agonist R1881 or the AR-antagonists AA or Cas for 2 h, harvested and lysed. CoIP was performed with specific antibody against AR. IgG was used as negative control. Immunoprecipitates were measured with SELDI-TOF-MS. Specific peak is indicated with the correspondent MW.

molecular weight (MW), and sequence tag (http://ca.expasy.org/tools/tagident.html).

5. Here the molecular weights of candidate proteins (data from MS) are "refined" with application of AR-characterizing keywords leading to a list of proteins that potentially interact with AR.
6. Specific data about these protein candidates were obtained via database search at UniProt (http://www.uniprot.org/) to specifically select putative AR-interacting proteins for

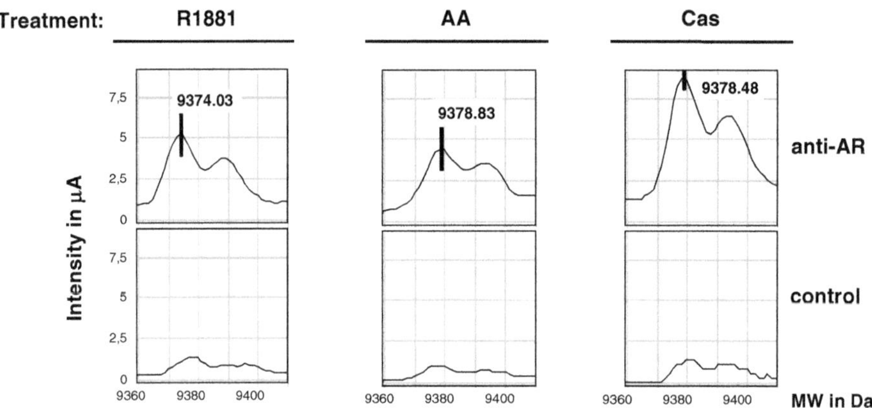

Fig. 14.4. AR-specific and ligand-independent MS signal was detected at 9.4 kDa. LNCaP cells were treated with the agonist R1881 or AR-antagonists AA or Cas for 2 h, harvested and lysed. CoIP was performed with specific antibody against AR. As control IgG was used. Immunoprecipitates were measured with SELDI-TOF-MS. Specific peaks are indicated with the corresponding MW.

further investigation, such as SDS-PAGE to separate AR-interacting proteins and Western blotting to specifically detect these proteins.

Thus, we successfully immunoprecipitated the androgen receptor (**Fig. 14.5**) and bound interacting proteins. Putative AR-interacting proteins were detected via SELDI-TOF-MS and a differentiation of the AR-specific SELDI signals suggesting AR-interacting proteins into ligand selective (**Figs. 14.2** and **14.3**) and non-ligand selective (**Fig. 14.4**) could be performed.

3.7. SDS-PAGE

SDS-PAGE is performed to separate different proteins according to their molecular weights. In a first step the proteins are exposed to SDS. When SDS binds to proteins they will be denatured and negatively charged. In a second step the proteins are separated by an electric field and migrate at different velocities depending on their molecular weight.

1. We used the Mini-Protean Tetra Cell system (Bio-Rad Laboratories^{TM}). Before usage, clean the glass surfaces with purified water and 70% (v/v) ethanol.
2. After assembling the glass chamber, fill in the separation gel with the calculated concentration of acrylamide and cover it carefully with 0.5 ml 2-propanol to obtain a smooth and even surface. The gel will polymerize sufficiently within 15 min.
3. Having poured away the 2-propanol, add the stacking gel solution onto the polymerized separation gel and insert the comb. After another 15 min the gel is ready for use.

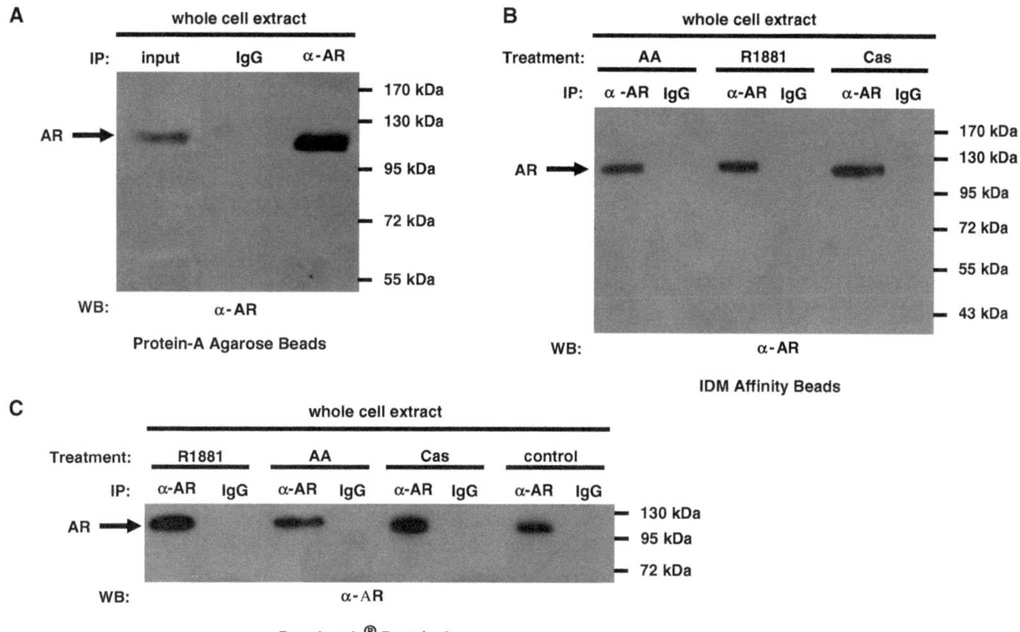

Fig. 14.5. Immunoprecipitation of AR using different affinity matrices. The affinity matrices used were the following: (**a**) Protein-A Agarose beads, (**b**) IDM Affinity Beads, (**c**) Dynabeads® Protein-A. (**a**) Untreated LNCaP cells, or (**b**) AA-, R1881-, or Cas-treated LNCaP cells were harvested and lysed. Experiment in (**c**) additionally includes a control. α-AR antibody was coupled to the beads followed by incubation with whole cell extracts. Proteins were separated by SDS-PAGE, followed by AR-detection via Western blotting (WB).

4. 10× stock solution running buffer: 250 mM Tris base, 1.91 M glycine, and 1% SDS, one part of this to be diluted in nine parts of purified water.
5. Put the gel in place into the chamber and fill in 1× running buffer in both parts of the chamber. Then carefully remove the comb from the gel and clean the wells with running buffer using a pipette or a syringe.
6. If the protein sample to be analyzed does not contain SDS loading buffer (already added after CoIP, *see* **Sections 3.5.1, 3.5.2**, and **3.5.3**) mix three parts of the protein sample (approx. 10–100 μg) and one part 4× SDS loading buffer and boil the mixture (up to 30 μl for the 1 mm 10-well comb) for 5 min at 95°C. After centrifuging for 1 min at 16,430×*g* fill the denatured protein solution into the wells. An adequate protein ladder running in a separate slot is necessary to estimate the molecular weight of the protein bands. We used a prestained protein marker (PeqGold prestained protein marker IV) that allows later on the detection on the membrane to ensure transfer of proteins to the membrane.
7. Close the lid and connect the apparatus to the power supply at 80 V until the bromophenol blue front reaches the edge of

the stacking gel (*see* **Note 16**). Afterward adjust the power supply to 200 V until the bromophenol blue front reaches the bottom of the gel (*see* **Note 17**). In this step, the cathode (minus) is connected to the upper part of the chamber where the samples have been loaded. The lower part that is in contact with the lower part of the gel is connected to the anode (plus).

3.8. Western Blot

In the subsequent Western blot, these separated proteins are transferred to a membrane. Now the proteins of interest can be bound by specific antibodies. After this step, these specific antibodies that remained on the membrane after washing steps are marked by a secondary AB that is coupled to an enzyme. In the final step, the enzymatic activity is measured, e.g., by luminescence, that can be detected and quantified.

1. Fill a tray up to about 3 cm above the bottom with transfer buffer and subsequently stack items in the following order to a cassette (**Fig. 14.6**) (later, the cathode will be placed here):

   Black plastic cover

   2× sponge (included in the Mini-Blot system)

   5× Whatman™ 3MM paper, soaked in transfer buffer

   Separating gel (handle with wet gloves to avoid adhesion), the stacking gel can be discarded

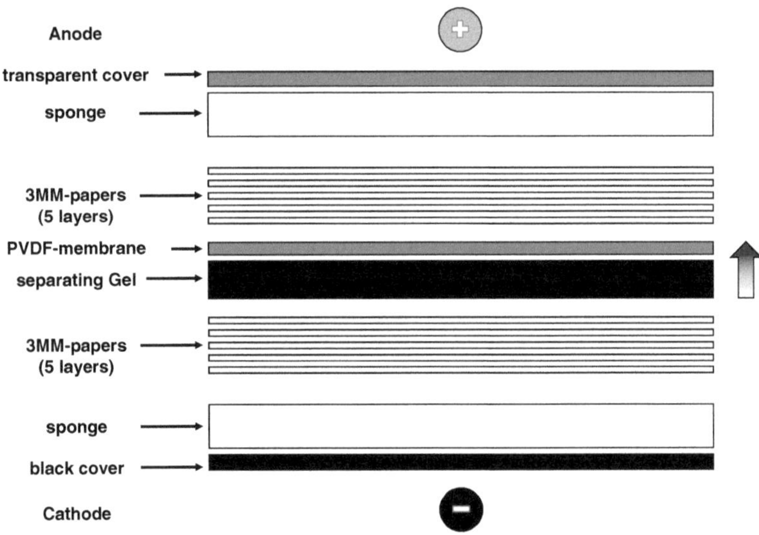

Fig. 14.6. Scheme of Western blotting. The order of the different layers in the Western transfer is shown. The methanol-activated membrane is stacked above the separation gel. Ensure to place the electric polarity in the correct way to ensure the proteins to migrate onto the PVDF membrane as indicated by the *arrow* at the right side.

Polyvinylidene fluoride (PVDF) membrane – activate it for 10 s in 100% methanol – like the Whatman paper the surface should cover the entire gel and avoid bubbles between the layers. The membrane should be handled only by forceps to avoid skin contact and thereby contamination with unwanted proteins

5× Whatman 3MM paper soaked in transfer buffer

2× sponge (included in the Mini-Blot system)

Transparent plastic cover (later, the anode will be placed here)

2. Now put the cassette into the apparatus ensuring that the PVDF membrane is between the separating gel and the anode. Otherwise the proteins will be lost in the transfer buffer.
3. After covering the cassette entirely with cooled transfer buffer (4°C) add a rotating magnetic stir bar and put the apparatus into the refrigerator to ensure a surrounding temperature of 4°C.
4. Connect the apparatus to the electric power supply at 400 mA without voltage restriction and let it run for 55 min (*see* **Note 18**). Cooling is necessary to avoid deterioration of the outcome.
5. Disassemble the cassette and indicate the orientation of the PVDF membrane by cutting an edge or marking it with a ball pen. The prestained marker should be visible on the membrane.
6. To detect the proper transfer of proteins to the membrane the membrane can be stained for 3–5 min in Ponceau red solution: 0.1% Ponceau red (w/v) in 5% acetic acid. After rinsing with water the proteins show a red staining. Note that Ponceau staining is reversible by rinsing with water.
7. Then shake the membrane for 60 min in blocking buffer (10% fat-free milk in TBS-T) to saturate it with unspecific proteins.
8. Remove the membrane from the blocking buffer and incubate it by shaking it for 3×5 min in TBS-T buffer on orbital shaker.
9. Incubate the membrane in a 50 ml falcon tube for 50 min with primary AB on an orbital shaker. Ensure that the surface of the membrane that faced the gel is orientated inward. The AB solution must wet the entire surface during rotation.

9a. For detection of ligand-selective AR-interacting proteins only:

To detect AR-interacting protein CDCA2 we used rabbit anti-CDCA2 as primary AB (*see* **Section 2.10**).

9b. For quantification of AR protein level only:
As primary AB to detect AR we used mouse anti-AR AB (Biogenex™). For loading control (to compare the total protein amounts loaded onto SDS-PAGE) we detected the house-keeping gene – product β-actin (constitutively expressed in all tissues) using rabbit anti-β-actin AB (*see* **Section 2.10**).

10. Remove unbound AB by washing the membrane 3×5 min on a shaker with TBS-T.

11. Incubate the membrane in a 50 ml falcon tube for 45 min with secondary AB on an orbital shaker.

11a. For detection of ligand-selective AR-interacting proteins only:
As secondary AB we used bovine anti-rabbit AB (*see* **Section 2.10**).

11b. For quantification of AR protein level only:
Goat anti-mouse AB was used as secondary AB to detect primary AB against AR. To detect primary AB against β-actin we used bovine anti-rabbit AB (*see* **Section 2.10**).

12. Remove unbound AB by washing the membrane 3×5 min on a shaker with 1× TBS-T.

13. The next steps are performed in a dark room. Take the ECL solution out of the refrigerator and let it adjust to RT.

14. Immediately before use mix 1 ml of each of the two components and apply it onto the membrane surface that was orientated to the inner part of the falcon tube.

15. Wait for 1 min, then remove the ECL solution and cover the membrane with a transparent thin plastic foil and put it into an x-ray film cassette.

16. Switch off the light and put a photographic film, e.g., GE Amersham hyperfilm ECL, onto the plastic foil. Ensure that you can reproduce the exact orientation after developing the film by, e.g., cutting an edge in the film. Fix the film with adhesive strips and incubate it in the closed cassette for various time periods starting with an exposure time of 1 min. Afterward, move the x-ray film into a developing machine (e.g., hyperprocessor, GE Healthcare™). Subsequently, repeat the same procedure with an incubation time of, e.g., 3 and 10 min.

17. If the signal is too low after 30 min exposure time, a more sensitive HRP-solution can be used (e.g., HRP-"juice" by

PJK^{TM}), the same membrane can be used directly without prior stripping. To allow quantification via densitometry programs the signal should not be saturated.

18. To estimate the molecular weight, label the marker bands of the membrane on the film by a pen by placing it in exactly the same position as it was placed in the x-ray cassette.

3.8.1. Stripping and Re-probing the Membrane

To detect other proteins on the membrane it is possible to remove the bound antibodies (stripping) and re-probe with other antibodies (*see* **Note 19**).

1. The stripping buffer contains 62.5 mM Tris–HCl, pH 6.8, 2% (w/v) SDS, and 100 mM β-mercaptoethanol. The buffer should be used for one time only. Store at 4°C.
2. For the stripping procedure, fill the buffer into a small plastic box and heat it in a water bath to 50°C. The buffer should cover the entire membrane in the next step.
3. Then place the membrane into the box with the buffer for an incubation time of 5–10 min maintaining the temperature of 50°C in the water bath. Incubation is performed in a shaking water bath. Note that the stripping procedure may also remove some of the membrane-bound proteins if longer incubation times are used. The prestained marker must therefore remain visible upon stripping; otherwise the incubation time must be reduced to avoid too much protein loss.
4. After this step, the membrane is washed 3×5 min with 1× TBS-T and blocked with the blocking buffer for 60 min. The membrane can then be handled as described from Step 8 onward as described in the previous section.

3.9. SDS-PAGE and WB Data Interpretation

3.9.1. For Detection of Ligand-Selective AR-Interacting Proteins

After examination of numerous hints of the SELDI-TOF-MS data we further focused on researching the CDCA2 performing immunological techniques such as CoIP, SDS-PAGE, and WB. Analyzing AR-co-immunoprecipitates via SDS-PAGE and WB using specific AB against CDCA2 we successfully detected CDCA2 in the presence of AR-antagonist bicalutamide in vivo (**Fig. 14.7**) verifying CDCA2 to be a ligand-selective AR-interacting protein.

3.9.2. For Quantification of AR Protein Level

We could show that the protein level of the AR (AR expression or AR protein stability) is decreased through treatment of LNCaP cells with the nonsteroidal anti-inflammatory drug FA (**Fig. 14.8**).

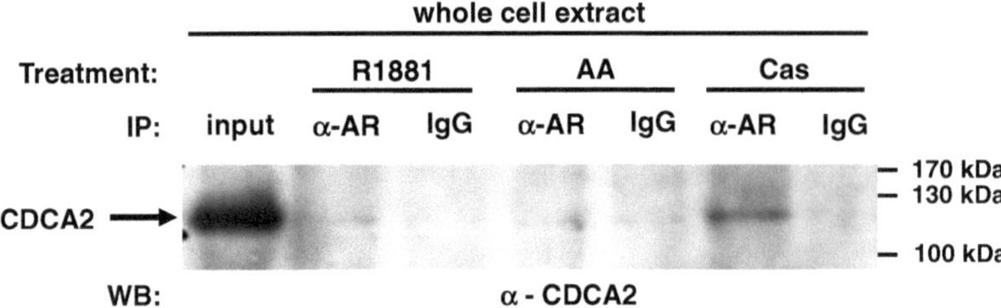

Fig. 14.7. Interaction of the cell division cycle-associated protein 2 (CDCA2) with AR exclusively in the presence of the AR-antagonist Cas. LNCaP cells were treated with the agonist R1881 or AR-antagonists AA or Cas for 2 h and then harvested and lysed. Co-immunoprecipitation (CoIP) was performed with specific antibody against AR. As negative control IgG was used. Immunoprecipitates were separated by SDS-PAGE followed by Western blot analysis using a specific antibody against CDCA2.

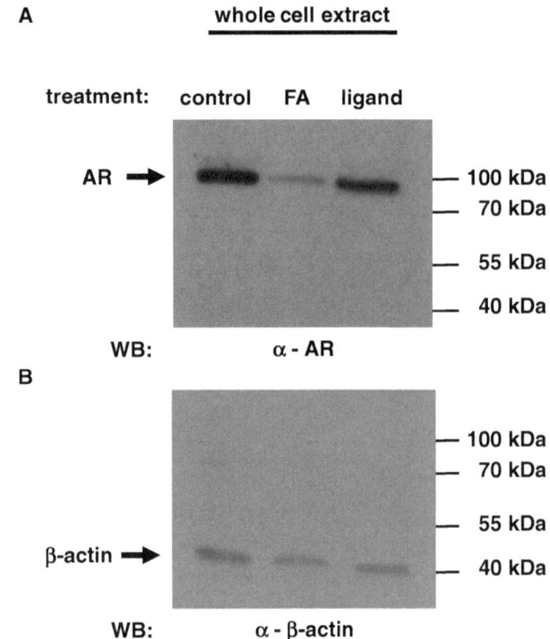

Fig. 14.8. Reduced AR protein level in LNCaP cells after treatment with FA. 6.4×10^4 LNCaP cells per well were cultured in 6-well plates and treated for 48 h with FA (10^{-7} M), AR-antagonist (ligand), or solvent DMSO (control) and harvested and lysed. Whole cell extracts were separated by SDS-PAGE and further analyzed by Western blotting using specific antibody against (**a**) AR and (**b**) β-actin.

4. Notes

1. To sustain ligand-dependent receptor–CoF binding during washing steps, all washing buffers should be supplemented with the cognate ligand consistent with the cell treatment.

2. To sustain ligand-dependent receptor–CoF binding during cell lysis, each lysis buffer should be supplemented with the ligand consistent with the cell treatment.
3. Washing buffers can be supplemented with different amounts (0.05–1%) of detergents, such as Triton X-100, to increase the stringency of the washing steps. You can check the strength of wash buffer via CoIP, SDS-PAGE, and WB; perform CoIP as described in **Section 3.5** with beads only (without antibodies) and separate precipitates in SDS-PAGE followed by Coomassie staining (see also Chapter 13, this volume). If proteins are visible you have to wash more stringently by increasing Triton X-100 concentration.
4. TEMED causes polymerization of the gels. Therefore, it should be added just before pouring it into the chamber.
5. It is recommended to cut the tip of the pipette to facilitate pipetting Tween 20.
6. To check if cells lost adherence upon trypsin addition, a light microscope can be used.
7. Working with proteins should always be performed on ice assuring low dissociation and degradation level of proteins.
8. Avoid storage of cell extracts longer than 7 days because AR activity, even when frozen, may decrease.
9. Thawing of deep frozen samples occurred always on ice assuring low dissociation and degradation level of proteins.
10. Resuspending the samples with lysis buffer must occur very gently to avoid dissociation of CoF from the AR caused by mechanical "forces." Therefore, gather the sample very gently and slowly with the pipette and transfer it back into the tube with very light pressure. Repeat this step twice.
11. Upon "shock freezing," degradation of proteins was avoided by gently thawing up the solution using a 37°C water bath and constantly and gently pivoting the tube to oppose radical temperature gradient inside (do not vortex).
12. During CoIP using the matrix only without coupling Protein-A and/or antibodies appears as an efficient negative control.
13. Cut off the very end of the pipette tip to facilitate adjusting the intended volume of beads.
14. The binding affinity of the applied antibodies to the different CoIP matrices can be checked by immunoblotting the eluate against these antibodies. This is necessary to ensure equally amounts of matrix-bound anti-antigen and negative antibodies.

15. The affinity of the applied antibodies to the antigen can be checked after the antigen coupling step using the supernatant for detection of the antigen by Western blotting against this antigen.
16. The edge between stacking and separation gels should be marked, but not on the glass chamber inside the apparatus.
17. Running the gel with high voltage results in fast protein separation but with a poor solution.
18. The blotting time depends on the size of proteins to be transferred. Small proteins transfer faster than larger proteins. Thus, use two PVDF membranes to avoid loss of smaller proteins due to longer blotting.
19. During the stripping procedure membrane-bound proteins may also be lost using longer incubation times. The prestained marker must therefore remain visible; otherwise the incubation time must be reduced to avoid too much protein loss. During the incubation, the buffer should be on a shaker.

References

1. Jemal, A. (2009) Cancer statistics. *CA Cancer J. Clin.* **59** (4), 225–249 (2009).
2. Andriole, G., Bruchovsky, N., Chung, L.W., Matsumoto, A.M., Rittmaster, R., Roehrborn, C., Russell, D., Tindall, D. (2004) Dihydrotestosterone and the prostate: the scientific rationale for 5alpha-reductase inhibitors in the treatment of benign prostatic hyperplasia. *J. Urol.* **172** (4 Pt 1), 1399–1403.
3. Ross, R., Bernstein, L., Judd, H., Hanisch, R., Pike, M., Henderson, B. (1986) Serum testosterone levels in healthy-young black-and-white men. *J. Natl. Cancer Inst.* **76**, 45–48.
4. Evans, R.M. (1988) The steroid and thyroid-hormone receptor superfamily. *Science* **240**, 889–895.
5. Culig, Z., Klocker, H., Bartsch, G., Hobisch, A. (2002) Androgen receptors in prostate cancer. *Endocr. Relat. Cancer* **9**, 155–170.
6. Papaioannou, M., Schleich, S., Prade, I., Degen, S., Roell, D., Schubert, U., Tanner, T., Claessens, F., Matusch, R., Baniahmad, A. (2009) The natural compound atraric acid is an antagonist of the human androgen receptor inhibiting cellular invasiveness and prostate cancer cell growth. *J. Cell. Mol. Med.* **13** (8B), 2210–2223.
7. Li, J., Al-Azzawi, F. (2009) Mechanism of androgen receptor action. *Maturitas* **63**, 142–148.
8. Kim, J., Coetzee, G.A. (2004) Prostate specific antigen gene regulation by androgen receptor. *J. Cell. Biochem.* **93** (2), 233–241.
9. Lemon, B.D., Freedman, L.P. (1999) Nuclear receptor CoF as chromatin remodelers. *Curr. Opin. Genet. Dev.* **9**, 499–504.
10. Xu, L., Glass, C.K., Rosenfeld, M.G. (1999) Coactivator and corepressor complexes in nuclear receptor function. *Curr. Opin. Genet, Dev.* **9**, 140–147.
11. Eisoldt, M., Asim, M., Eskelinen, H., Linke, T., Baniahmad, A. (2009) Inhibition of MAPK-signaling pathway promotes the interaction of the corepressor SMRT with the human androgen receptor and mediates repression of prostate cancer cell growth in the presence of antiandrogens. *J. Mol. Endocrinol.* **42**, 429–435.
12. Glass, C.K., Rosenfeld, M.G. (2000) The coregulator exchange in transcriptional functions of nuclear receptors. *Genes Dev.* **14**, 121–141.
13. Farla, P., Hersmus, R., Trapman, J., Houtsmuller, A.B. (2005) Antiandrogens prevent stable DNA-binding of the androgen receptor. *J. Cell. Sci.* **118**, 4187–4198.

14. Hodgson, M.C., Astapova, I., Cheng, S., Lee, L.J., Verhoeven, M.C., Choi, E., Balk, S.P., Hollenberg, A.N. (2005) The androgen receptor recruits nuclear receptor CoRepressor (N-CoR) in the presence of mifepristone via its N and C termini revealing a novel molecular mechanism for androgen receptor antagonists. *J. Biol. Chem.* **280**, 6511–6519.
15. Masiello, D., Cheng, S., Bubley, G.J., Lu, M.L., Balk, S.P. (2002) Bicalutamide functions as an androgen receptor antagonist by assembly of a transcriptionally inactive receptor. *J. Biol. Chem.* **277**, 26321–26326.
16. Zhu, W., Smith, A., Young, C.Y. (1999) A nonsteroidal anti-inflammatory drug, flufenamic acid, inhibits the expression of the androgen receptor in LNCaP cells. *Endocrinology* **140** (11), 5451–5454.
17. von Eggeling, F., Junker, K., Fiedle, W., Wollscheid, V., Dürst, M., Claussen, U., Ernst, G. (2001) Mass spectrometry meets chip technology: a new proteomic tool in cancer research? *Electrophoresis* **22** (14), 2898–2902.
18. Abramovitz, M., Leyland-Jones, B. (2006) A systems approach to clinical oncology: focus on breast cancer. *Proteome Sci.* **4**, 4–5.
19. Miyamae, T., Malehorn, D.E., Lemster, B., Mori, M., Imagawa, T., Yokota, S., Bigbee, W.L., Welsh, M., Klarskov, K., Nishomoto, N., Vallejo, A.N., Hirsch, R. (2005) Serum protein profile in systemic-onset juvenile idiopathic arthritis differentiates response versus nonresponse to therapy. *Arthritis Res. Ther.* 7 (4), R746–R755.

Section V

Recent Advances in Studying Androgen Receptor Signalling

Chapter 15

Global Identification of Androgen Response Elements

Charles E. Massie and Ian G. Mills

Abstract

Chromatin immunoprecipitation (ChIP) is an invaluable tool in the study of transcriptional regulation. ChIP methods require both a priori knowledge of the transcriptional regulators which are important for a given biological system and high-quality specific antibodies for these targets. The androgen receptor (AR) is known to play essential roles in male sexual development, in prostate cancer and in the function of many other AR-expressing cell types (e.g. neurons and myocytes). As a ligand-activated transcription factor the AR also represents an endogenous, inducible system to study transcriptional biology. Therefore, ChIP studies of the AR can make use of treatment contrast experiments to define its transcriptional targets. To date several studies have mapped AR binding sites using ChIP in combination with genome tiling microarrays (ChIP-chip) or direct sequencing (ChIP-seq), mainly in prostate cancer cell lines and with varying degrees of genomic coverage. These studies have provided new insights into the DNA sequences to which the AR can bind, identified AR cooperating transcription factors, mapped thousands of potential AR regulated genes and provided insights into the biological processes regulated by the AR. However, further ChIP studies will be required to fully characterise the dynamics of the AR-regulated transcriptional programme, to map the occupancy of different AR transcriptional complexes which result in different transcriptional output and to delineate the transcriptional networks downstream of the AR.

Key words: Nuclear hormone receptor, androgen receptor, chromatin immunoprecipitation (ChIP), high-throughput sequencing, genomic tiling microarray, transcription, hormone, cancer.

1. Introduction

Chromatin immunoprecipitation (ChIP) has been used to map sites of DNA methylation (1), histone modifications (2), and sites of protein occupancy (3–5), utilising specific antibodies to allow enrichment of the genomic regions associated with these epitopes (**Fig. 15.1a**). To identify transcription factor binding sites, protein–DNA interactions are cross-linked using formaldehyde

F. Saatcioglu (ed.), *Androgen Action*, Methods in Molecular Biology 776,
DOI 10.1007/978-1-61779-243-4_15,

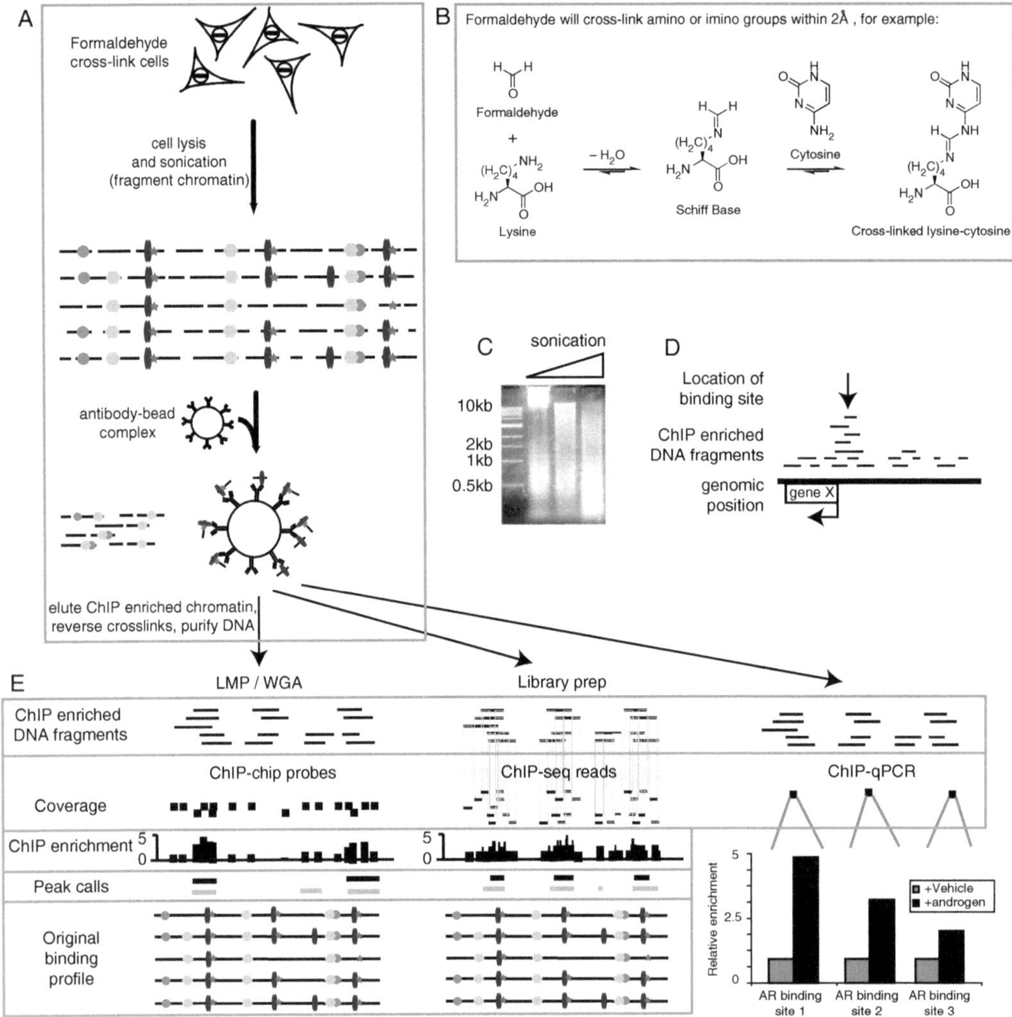

Fig. 15.1. Overview of chromatin immunoprecipitation combined with array detection (ChIP-chip) or direct sequencing (ChIP-seq) of enriched DNA fragments. (**a**) Schematic summary of chromatin immunoprecipitation (ChIP) method. (**b**) Formaldehyde cross-linking chemistry. (**c**) Example of gel electrophoresis to visualise chromatin shearing after sonication. (**d**) Schematic representation of DNA fragment enrichment using the ChIP method (*arrow* indicates site of protein–DNA interaction). (**e**) Schematic comparing ChIP-chip, ChIP-seq and ChIP-qPCR methods to assess enrichment of DNA binding sites.

(**Fig. 15.1b**), stabilising these transient interactions and allowing their isolation. Chromatin is then fragmented (**Fig. 15.1c**) to allow separation of genomic fragments bound by the transcription factor of interest away from those which are not bound. Following antibody enrichment (immunoprecipitation), formaldehyde cross-links are reversed and enriched DNA fragments are then purified (**Fig. 15.1d**). ChIP-enriched genomic DNA fragments can be mapped to the genome and quantified using genomic microarrays (ChIP-chip), direct sequencing (ChIP-seq) or quantitative PCR (**Fig. 15.1e**). Each of these approaches offers specific advantages and disadvantages (**Table 15.1**). For example, the genomic binding sites of a transcription factor must be predicted or mapped prior to the application of quantitative PCR to assess DNA fragments enriched using ChIP sites (**Fig. 15.1e**). Genomic tiling microarrays have been widely used to map transcription factor binding sites (ChIP-chip), with genomic coverage targeted to specific regions of interest (6), gene promoter regions (4) or whole non-repetitive genomes (5). ChIP-chip approaches have greatly advanced our understanding of transcriptional networks in many areas of biology, allowing assessment of histone marks in as few as 10^3 cells (7) and transcription factor binding from $>10^4$ cells (8). However, ChIP-chip approaches require many rounds of amplification to yield sufficient DNA quantities for microarray hybridisation and are limited by the restraints of microarray probe design, resulting in uneven genomic coverage and low resolution of DNA binding sites (**Fig. 15.1e** and **Table 15.1**). Given the large size of higher eukaryote genomes (e.g. human genome $\sim 3 \times 10^9$ bp), genome tiling microarrays covering the whole non-repetitive genome are printed on 7–38 individual microarrays (depending on the platform and probe spacing); thus the amount of ChIP DNA required and the cost of each replicate is high (**Table 15.1**). The combination of ChIP with second-generation sequencing technologies (ChIP-seq) offered an attractive alternative to ChIP-chip, requiring fewer amplification steps, providing more complete genome coverage, increasing the resolution of DNA binding sites and reducing the cost of whole genome coverage (**Table 15.1**) (9). However, ChIP-seq approaches require access to specialist equipment and technical expertise and more starting material compared to ChIP-chip, and importantly, the analysis of ChIP-seq data remains less well defined (**Table 15.1**) (10). These limitations are related to the relatively recent development of second-generation sequencing technologies and are being addressed by the research community. For example, a recent community challenge reported a detailed comparison of 11 currently available ChIP-seq analysis packages (10) and a separate study reported a new ChIP-seq method using the Heliscope single molecule sequencing platform (Helicos) allowing successful genome-wide coverage from low cell numbers, without the need for amplification steps (**Table 15.1**). However, a more fundamental question remains with regard to genome-wide coverage using either ChIP-chip or ChIP-seq, since most studies resort to taking arbitrary windows around genes of 10–100 kb to attempt to link transcription factor binding sites to their functional targets. Therefore, most studies could provide the same level of utilised data using genomic tiling microarrays focussed on protein coding genes and

Table 15.1
Comparison of ChIP-chip and ChIP-seq approaches

	ChIP-chip	ChIP-seq
Starting material	10^2–10^7 cells for histone marks or RNAP II 10^6–10^7 cells for transcription factors	10^7–10^8 cells for Illumina sequencing 5×10^4–10^6 cells for Helicos sequencing
Sample preparation	Pooling replicate ChIP DNA or amplification using WGA (14 + 14 cycles), LMP (15 + 25 cycles) and random priming (20 + 15 cycles)	Single-end Illumina sequencing library prep (18 cycles)
Amount of ChIP DNA required	>2–5 μg	>5–10 ng for Solexa GA (Illumina) 50 pg–3 ng for Heliscope (Helicos)
Coverage	Custom design, promoter or whole genome (non-repetitive)	Whole genome (including uniquely mapping repeat sequences)
Amplification bias	Likely, given amplification steps. Enrichment can be measured by qPCR, but amplification bias is not assessed and could lead to over- or underrepresentation of ChIP-enriched regions	Less likely given fewer amplification steps. Amplification bias can be removed during sequence tag alignment (i.e. as non-unique sequence reads), although some evidence of CG bias during library preparation (e.g. Illumina)
Binding site resolution	Set by array coverage and analysis (typically 500–1000 bp)	Set by DNA binding profile of ChIP-target and fragment size selected for sequencing (typically ~50–300 bp)
Analysis	Well-defined methods available for peak calling and FDR calculation	Many tools available, although significant differences in peak calling[a]
Equipment	Microarray hybridisation/processing equipment and scanner	Illumina Genome Analyser, SOLiD System, 454 Genome Sequencer or HeliScope Sequencer
Cost (per whole genome replicate)	Roche Nimblegen 10 array set = £4709 Affymetrix 7 array set	Single-end library preparation and 36 bp Illumina (Solexa) sequencing lane currently = £570
Sensitivity	Subject to microarray artefacts: individual probe performance, genomic coverage, hybridisation efficiency, spatial effects	Digital readout, sensitivity defined by sequencing depth (related to number of unique sequence tags)
Other	May allow more sensitive detection of ChIP enrichment at certain genomic regions	Sequence level data allow SNP calling and mutation analysis of DNA binding sites

[a]ChIP-seq community challenge (http://sourceforge.net/projects/useq/files/CommunityChIPSeqChallenge/)

miRNA loci. This situation is likely to continue until information about the three-dimensional structure of chromatin in the nucleus is mapped for a given cell type.

2. Materials

The materials and methods described below are adapted from previously published protocols (3, 4, 11) and provide a focused description of positive and negative control experiments to measure androgen-stimulated AR binding using ChIP in combination with Roche Nimblegen genome tiling microarrays or direct Solexa sequencing (Illumina). However, these methods are also more generally applicable to the study of AR in other contexts and also the study of other chromatin bound factors.

2.1. Cell Culture and Cross-Linking

1. RPMI (Invitrogen) supplemented with 10% foetal bovine serum (FBS, HyClone)
2. Phenol red-free RPMI (Invitrogen) supplemented with 10% Charcoal dextran stripped FBS (HyClone)
3. AR ligands (e.g. DHT or the synthetic androgen R1881)
4. 11% formaldehyde in 50 mM HEPES–KOH (pH 7.5), 100 mM NaCl, 1 mM EDTA and 0.5 mM EGTA (*see* **Note 1**)
5. Formaldehyde quenching solution of 2.5 M glycine

2.2. Harvesting Cells and Sonication

1. Cell scrapers (Corning)
2. Phosphate-buffered saline (PBS) supplemented with Complete protease inhibitors (Roche)
3. Rotating tube mixer at 4°C
4. ChIP cell lysis buffer: 50 mM HEPES–KOH (pH 7.5), 140 mM NaCl, 1 mM EDTA, 10% glycerol, 0.5% Igepal (NP-40), 0.25% Triton X-100 and 1× SIGMAFAST protease inhibitors (Sigma)
5. ChIP nuclei wash buffer: 10 mM Tris (pH 7.5), 200 mM NaCl, 1 mM EDTA and 0.5 mM EGTA
6. ChIP nuclear lysis buffer: 10 mM Tris (pH 7.5), 100 mM NaCl, 1 mM EDTA, 0.5 mM EGTA, 0.1% sodium deoxycholate and 0.5% SDS (*see* **Note 2**)
7. 10% Triton X-100
8. Sonicating water bath (e.g. Diagenode Bioruptor) or probe sonicator (*see* **Note 3**)

9. Benchtop centrifuge
10. Agarose, TBE and electrophoresis equipment

2.3. Immunoprecipitation

1. Protein-A/G Dynal magnetic beads (Invitrogen) and magnetic tube rack (*see* **Note 4**)
2. Antibodies to target proteins (e.g. rabbit anti-AR N20, Santa Cruz) (*see* **Note 5** for resources listing validated ChIP-grade antibodies)
3. PBS supplemented with 0.5% BSA
3. Rotating tube mixer at 4°C
4. RIPA ChIP wash buffer: 50 mM HEPES (pH 7.6), 1 mM EDTA, 0.5 M LiCl, 1% Igepal (NP-40), 0.7% sodium deoxycholate
5. TE with 50 mM NaCl: 10 mM Tris–HCl (pH 8), 1 mM EDTA and 50 mM NaCl
6. Elution buffer: 1% SDS and 0.1 M $NaHCO_3$

2.4. DNA Isolation

1. TE: 10 mM Tris–HCl (pH 8.0) and 1 mM EDTA
2. RNase A, 1 mg/ml (DNase free)
3. Proteinase K, 20 mg/ml
4. Sodium chloride: 5 M NaCl
5. Glycogen (Roche) or suitable carrier for precipitation
6. Benchtop centrifuge
7 Phenol:chloroform:isoamyl alcohol (25:24:1)
8. Isopropanol
9. Ethanol, 75%
10. Tris pH 8, 10 mM

2.5. Analysis of Enrichment

2.5.1. Real-Time PCR

1. Real-time PCR instrument (e.g. ABI 7900)
2. Oligonucleotide primers to genomic regions of interest
3. SYBR Green PCR Master Mix (Applied Biosystems)
4. Optical PCR plates and adhesive covers compatible with the real-time PCR instrument

2.5.2. ChIP-chip

1. Nanodrop spectrophotometer or Qubit fluorometer
2. Linker oligonucleotides (P1: GCGGTGACCCGGGA-GATCTGAATTC and P2: GAATTCAGATC)
3. T4 DNA polymerase
4. T4 DNA ligase
5. Taq polymerase
6. dNTP mix

7. BioPrime array CGH labelling kit (Invitrogen)
8. Cy3-dUTP and Cy5-dUTP
9. Sephadex G-50 columns
10. Microarray hybridisation apparatus (e.g. Roche Nimblegen Hybridisation System and Nimblegen Hybridisation Kit)
11. Microarray scanner (e.g. GenePix 4000)
12. Image analysis software (e.g. NimbleScan)
13. Bioinformatics support for analysis of ChIP-chip data (e.g. implementing Ringo and limma analysis packages in the R statistical program (*see* **Note 6**).

2.5.3. ChIP-seq

1. T4 DNA polymerase
2. Klenow DNA polymerase
3. T4 polynucleotide kinase
4. dNTP mix (10 mM)
5. DNA Clean and Concentrator-5 Kit (Zymo Research)
6. Klenow fragment (3′5′ exo-minus, 5 U/μl)
7. dATP (1 mM)
8. Illumina oligonucleotide adapters
9. T4 DNA ligase
10. Phusion DNA polymerase
11. Illumina oligonucleotide primers 1.1 and 2.1
12. Dedicated 'clean' electrophoresis equipment
13. High-purity agarose (e.g. Low Range Ultra Agarose, Bio-Rad)
14. SYBR Safe DNA stain
15. Dark reader transilluminator (Clare Chemical)
16. Qiagen MinElute Gel Extraction Kit
17. Agilent Bioanalyser (*see* **Note 7**)

3. Methods

The ChIP-chip and ChIP-seq methods described below have been successfully used to map AR binding sites; however, they have also been used to map binding sites for other transcriptional regulators and RNAP II occupancy. Therefore, the ChIP methods presented could be used to address the outstanding questions in AR biology noted above or applied to the study of other factors. The AR is activated by androgen stimulation, allowing the same antibody to be used in both positive and negative control conditions, thus providing an ideal control for antibody specificity (*see* **Note 8**).

3.1. Cell Culture and Cross-Linking

1. Maintain LNCaP cells in RPMI supplemented with 10% FBS in cell culture incubators (5% CO_2 at 37°C) and passage at a dilution of 1:3 when approaching confluence with trypsin/EDTA. For ChIP-chip assays we use 5×10^6 cells for each ChIP reaction and for ChIP-seq assays we use 10^7 cells for each ChIP reaction.
2. When cells are ~70% confluent aspirate media from culture flasks, wash cells with 1× PBS and replace media with phenol red-free RPMI supplemented with 10% charcoal dextran stripped FBS.
3. After 72 h replace cell culture media with media supplemented with androgens (e.g. 1 mM R1881) or an equal volume of ethanol (vehicle) and return cells to the incubator for 4 h.
4. For every 10 ml of culture media add 1/10th volume of 11% formaldehyde solution to cell culture media. Incubate flasks at room temperature for 10 min (*see* **Note 1**).
5. Quench the formaldehyde cross-linking reaction by adding 1/20th volume of 2.5 M glycine solution directly to culture media (to give a final concentration of 125 mM) and incubate for 5 min at room temperature.

3.2. Harvesting Cells and Sonication

1. Transfer flasks to ice, remove media and wash cells twice with ice-cold 1× PBS supplemented with protease inhibitors.
2. Aspirate PBS from cells and harvest cells using a cell scraper. Transfer cells to a 15 ml tube using a wide-bore pipette tip and centrifuge cells at 1000×g for 3 min at 4°C.
3. Aspirate residual PBS and add 5 ml of ChIP cell lysis buffer per 5×10^6 cells. Incubate on a rotary tube mixer at low speed for 10 min at 4°C. Centrifuge at 1200×g for 5 min in a benchtop centrifuge and discard the supernatant.
4. Resuspend pellet in 5 ml of ChIP nuclei wash buffer and incubate on a rotary tube mixer at low speed for 5 min at 4°C. Centrifuge in at 1200×g for 5 min in a benchtop centrifuge and discard the supernatant.
5. Resuspend pellet in 1 ml of ChIP nuclear lysis buffer. Split nuclear lysate into 4 × 250 μl aliquots and sonicate for 15 min at maximum power in a Bioruptor sonicator (Diagenode) to fragment chromatin to an average length of 500 bp (*see* **Note 3**). Re-pool sonicated lysates, add 100 μl of 10% Triton X-100 and centrifuge in a benchtop microfuge

at 14,000×g for 10 min at 4°C. Transfer the supernatant to a 15 ml tube and add 2 ml of ChIP nuclear lysis buffer and 200 μl of 10% Triton X-100.

6. Take 50 μl from each sample for the total genomic input control. Assess the extent of sonication by electrophoresis on a 1% agarose gel, after reversing formaldehyde cross-links (see below). A smear should be visible, with the majority of fragments between 250 bp and 1 kb (*see* **Fig. 15.1c**).

3.3. Immunoprecipitation

1. Aliquot 100 μl of protein-A Dynal beads per ChIP reaction and wash three times with 1 ml PBS-BSA (0.5%), collecting beads on a magnetic rack in between washes. Resuspend beads in 250 μl of PBS-BSA, add 7.5 μg of AR N20 antibody (Santa Cruz) and incubate overnight on a rotary mixer at 4°C. Wash antibody-bead complexes three times with 1 ml PBS-BSA, using a magnetic rack, and resuspend in 100 μl PBS-BSA.
2. Add 100 μl of antibody–bead complexes to the 3 ml of sonicated lysates in 15 ml tubes and incubate overnight at 4°C on a rotary mixer at low speed.
3. Working in a cold room at 4°C transfer chromatin–antibody–bead complexes to 1.5 ml centrifuge tubes, by sequentially adding 1 ml of the mixture to tubes on a magnetic rack and discarding the supernatant.
4. Wash beads six times with 1 ml of ice-cold RIPA ChIP wash buffer, taking care to fully resuspend beads during washes and using a magnetic rack to immobilise beads between washes.
5. Wash beads once with 1 ml ice-cold TE supplemented with 50 mM NaCl, immobilise beads using a magnetic rack and discard the supernatant. Centrifuge at 3000×g in a benchtop microfuge for 2 min, at 4°C, and discard the residual supernatant.
6. Elute ChIP material by adding 200 μl of ChIP elution buffer and incubating at room temperature for 15 min with vigorous mixing. Centrifuge samples at 300×g for 3 min to pellet beads and transfer the supernatant to a fresh tube. Repeat the elution and combine eluates.

3.4. DNA Isolation

1. Reverse protein–DNA cross-links of total genomic input control and ChIP samples by adding NaCl to a final concentration of 200 mM and incubate at 65°C overnight.
2. Digest contaminating RNA by adding RNase A (20 μg/ml final concentration). Incubate at 37°C for 30 min.
3. Digest proteins by adding EDTA (10 mM final concentration), Tris–HCl pH 6.7 (20 mM final concentration) and

Proteinase K (80 μg/ml final concentration). Incubate at 55°C for 1 h.

4. Recover DNA by adding an equal volume (~500 μl) of phenol:chloroform:isoamyl alcohol, vortex vigorously and centrifuge at 13,000×g for 15 min. Carefully transfer the aqueous phase to a fresh tube, add 20 μg glycogen and an equal volume of isopropanol. Vortex vigorously and centrifuge at 13,000×g for 15 min. Discard supernatant and add 1 volume of 75% ethanol per volume of isopropanol used. Centrifuge at 7000×g for 8 min, discard supernatant and air-dry pellet. Resuspend pellet in 60 μl of 10 mM Tris, pH 8.

3.5. Analysis of Enrichment

Three alternative methods for quantifying ChIP enrichment are described below: ChIP-quatitative PCR, ChIP-chip and ChIP-seq. When designing and analysing ChIP experiments, it is important to bear in mind that ChIP enriches rather than isolates genomic targets. Therefore, while the enriched protein-bound DNA fragments will be highly enriched, the majority of DNA isolated from a ChIP reaction is likely to comprise fragments not bound by the protein of interest, purely because these DNA fragments constitute such a large proportion of the genome (**Fig. 15.1d**).

3.5.1. Real-Time PCR

When analysing ChIP enrichment using real-time PCR it is necessary to compare the test ChIP with a control ChIP (*see* **Note 8**) to assess specific enrichment over background (**Fig. 15.1e**). It is also necessary to compare the candidate genomic region (i.e. the region believed to be bound by the protein of interest) with a control genomic region which is not bound by the protein. The control genomic region allows an assessment of the non-specific DNA from each ChIP and can be used to normalise between the test and control ChIPs. Finally, in order to avoid bias caused by differences in PCR efficiency between test and control PCR reactions it is advisable to use a serial dilution of input material as a standard curve for each PCR reaction (e.g. 1× to 1/128).

1. Into an optical PCR plate aliquot 1 μl of ChIP DNA and standard curve samples in triplicate for each genomic region to be analysed by real-time PCR (usually a minimum of two PCR reactions, for the candidate region and the control region).
2. Mix 10 pmol of each primer with water and SYBR Green Master Mix to a 1× final concentration in a final volume of 10 μl.
3. Aliquot PCR mix onto ChIP DNA, seal plates with adhesive covers and centrifuge briefly.
4. Use the PCR conditions suggested for use with the SYBR Green Mix used (e.g. hot-start: 50°C 2 min, 95°C 10 min

[95°C 15 s, 60°C 1 min], repeat 40 times). The addition of a dissociation curve at the end of the PCR reaction allows an assessment of the specificity of the PCR.

5. Specific enrichment by ChIP can be assessed using the equation

$$\text{Relative enrichment} = \text{TestEff}^{(\text{control sample Ct} - \text{test sample Ct})} / \text{ControlEff}^{(\text{control sample Ct} - \text{test sample Ct})}$$

where 'ControlEff' is the efficiency of the control PCR and 'TestEff' is the efficiency of the test PCR, both calculated using the formula $10^{(1/-\text{slope of standard curve})}$. 'Control sample Ct' and 'test sample Ct' are the cycle thresholds (Ct) at which the PCR reactions for control or test samples become exponential.

3.5.2. Array Detection (ChIP-chip)

Microarray detection of ChIP DNA has been widely used to allow de novo identification of genomic regions enriched by ChIP (3, 4, 6, 12, 13). The advantages and limitations of ChIP-chip approaches have been outlined above and summarised in **Table 15.1**. However, two key considerations are the method of amplification of ChIP material and the choice of microarray platform. ChIP yields small amounts of DNA (typically in the nanogram range); therefore, either multiple replicate ChIP reactions must be pooled or several rounds of amplification are required to yield sufficient material for microarray hybridisation. Most ChIP-chip studies to date have used random priming, linker-mediated PCR (LMP) amplification or proprietary DNA amplification kits (e.g. Sigma WGA). Direct comparison of these amplification methods suggests that low cycle LMP (method described below) and WGA (Sigma) have similar performance (14). A recent study compared multiple genomic tiling arrays and analysis methods, finding that platforms with longer microarray probe sequences (e.g. Nimblegen and Agilent) have the highest reproducibility and sensitivity (15).

1. Blunt ends of 50 μl ChIP DNA and 2 μl total input control (200 ng) with T4 DNA polymerase (1 U) in 1× T4 DNA polymerase reaction buffer supplemented with 45 ng/μl BSA and 90 μM dNTP mix. Incubate at 12°C for 20 min. Purify blunted ChIP DNA by adding 0.1 volume of 3 M NaOAc and 10 μg glycogen. Mix and then add 1 volume of phenol:chloroform:isoamyl alcohol (25:24:1), vortex and centrifuge for 15 min at 13,000×g. Transfer aqueous phase to a fresh tube and add 2.5 volumes of 100% ethanol, vortex and centrifuge for 15 min at 13,000×g. Discard supernatant and wash pellet with 2.5 volumes of 70% ethanol. Centrifuge

for 7.5 min at 13,000×g. Discard supernatant, air-dry pellet and resuspend in 25 μl UP water.

2. Anneal unidirectional linkers (P1: GCGGTGACCCGGAGATCTGAATTC and P2: GAATTCAGATC) by mixing 15 μmol of each primer in 250 mM Tris–HCl, pH 8, heating to 95°C for 5 min then cooling slowly (<1°C/min). Ligate 6.7 μl annealed linkers to 25 μl blunt-ended ChIP and input DNA with 5 U T4 DNA ligase in 1× T4 DNA ligase buffer for 16 h at 16°C. Precipitate linker-ligated ChIP DNA by adding 0.1 volume 3 M NaOAc and 2.5 volumes 100% ethanol. Vortex and centrifuge for 15 min at 13,000×g. Discard supernatant and wash pellet with 2.5 volumes of 70% ethanol. Air-dry pellet and resuspend in 35 μl UP water.
3. Amplify 35 μl linker-ligated ChIP DNA and 1 μl total input DNA for each planned array in 1× PCR buffer supplemented with 2 mM $MgCl_2$, 0.5 mM dNTP mix, 1 μM linker primer P1 and 5 U Taq polymerase. Using the PCR programme 55°C 2 min, 72°C 5 min, 95°C 2 min [95° 30 s, 55°C 30 s, 72°C 1min], repeat 15 times (*see* **Note 9**) and 72°C 2 min.
4. Dilute the 50 μl of LMP-15-amplified ChIP and input DNA 1-in-10 with UP water, take 10 μl for a further round of amplification in 1× PCR buffer supplemented with 2 mM $MgCl_2$, 0.5 mM dNTP mix, 1 μM linker primer P1 and 5 U Taq polymerase. Using the PCR programme 55°C 2 min, 72°C 5 min, 95°C 2 min [95°C 30 s, 55°C 30 s, 72°C 1min], repeat 25 times (*see* **Note 9**) and 72°C 2 min.
5. Visualise amplification by loading 0.1 volume of the PCR reaction on a 1.5% gel. Successful amplification results in a smear from 200 to 600 bp (i.e. slightly shorter fragments than the smear before amplification). Clean up amplified ChIP and input material using a Qiagen PCR purification kit and quantify DNA yields using a Nanodrop spectrophotometer.
6. Label 1 μg of amplified ChIP DNA with Cy5-dUTP and 1 μg of amplified total input DNA with Cy3-dUTP using the BioPrime array CGH labelling kit, according to the manufacturer's instructions (*see* **Note 10**). Purify labelled samples using Sephadex G-50 columns and measure Cy dye incorporation using a Nanodrop spectrophotometer.
7. Hybridise 5 μg of Cy5-labelled ChIP DNA together with 5 μg of Cy3-labelled total input DNA. Hybridisation and array washing conditions vary with the type of array used, the array manufacturer and the hybridisation apparatus used. Refer to the microarray or hybridisation platform manufacturer's guidelines for detailed protocols. For example, for

hybridisation to Nimblegen oligo arrays use a Nimblegen Hybridisation System with NimbleChip Hybridisation Mixers at 42°C for 16–20 h in Nimblegen Hybridisation Buffer. Wash Nimblegen arrays at room temperature with vigorous agitation in Nimblegen Wash Buffer I supplemented with 100 μM DTT for 2 min, transfer slides to Nimblegen Wash Buffer II supplemented with 100 μM DTT for 1 min and finally Nimblegen Wash Buffer III supplemented with 100 μM DTT for 15 s. Dry Nimblegen arrays in a Microarray High-Speed Centrifuge.

8. Scan arrays using a microarray scanner (e.g. GenePix 4000 scanner) and extract numeric data from scanned images using image analysis software (e.g. NimbleScan). It is possible to perform preliminary data analysis using image analysis software; however detailed downstream analysis of ChIP-chip data represents a significant bioinformatics challenge. Several analysis methodologies for ChIP-chip data have been published (16–19) (*see* **Note 6**) and detailed workflows for the analysis of ChIP-chip data have been published (20).

3.5.3. Direct Sequencing

The combination of sequencing-based approaches with ChIP circumvents many of the problems associated with ChIP-chip (e.g. probe design, probe specificity, genome coverage and bias introduced by amplification). Comparison of ChIP-seq and ChIP-chip for the STAT1 transcription factor revealed a 64–71% overlap between the binding sites identified by both techniques, although ChIP-seq found 3.8-fold more binding sites in total suggesting that it is the more sensitive method (21). However, this relatively new technology is not without its problems (outlined above and in **Table 15.1**), not least the larger number of cells required and the poor concordance between different analysis packages. The emergence of third-generation single molecule sequencing platforms such as the Heliscope system (Helicos) and the promise of Nanopore technology (22) suggest that in the near future the amount of starting material required will be greatly reduced and post-ChIP amplification steps may no longer be required (**Table 15.1**). However, for most applications where starting material is not severely limited (e.g. larger tissue samples or cell lines), the ever-increasing capacity and falling cost of second-generation sequencing technologies suggests that they will remain the platform of choice for ChIP-seq studies. The method outlined below describes the ChIP-seq library preparation for the second-generation Solexa (Illumina) sequencing platform.

1. Blunt DNA fragments from 50 μl of the ChIP material and 50 ng of total genomic input control using 15 U T4 DNA polymerase, 5 U Klenow DNA polymerase and 50 U T4 polynucleotide kinase in 1× T4 DNA ligase buffer. Incubate

for 30 min at 20°C. Clean up DNA using the Zymogen DNA Clean and Concentrate-5 kit, eluting in 32 μl of preheated EB.

2. Add A-overhangs to 32 μl of blunted ChIP and input DNA samples using 15 U Klenow 3′→5′ exo-minus, with 200 μM dATP in Klenow buffer. Incubate at 37°C for 30 min. Clean up DNA using the Zymogen DNA Clean and Concentrate-5 kit, eluting in 8 μl of preheated EB.
3. Ligate Illumina adapters to 8 μl of ChIP and input DNA using 2 μl of adapters and 2.5 μl of Quick T4 DNA ligase (NEB) in 1× Quick DNA ligase buffer. Incubate at room temperature for 15 min. Clean up DNA using the Zymogen DNA Clean and Concentrate-5 kit, eluting in 23 μl of preheated EB.
4. Amplify and enrich adapter-ligated DNA fragments by PCR using 23 μl of ChIP and input DNA, 1 μl Illumina PCR primer 1.1, 1 μl Illumina PCR primer 2.1, 0.5 μl Phusion DNA polymerase, 1× high-fidelity PCR buffer and 200 μM dNTP. Amplify using the following conditions: 98°C 30 s [98°C 10 s, 65°C 30 s, 72°C 30 s], repeat 17 times, 72°C 5 min. Clean up amplified material using a Qiagen PCR purification kit, eluting DNA in 30 μl of preheated EB.
5. Size select ChIP and total genomic input material using agarose gel electrophoresis. Pour a 2% TAE agarose gel with 1× SYBR Safe DNA stain. Load DNA ladder, ChIP and input samples using only glycerol (12% final glycerol concentration). Use specific electrophoresis equipment for library preparation and run only one library per gel. Run gel at 120 V for 45 min, visualise on a Dark Reader transilluminator and excise the 200–300 bp part of the DNA smear. Purify the DNA using a Qiagen MiniElute Gel Extraction Kit, eluting in 15 μl of preheated EB (*see* **Note 11**).
6. Measure DNA concentration on an Agilent Bioanalyser chip and proceed with sequencing on an Illumina Genome Analyser.
7. Base calling from raw image files, quality control of sequence reads, alignment of short sequence reads to the reference genome, removal of exact duplicate reads and peak calling require the implementation of an Illumina bioinformatics pipeline in combination with other bioinformatics programmes (e.g. MAQ and MACS), requiring significant bioinformatics support (*see* **Note 12**). With regard to the analysis of filtered and aligned data, many ChIP-seq peak calling packages have been directly compared (10) and careful implementation of the best performing analysis packages in these comparisons (i.e. MACS 1.3.5, USeq, Partek,

SWEMBL or BPC) alone or in combination should provide the most reliable peak calls. These analysis tools require specialist bioinformatics support and so simplified web-based analysis tools offer a useful alternative to wet lab scientists (e.g. Cistrome or Sole Search) (23, 24), although it is likely that these tools may be most useful for first pass analysis.

4. Notes

1. Cross-linking of protein–DNA interactions is commonly used when studying transcription factor binding. Formaldehyde is most widely used for this purpose and will produce covalent cross-links between amino or imino groups which are within 2 Å from each other (**Fig. 15.1b**). It is also possible to use other cross-linking agents such as imidoesters or NHD esters (e.g. DMP [Dimethyl pimelimidate] or DSG [disuccinimidyl glutarate]) in combination with formaldehyde to increase the efficiency of cross-linking, which may be most applicable to low-abundance DNA binding proteins (25). The use of imidoesters or NHS esters as cross-linkers also provides an opportunity to alter the resolution from 2 to 20 Å, depending on the spacer length of the ester used (25). It is possible to perform ChIP without cross-linking (i.e. native ChIP); however, this is only suitable for proteins which bind stably to DNA and is mainly used in ChIP assays for histones (26).
2. Sodium dodecyl sulphate (SDS) is liable to precipitate and can cause foaming when used with probe-based sonicators. Therefore, *N*-lauroyl sarcosine can be substituted with SDS to avoid these problems. However, in our experience when using a waterbath sonicator SDS or *N*-lauroyl sarcosine works equally well.
3. It is essential to optimise sonication conditions for each cell type and sonicator, since this step defines the resolution of ChIP. It is advisable to test a range of conditions including length of pulse, number of pulses and amplitude of sonication. The efficiency of sonication should be assessed by resolving a sample of total chromatin after sonication and decross-linking using agarose gel electrophoresis (*see* **Fig. 15.1c**).
4. Dynal magnetic protein-A/G beads have lower non-specific DNA binding, reducing background and increasing specific enrichment compared to either agarose or sepharose beads. In general protein-A or protein-G beads

may be used for ChIP using rabbit antibodies and protein-G may be used for ChIP using mouse, sheep and goat IgG1 antibodies. However, we use an equal mixture of protein-A and protein-G beads to allow comparison of ChIP using antibodies from different species.

5. A number of research groups and companies provide searchable databases and compendia of validated ChIP-grade antibodies (27–33).
6. Many software packages have been published for the analysis of ChIP-chip data. Many of these packages are platform-specific analysis tools and large studies have been undertaken to compare the performance of these different analysis tools (15) and to provide detailed workflows for ChIP-chip analysis (20).
7. Quantitation of adapter-ligated ChIP DNA in sequencing libraries is an essential step to determine the loading on sequencing flow cells. It is common to use the Agilent Bioanalyser system to quantify ChIP-seq libraries following limited PCR enrichment of adapter-ligated DNA fragments. However, quantitative PCR-based methods have also been described, which utilise the adapter sequences to allow accurate quantitation of only DNA fragments which are attached to sequencing adapters (34).
8. There are many control ChIP experiments which can be used as a reference to assess specific enrichment in the test ChIP. The choice of which control to use will depend on the system under investigation and the question to be addressed. Many studies use an IgG ChIP control to assess non-specific enrichment caused by protein–DNA complexes binding to beads or IgG. However, this control does not account for any 'off-target' binding of the specific antibody used for ChIP. In order to assess the specific enrichment by a ChIP antibody it is necessary to compare isogenic cells in which the protein of interest is not bound to DNA or alternatively lacks the target protein completely. In the case of NHRs it is possible to compare hormone-deprived cells to cells stimulated with the specific NHR ligand resulting in nuclear translocation and DNA binding (e.g. androgen treatment to activate the AR). Where possible the best controls for ChIP may be isogenic cells which are null for the target protein (e.g. have targeted deletions of the gene encoding the target protein) or alternatively RNAi 'knockdown' of the target protein.
9. When using LMP to amplify ChIP material fewer rounds of amplification should introduce less bias; therefore, it is recommended that a range of PCR cycles (e.g. 28–35 cycles)

are tested. The lowest cycle number to yield the required amount of DNA should be used for downstream applications. Although it is possible to perform multiple rounds of LMP on the same sample it should be noted that this may introduce more bias into the ChIP-enriched DNA.

10. Several methods can be employed to label ChIP material for hybridisation on microarrays. These include direct incorporation as described above (e.g. Bioprime kit), indirect labelling by aa-dUTP incorporation during LMP amplification and subsequent coupling with monoreactive ester Cy dyes or finally by using Cy dye labelled random 7/9mer oligonucleotide primers.

11. A recent methods paper has comprehensively described improvements to the preparation of Illumina sequencing libraries (34). This included a suggestion that heating dsDNA during library preparation may have an impact on GC bias in the resultant library. Therefore, the increased yields afforded by preheating elution buffers in DNA purification columns and gel extraction kits may be counterbalanced by the cost of introducing experimental noise into this sensitive system, with implications for the fidelity of ChIP-seq libraries.

12. A recent review discussed the characteristics of ChIP-seq analysis software packages (35) and many of the currently available analysis tools were compared side-by-side in a ChIP-seq community challenge (10) The top scoring analysis tools were MACS 1.3.5, USeq, Partek, SWEMBL and BPC, which when used alone or in combination should provide the most reliable peak calls. However, the published comparisons may not accurately model all types of ChIP-seq data sets and it remains possible that some tools may perform better than others for certain ChIP-seq profiles.

Acknowledgements

C.E.M. is a postdoctoral researcher funded by a Cancer Research UK Programme Grant and I.G.M is a CRUK core-funded associate scientist. The authors are grateful and thank Samantha Cheung for help with illustrations and use of the ChemDraw application.

References

1. Cheung, H. H., Lee, T. L., Davis, A. J., Taft, D. H., Rennert, O. M., and Chan, W. Y. (2010) Genome-wide DNA methylation profiling reveals novel epigenetically regulated genes and non-coding RNAs in human testicular cancer, *Br J Cancer* ***102***, 419–427.
2. Barski, A., Cuddapah, S., Cui, K., Roh, T. Y., Schones, D. E., Wang, Z., Wei, G., Chepelev, I., and Zhao, K. (2007) High-resolution profiling of histone methylations in the human genome, *Cell* ***129***, 823–837.
3. Carroll, J. S., Meyer, C. A., Song, J., Li, W., Geistlinger, T. R., Eeckhoute, J., Brodsky, A. S., Keeton, E. K., Fertuck, K. C., Hall, G. F., Wang, Q., Bekiranov, S., Sementchenko, V., Fox, E. A., Silver, P. A., Gingeras, T. R., Liu, X. S., and Brown, M. (2006) Genome-wide analysis of estrogen receptor binding sites, *Nat Genet* ***38***, 1289–1297.
4. Massie, C. E., Adryan, B., Barbosa-Morais, N. L., Lynch, A. G., Tran, M. G., Neal, D. E., and Mills, I. G. (2007) New androgen receptor genomic targets show an interaction with the ETS1 transcription factor, *EMBO Rep* ***8***, 871–878.
5. Wang, Q., Li, W., Zhang, Y., Yuan, X., Xu, K., Yu, J., Chen, Z., Beroukhim, R., Wang, H., Lupien, M., Wu, T., Regan, M. M., Meyer, C. A., Carroll, J. S., Manrai, A. K., Janne, O. A., Balk, S. P., Mehra, R., Han, B., Chinnaiyan, A. M., Rubin, M. A., True, L., Fiorentino, M., Fiore, C., Loda, M., Kantoff, P. W., Liu, X. S., and Brown, M. (2009) Androgen receptor regulates a distinct transcription program in androgen-independent prostate cancer, *Cell* ***138***, 245–256.
6. Bolton, E. C., So, A. Y., Chaivorapol, C., Haqq, C. M., Li, H., and Yamamoto, K. R. (2007) Cell- and gene-specific regulation of primary target genes by the androgen receptor, *Genes Dev* ***21***, 2005–2017.
7. Dahl, J. A., Reiner, A. H., and Collas, P. (2009) Fast genomic muChIP-chip from 1,000 cells, *Genome Biol* ***10***, R13.
8. Acevedo, L. G., Iniguez, A. L., Holster, H. L., Zhang, X., Green, R., and Farnham, P. J. (2007) Genome-scale ChIP-chip analysis using 10,000 human cells, *BioTechniques* ***43***, 791–797.
9. Johnson, D. S., Mortazavi, A., Myers, R. M., and Wold, B. (2007) Genome-wide mapping of in vivo protein-DNA interactions, *Science* ***316***, 1497–1502.
10. http://sourceforge.net/projects/useq/files/CommunityChIPSeqChallenge/. ChIP-seq community challenge.
11. Schmidt, D., Wilson, M. D., Spyrou, C., Brown, G. D., Hadfield, J., and Odom, D. T. (2009) ChIP-seq: using high-throughput sequencing to discover protein-DNA interactions, *Methods (San Diego, Calif)* ***48***, 240–248.
12. Takayama, K., Kaneshiro, K., Tsutsumi, S., Horie-Inoue, K., Ikeda, K., Urano, T., Ijichi, N., Ouchi, Y., Shirahige, K., Aburatani, H., and Inoue, S. (2007) Identification of novel androgen response genes in prostate cancer cells by coupling chromatin immunoprecipitation and genomic microarray analysis, *Oncogene* ***26***, 4453–4463.
13. Wang, Q., Li, W., Liu, X. S., Carroll, J. S., Janne, O. A., Keeton, E. K., Chinnaiyan, A. M., Pienta, K. J., and Brown, M. (2007) A hierarchical network of transcription factors governs androgen receptor-dependent prostate cancer growth, *Mol Cell* ***27***, 380–392.
14. Ponzielli, R., Boutros, P. C., Katz, S., Stojanova, A., Hanley, A. P., Khosravi, F., Bros, C., Jurisica, I., and Penn, L. Z. (2008) Optimization of experimental design parameters for high-throughput chromatin immunoprecipitation studies, *Nucleic Acids Res* ***36***, e144.
15. Johnson, D. S., Li, W., Gordon, D. B., Bhattacharjee, A., Curry, B., Ghosh, J., Brizuela, L., Carroll, J. S., Brown, M., Flicek, P., Koch, C. M., Dunham, I., Bieda, M., Xu, X., Farnham, P. J., Kapranov, P., Nix, D. A., Gingeras, T. R., Zhang, X., Holster, H., Jiang, N., Green, R. D., Song, J. S., McCuine, S. A., Anton, E., Nguyen, L., Trinklein, N. D., Ye, Z., Ching, K., Hawkins, D., Ren, B., Scacheri, P. C., Rozowsky, J., Karpikov, A., Euskirchen, G., Weissman, S., Gerstein, M., Snyder, M., Yang, A., Moqtaderi, Z., Hirsch, H., Shulha, H. P., Fu, Y., Weng, Z., Struhl, K., Myers, R. M., Lieb, J. D., and Liu, X. S. (2008) Systematic evaluation of variability in ChIP-chip experiments using predefined DNA targets, *Genome Res* ***18***, 393–403.
16. Ji, X., Li, W., Song, J., Wei, L., and Liu, X. S. (2006) CEAS: cis-regulatory element annotation system, *Nucleic Acids Res* ***34***, W551–W554.
17. Peng, S., Alekseyenko, A. A., Larschan, E., Kuroda, M. I., and Park, P. J. (2007) Normalization and experimental design for ChIP-chip data, *BMC Bioinformatics* ***8***, 219.
18. Scacheri, P. C., Crawford, G. E., and Davis, S. (2006) Statistics for ChIP-chip and DNase hypersensitivity experiments on NimbleGen arrays, *Methods Enzymol* ***411***, 270–282.

19. Toedling, J., Sklyar, O., and Huber, W. (2007) Ringo–an R/Bioconductor package for analyzing ChIP-chip readouts, *BMC Bioinformatics* *8*, 221.
20. Toedling, J., and Huber, W. (2008) Analyzing ChIP-chip data using bioconductor, *PLoS Comput Biol* *4*, e1000227.
21. Robertson, G., Hirst, M., Bainbridge, M., Bilenky, M., Zhao, Y., Zeng, T., Euskirchen, G., Bernier, B., Varhol, R., Delaney, A., Thiessen, N., Griffith, O. L., He, A., Marra, M., Snyder, M., and Jones, S. (2007) Genome-wide profiles of STAT1 DNA association using chromatin immunoprecipitation and massively parallel sequencing, *Nat Methods* *4*, 651–657.
22. http://www.nanoporetech.com/. Nanopore.
23. http://cistrome.dfci.harvard.edu/ap/. Cistrome.
24. http://chipseq.genomecenter.ucdavis.edu/cgi-bin/chipseq.cgi. Sole Search.
25. Nowak, D. E., Tian, B., and Brasier, A. R. (2005) Two-step cross-linking method for identification of NF-kappaB gene network by chromatin immunoprecipitation, *BioTechniques* *39*, 715–725.
26. West, A. G., Huang, S., Gaszner, M., Litt, M. D., and Felsenfeld, G. (2004) Recruitment of histone modifications by USF proteins at a vertebrate barrier element, *Mol Cell* *16*, 453–463.
27. http://www.chiponchip.org/Antibody/chip.html. Compendium of ChIP grade antibodies.
28. http://www.abcam.com/index.html?c=917. Abcam ChIP grade antibodies.
29. http://www.diagenode.com/en/topics/antibodies/antibodies.php. Diagenode ChIP antibodies.
30. http://www.cellsignal.com/technologies/chip.html. Cell Signalling ChIP antibodies.
31. http://www.activemotif.com/catalog/18/chip-validated-antibodies.html. Active motif ChIP antibodies.
32. http://www.millipore.com/microsites/search.do?q=&filterProductTypes=taxonomy%3a%5e73UUAR%2f73UUB8&module=antibody&tabValue=ANTIBODY&filter=61964%3a%5e%22Epigenetics+%26+Nuclear+Function%22%24&filter=61965%3a%5e%22Chromatin+Biology%22%24&filter=60679%3a%5e%22Chromatin+Immunoprecipitation+(ChIP)%22%24show=10#0:0. Millipore ChIP antibodies.
33. http://www.invitrogen.com/site/us/en/home/Products-and-Services/Applications/RNAi-Epigenetics-and-Gene-Regulation/Chromatin-Remodeling/Chromatin-Immunoprecipitation-ChIP/antibodies-for-chip.html. Invitrogen ChIP antibodies.
34. Quail, M. A., Kozarewa, I., Smith, F., Scally, A., Stephens, P. J., Durbin, R., Swerdlow, H., and Turner, D. J. (2008) A large genome center's improvements to the Illumina sequencing system, *Nat Methods* *5*, 1005–1010.
35. Pepke, S., Wold, B., and Mortazavi, A. (2009) Computation for ChIP-seq and RNA-seq studies, *Nat Methods* *6*, S22–S32.

Chapter 16

Tissue-Specific Knockout of Androgen Receptor in Mice

Tzu-hua Lin, Shuyuan Yeh, and Chawnshang Chang

Abstract

Androgen acting through the androgen receptor (AR) is known to be essential for male sexual differentiation and development. Using *Cre*-lox technology, we have generated the floxed AR mice, which have been bred with general or tissue-specific *Cre* expressing transgenic mice to knock out the AR gene in specific target cells. Our findings indicated that AR is required for sexual development and that loss of AR can have significant effects on many aspects of physiological functions and disease progression, such as immune function, metabolism, and tumorigenesis. Furthermore, our strategy can generate AR knockout (ARKO) in female mice, which allows researchers to study the AR function in the female. In brief, our floxed AR mouse model provides a powerful tool to study in vivo AR functions in selective tissues and cell types and has made possible several research breakthroughs in the field of endocrinology.

Key words: Androgen receptor, *Cre*-lox system, transgenic mice, androgen receptor knockout mice, tissue-specific knockdown.

1. Introduction

The androgen receptor (AR), a member of the nuclear receptor superfamily, was first cloned in 1988 (1). It contains an N-terminal transactivation domain, a central DNA-binding domain, and a C-terminal ligand-binding domain. Once the androgen binds to the ligand-binding domain, AR could undergo a conformational change to be released from the heat shock protein, form a dimer, and then be transported into the nucleus (2). AR functions as a transcription factor by binding to its response element through the DNA-binding domain encoded by exons 2 and 3 of the AR gene. The androgen/AR signaling is well known to be important in male development and aged-related diseases. Recent findings also indicate that it may play essential

F. Saatcioglu (ed.), *Androgen Action*, Methods in Molecular Biology 776,
DOI 10.1007/978-1-61779-243-4_16, © Springer Science+Business Media, LLC 2011

roles in female physiological processes, including the folliculogenesis (3), bone metabolism (4), and several female cancers (5–7). In order to study the functions and detailed mechanisms of AR in vivo, we have generated the AR knockout (ARKO) mice through use of the *Cre*-lox technology

To specifically knock out the AR in the desired cells, our floxed AR mice have been bred with different tissue-specific *Cre* mice. Since the AR gene is located on the X-chromosome and AR is critical for male fertility, it has been difficult to produce female ARKO mice using the conventional gene knockout strategy. But, by using our powerful in vivo model (8, 9), we were able to generate the female ARKO mice and study the AR roles in mammary gland development and cancer progression (10, 11). To date, we have clarified specific AR roles in male fertility (12–15), prostate development (9, 16), metabolism and diabetes (17–19), immune functions (20–23), bone metabolism (24), and several cancers (25–28). Our findings (see **Table 16.1** for a summary of ARKO mice studies from the Chang Lab) have indicated the great potential of AR-targeting therapy in future clinical application.

Table 16.1
Summary of general or tissue-specific ARKO studies using the AR exon2 floxed mice (Chang Lab)

Name	*Cre* mice	Target cells	Phenotypes
Prostate development			
G-ARKO	ACTB-Cre (FVB)	All	No prostate developed (9)
pes-ARKO	Pb-Cre (C57-B6)	Prostate epithelial cells	Increased prostate cell proliferation and loss of cell differentiation (16)
FSP-ARKO	FSP1-Cre (C57-B6)	Fibroblast	Partial loss of differentiation in ventral prostate[c]
Tgln-ARKO	Tgln-Cre (C57-B6)	Smooth muscle	Abnormal development with loss of folding structure in anterior prostate (36)
d-ARKO	FSP1 and Tgln-Cre	Fibroblast/ smooth muscle	Impaired folding structure and branching morphogenesis and decreased number in basal and luminal epithelial cells in anterior prostate[a]
K-ARKO	K5-Cre (FVB)	Epithelial (basal) CK5 cells	Partial detachment of epithelial cells from basement membrane and increased basal cell proliferation before puberty[b]
Male fertility			
G-ARKO	ACTB-Cre (FVB)	All	Female like appearance with shrinkage of testis and sex accessory organs, lower androgen level, and arrested spermatogenesis (9)
Tgln-ARKO (PM-ARKO)	Tgln-Cre (C57-B6)	Smooth muscle (peri-tubular myoid cells)	Normal fertility but has smaller testis and oligozoospermia in epididymis (15)

Table 16.1 (continued)

Name	*Cre* mice	Target cells	Phenotypes
S-ARKO	AMH-Cre (C57-B6)	Sertoli cells	Infertility with defective spermatogenesis and Hypotestosteronemia (12–14)
Ley-ARKO	AMHRII-Cre (C57-B6/ 129seve)	Leydig cells and partial in sertoli cells	Infertility with defective spermatogenesis and Hypotestosteronemia (13)
Germ-ARKO	Sycp-Cre (C57-B6)	Germ cells	Normal fertility and sperm count (13)
A-ARKO	aP2-Cre (C57-B6)	Adipose tissue	Sub-fertility with reduced testis size and sperm number[b]
E-ARKO	GPX5-Cre (C57-B6)	Epididymus	Normal fertility
Diabetes and metabolism			
G-ARKO	ACTB-Cre (FVB)	All	Reduced body weight in young mice but gain obesity when aged (9, 17) Insulin and leptin resistance with hyperleptinemia at advanced age (17)
A-ARKO	aP2-Cre (C57-B6)	Adipose tissue	Hyperleptinemia, hypotriglyceridemia, and hypocholesterolemia. No leptin resistance and obesity (19)
L-ARKO (H-ARKO)	Alb-Cre (C57-B6)	Hepatocyte	Increased hepatic steatosis and insulin resistance in H-ARKO male mice when fed with high fat diet (18)
N-ARKO	Synapsin I-Cre (C57-B6)	Neuron	Insulin resistance and hyperleptinemia, increased hepatic steatosis[a]
HP-ARKO	Insulin-Cre (C57-B6)	Hypothalamic neuron Pancreas β-cell	Mild insulin resistance and hyperleptinemia[a]
Immune function and wound healing			
G-ARKO	ACTB-Cre (FVB)	All	Neutropenia and susceptible to acute bacterial infection (21) Accelerated cutaneous wound healing (22) Increased immature B cell number in peripheral blood and bone marrow (20) Increased thymus size and thymocyte number[a]
M-ARKO	Lyz-Cre (C57-B6)	Myeloid cells	Acceleration of cutaneous wound healing by impaired local TNF-alpha production (22)
B-ARKO	CD19-Cre (C57-B6)	B lymphocyte	Increased immature B cell number in peripheral blood and bone marrow (20) Apoptosis resistance and increased proliferation of B cell precursor in bone marrow (20)
T-ARKO	Lck-Cre (C57-B6)	T lymphocyte	No effect on thymus size and thymocyte numbers[a]

Table 16.1 (continued)

Name	*Cre* mice	Target cells	Phenotypes
K-ARKO	K5-Cre (FVB)	(basal) Epithelial cells	Reduced re-epithelialization after cutaneous wound injury (22) Increased thymus size and thymocyte number[a]
FSP-ARKO	FSP1-Cre (C57-B6)	Fibroblast	Enhanced re-epithelialization after cutaneous wound injury (22) No effect on thymus size and thymocyte number[a]
Cardiovescular disease			
G-ARKO	ACTB-Cre (FVB)	All	Research on mesenchymal stem cell transplantation on myocardial infarction model in progress
G-ARKO	E2A-Cre (C57-B6)	All	Research on arthrosclerosis and abdominal aorta aneurysm (AAA) in progress
M-ARKO	Lyz-Cre (C57-B6)	Myeloid cells	Research on arthrosclerosis and AAA in progress
Tgln-ARKO	Tgln-Cre (C57-B6)	Smooth muscle	Research on arthrosclerosis and AAA in progress
Prostate cancer			
Pes-ARKO	Pb-Cre (C57-B6)	Prostate epithelial cells	Developed larger and more invasive prostate cancer in transgenic mice model (TRAMP) (26, 27)
Ind-ARKO	Mx1-Cre (C57-B6)	All (induced by polyI C)	Reduced tumorigenesis at early stage but no effect at late stage in TRAMP model (27)
Bladder cancer			
G-ARKO	ACTB-Cre (FVB)	All	Reduced cancer incidence in BBN-induced bladder cancer model (25)
UP-ARKO	UPII-Cre (FVB)		Reduced cancer incidence in SV40 T antigen or BBN-induced bladder cancer model[b]
Liver cancer			
G-ARKO	ACTB-Cre (FVB)	All	Reduced cancer incidence in DEN-induced liver cancer model (24)
L-ARKO	Albumin-Cre (C57-B6)	Hepatocyte	Reduced cancer incidence but enhanced tumor progression in DEN-induced liver cancer model Reduced cancer incidence in HBV transgenic liver cancer model (28)
Others			
G-ARKO	ACTB-Cre (FVB)	All	Abnormal mammary gland development and growth retardation (10) Sub-fertility and defective folliculogenesis (11) Reduced mineralization by diminished osteoblast activity (23)
Ind-ARKO	Mx1-Cre (C57-B6)	All (induced by polyI C)	Induced bone loss after peak bone mass formation[a]

[a] Paper submitted
[b] Chang et al. (in preparation)
[c] Accepted in 2011

2. Materials

2.1. Construction of Targeting Vectors

1. ES129/SVJ bacteriophage library (Stratagene, Santa Clara, CA) is kept at 4°C for short-term storage (<2 months) or aliquoted into small amounts and kept in –80°C for long-term storage (>12 months).
2. *Escherichia coli* (*E. coli*) is stored at –80°C with glycerol for long-term storage.
3. Luria-Bertani (LB) plate with ampicillin is stored at 4°C.
4. Nitrocellulose membrane (Invitrogen, Carlsbad, CA) is stored at room temperature.
5. AR exon 2 DNA probe labeled with α^{32}P-dCTP is generated by PCR. Stored at –20°C and handled carefully to avoid radiation exposure and within 2–3 weeks to avoid decay.
6. High-fidelity DNA polymerase and reaction buffer are stored at –20°C and used in the indicated buffer conditions.
7. DNA recovery system (for example, the Qiaquick gel extraction kit, Qiagen, Valencia, CA) is stored at room temperature.
8. Klenow fragment (New England Biolabs, Ipswich, MA) is stored at –20°C. The restriction enzyme buffer number 1–4 supplied by New England Biolabs can all be used with Klenow.
9. PKI vector is modified from pBluescript vector (29) and stored at 4°C.
10. *Xho*I, *Kpn*I, *and Not*I restriction endonucleases (New England Biolabs, Ipswich, MA) are stored at –20°C and used with the recommended reaction buffer and bovine serum albumin (BSA) conditions as indicated by the manufacturer.
11. T4 DNA ligase and the 10X ligation buffer (New England Biolabs, Ipswich, MA) are stored at –20°C.

2.2. Generation of Chimera Founder Mice

1. ES129/SEVE cell line (Stratagene, Santa Clara, CA).
2. STO cell lines (CRL-1503, ATCC).
3. Dulbecco's modified Eagle's medium (DMEM, Gibco/Invitrogen, Carlsbad, CA) supplied with 15% embryonic specific-fetal bovine serum (ES-FBS, Gibco/Invitrogen, Carlsbad, CA) for ES129 cells or 10% FBS for STO cells. The other supplies include 3.7 g/l sodium bicarbonate ($NaHCO_3$, Sigma-Aldrich, St. Louis, MO),

4 mM L-glutamine (Gibco/Invitrogen, Carlsbad, CA), 4.5 g/l Glucose (Sigma-Aldrich, St. Louis MO), 0.1 mM beta-mercaptoethanol (Sigma-Aldrich, St. Louis MO), 1% penicillin/streptomycin stock for cell culture (Gibco/Invitrogen, Carlsbad, CA), 1% modified Eagle's medium – non-essential amino acid mix (MEM-NEAA, Gibco/Invitrogen, Carlsbad, CA), and 1000 U/ml of leukemia inhibition factor (LIF, Sigma-Aldrich, St. Louis MO). All reagents are combined and filtered through 0.2 μm filter units.

4. 0.1% Gelatin (Sigma-Aldrich, St. Louis MO) in sterile ddH_2O is stored at 4°C.
5. Mitomycin C (Sigma-Aldrich, St. Louis MO) 0.5 mg/ml in PBS is stored at 4°C.
6. Gene pulsar II system (Bio-Rad/Life Science Research, Hercules, CA).
7. G418 (Sigma-Aldrich, St. Louis MO) stock solution at 50 mg/ml in ddH_2O and aliquoted in 1.5 ml tubes and stored at –20°C.
8. *Kpn*I restriction endonuclease (New England Biolab, Ipswich, MA), as above.
9. 0.7% agarose gel (Sigma-Aldrich, St. Louis MO) with ethidium bromide (Sigma-Aldrich, St. Louis MO) (*see* **Note 1**): to make 0.7% agarose gel mix, add 0.35 g of agarose powder with 50 ml 0.5X TBE buffer in 250 ml flask and heat in a microwave oven to dissolve the agarose. While the gel cools down to 60°C, 2.5 μl of ethidium bromide (10 mg/ml) is added and gently mixed for a final concentration of 0.5 μl/ml. The gel is then poured into the gel rack, the comb inserted, and solidified at room temperature.
10. Buffer for DNA transfer
    a. Depurination buffer: 0.25 M HCl.
    b. Denaturation buffer: 0.5 M NaOH, 1.5 M NaCl.
    c. Neutralization buffer: 0.5 M Tris pH 7.0, 3.0 M NaCl.
11. Nylon membrane (Millipore, Billerica, MA).
12. Blotting paper (for example, Whatman 3MM paper) and paper towels.
13. Buffer for southern blot hybridization.
    a. Transfer buffer (6X SSC) is diluted from 20X SSC buffer: 3 M NaCl, 0.3 M Na-citrate, pH 7.0.
    b. Hybridization buffer: 5X SSC, 3% bovine serum albumin, 0.1% *N*-lauroylsarcosine, Na-salt 0.02% SDS, 50% formamide.

c. Wash buffer 1: 2X SSC, 0.1% SDS.
d. Wash buffer 2: 0.2X SSC, 0.1% SDS.
e. 2X SSC buffer.

14. pCMV-*Cre* vector.
15. Female mice at 12–24 weeks of age.

2.3. Breeding with Tissue-Specific Cre Mice

1. Male breeder: transgenic male mice with *Cre* recombinase expression under the control of general or tissue-specific promoter. (*See* **Notes 2** and **3**).
2. Female breeder: transgenic female mice with floxed AR gene inserted in one or both alleles.
3. Ear punch.

2.4. Primer Design and Genotyping of AR Knockout Mice

1. Direct PCR tail buffer (Viagen Biotech, Los Angeles, CA) is stored at 4°C.
2. Proteinase K (Promega, Madison, WI): stored at –20°C.
3. Primer sequence: (1) floxed AR genotyping: "select": 5′-GTTGATACCTTAACCTCTGC-3′; "2–9": 5′-CCTACATGTACTGTGAGAGG-3′. (2) *Cre* recombinase: forward: 5′-GCG GTC TGG CAG TAA AAA CTA TC-3′; reverse: 5′-GTG AAA CAG CAT TGC TGT CAC TT-3′. (3) IL-2: forward: 5′-CTA GGC CAC AGA ATT GAA AGA TCT-3′; reverse: 5′-GTA GGT GGA AAT TCT AGC ATC ATC C-3′. Store at –20°C in 100 mM for stock concentration and 10 mM for use.
4. *Taq* DNA polymerase and related polymerase chain reaction (PCR) buffer (e.g., GoTaq DNA polymerase, Promega, Madison, WI). Store at –20°C.
5. Deoxynucleotide (dNTP) mixture including dATP, dTTP, dCTP, dGTP in 10 mM concentration (Fermentas, Ontario, Canada). Store at –20°C.
6. 1.2% agarose gel (Sigma-Aldrich, St. Louis MO) with ethidium bromide (Sigma-Aldrich, St. Louis MO). Prepare as described above adjusting the agarose amount.
7. 6X DNA loading dye: 25 mg bromophenol blue, 25 mg xylene xyanol, 4 g sucrose, adjust volume to 10 ml with H_2O. Store at room temperature.
8. Tris/borate/EDTA (TBE) buffer: prepared in 5X stock concentration by mixing 53 g Tris base, 27.5 g boric acid, and 20 ml 0.5 M EDTA in 1 l of water. It should be diluted to 0.5X before use. Store at room temperature.

3. Methods

The process of generating the tissue-specific ARKO in mice, including the generation of the floxed AR mice, the breeding strategy, and genotyping of the mice, is illustrated in the **Fig. 16.1**. To design the construction of the targeted gene with two loxP sites may be one of the most critical steps for the successful establishment of floxed AR mice. The exon encoding the functional domain, which is essential for its protein function, should be selected as a target to be knocked out by *Cre*-mediated recombination. To establish the floxed AR mouse, the AR DNA-binding domain (DBD), encoded by exons 2 and 3, which is required for the binding to and activation from the AR response element in the target gene promoter, is targeted. Deletion or mutation of the DBD will make AR lose its DNA-binding capacity and functions. To disrupt the AR function by deleting the exon 2, two loxP sites were designed to be inserted into the intron regions surrounding both sides of AR exon 2 (*see* **Note 4**). The *Cre* recombinase can then delete the AR exon 2 efficiently and the target cells will then lose the expression of a functional AR.

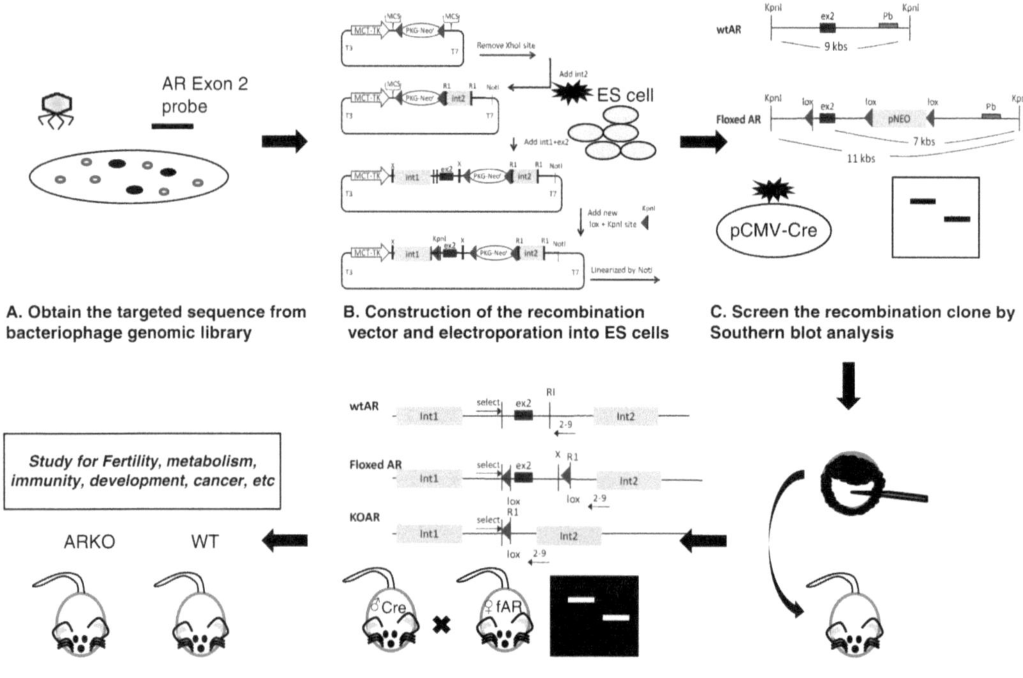

Fig. 16.1. The strategy of generating the floxed AR mice.

Since the general ARKO in male and female mice is not life threatening, it is not expected that tissue-specific ARKO in mice would be lethal. The AR gene is located on the X-chromosome, thus the loss of AR in testis, ovary, or sex accessory organs can cause infertility in male and female mice (*see* **Note 5**). Breeding the mice is a time-consuming process; thus, the optimized conditions for genotyping with suitable controls to interpret the results are essential to obtain the desired tissue-specific ARKO mice.

3.1. Construction of Targeting Vectors

1. The following instructions are based on the use of a bacteriophage genomic library to clone the AR genomic DNA for the construction of floxAR recombinant plasmid (*see* **Note 6**). First, *E. coli* infected with the bacteriophage DNA library is plated onto LB agarose plates (*see* **Note 7**). After the infected lytic bacteria plaques are clearly observed on the plate, the released bacteriophage clones are transferred onto nitrocellulose membranes and used for further analysis.
2. The membranes are then hybridized with the AR exon 2 as the probe by labeling with a α^{32}P-dCTP. After an extensive washing step to remove the excess unbound DNA probe, the membranes are imaged by exposure to X-ray film. The spots on the membrane with strong positive signals are then matched to the plate with the lytic bacteriophage plaques. The bacteriophage plaques harboring genomic DNA containing the AR exon 2 sequence are harvested and re-infected into *E. coli* (*see* **Note 8**). The candidate phage-infected bacteria are again plated, the lytic phage plaques then transferred onto a nitrocellulose membrane, and again analyzed by the AR exon 2 DNA probe to separate the specific clones with 10–25 kb of intron genomic DNA surrounding the AR exon 2.
3. The isolated bacteriophage clones are amplified, DNA is purified, and then sequenced to identify the coverage of sequences in each clone (*see* **Note 9**). The primer reading should start from the AR exon 2 at both the 5′ and the 3′ directions, and the reading should be continuous with designed primers until it reaches the sequence on the bacteriophage genome. The clones with AR genome sequence of at least 3–5 kb for recombination at either side of exon 2 are then used as templates for further PCR amplification and modification.
4. Generation of the floxed AR exon 2 recombination plasmid. The PKI is used as the vector to construct the floxed AR exon 2 with two flanking arms for the genomic recombination. PKI vector is modified from the pBluescript plasmid. It contains a T7 promoter at the 3′ end, a T3 promoter at the 5′ end, two multiple cloning sites (MCS), two loxP sites,

a positive Neo selective marker (PKG-Neo^r), and a negative thymidine kinase selective marker (MCT-TK). For the cloning, the *Xho*I site at the 5′ end MCS was first destroyed. The vector is digested by *Xho*I in the indicated buffer condition and incubated at 37°C for 1 h. The Klenow fragment and dNTP are then added into the reaction mixture to fill in the cutting site to generate a blunt end. After that, the linear form vector is recovered and then the T4 DNA ligase with its buffer is added and the reaction mixture is incubated overnight at 16°C for re-ligation of the vector.

5. A 3 kb intron 2 fragment with two *Eco*RI sites at each end is generated by Hercules high-fidelity DNA polymerase from bacteriophage genomic library, sequenced, and then introduced into the 3′ *EcoR1* cloning site (R1). Then a fragment containing intron 1, exon 2, and a small fragment of intron 2 sequences with *Xba*I site at each end is again generated by PCR and inserted into 5′ *Xba*I site (X). Both of the restriction enzymes are incubated at 37°C for 2 h. The DNA recovery and ligation steps are performed as described.
6. A loxP sequence plus an artificial *Kpn*I site (*see* **Note 10**) are engineered adjacent to exon 2 by a synthetic hanging primer and PCR filled-in. The product DNA is finally inserted into the *Xho*I site shortly 5′ to the beginning of exon 2. The constructed plasmid is linearized by *Not*I before being electroporated into ES cells.

3.2. Generation of Chimera Founder Mice

1. For ES cell culture, all the culture dishes need to be gelatinized before use. The 0.1% gelatin is applied to the culture dishes and incubated at room temperature for 20 min, then the excess gelatin is aspirated. The plates are then air dried in the hood.
2. To establish the feeder cell layer, 4 × 10^6 STO cells are seeded on 100 cm^2 dish. STO cells are inactivated by adding 0.2 ml mitomycin (0.5 mg/ml) into 10 ml of STO cell culture medium, and incubated at 37°C for 2 h. The mitomycin media are aspirated and plates rinsed by PBS twice, then 5 × 10^6 of ES 129/SEVE cells are seeded on the plates with feeder cell layers and grown at 37°C.
3. For the electroporation of floxed AR exon 2 DNA, 40 μg of the linearized DNA is suspended together with 10^9 ES cells in 1 ml of DMEM. The DNA electroporation is conducted at 300 F, 0.4 ms using the Gene Pulsar II system (Bio-Rad).
4. After the electroporation, the ES cells are mixed with 5 ml culture medium and then seeded on the plate with

feeder layer and cultured at 37°C. The neomycin-resistant colonies are selected in the presence of 300 μg/ml G418. After 1–2 weeks the remaining colonies of cells are isolated and grown separately.

5. The clones with homologous recombination are then screened by Southern blot hybridization (30) with *Kpn*I digestion (*see* **Note 11**).
6. First, the genomic DNA harvested from each clone is digested by *Kpn*I at 37°C for 1 h, then subjected to electrophoresis in a 0.7% agarose gel with 80 V for 2 h. Second, the gel with digested genomic DNA is then depurinated in 0.25 M HCl for 5 min, then immersed in denaturation buffer for 20 min twice, and then immersed in neutralization buffer for 20 min twice.
7. The nylon membrane and four pieces of blotting paper are soaked with distilled water and then immersed into 6X SSC.
8. Assemble the transfer stack in the following order: a sponge, one piece of blotting paper soaked in 6X SSC, the gel (put the wells-side down), the nylon membrane, two pieces of blotting paper, and 4 cm stack of paper towel. Wrap whole transfer stack in plastic wrap. Wet each layer with 6X SSC buffer by glass pipette and remove the trapped air during the process. Place glass plate and weight on top and leave overnight. After the transfer is completed, mark the gel position and orientation on the nylon membrane and place it on blotting paper to dry.
9. Before hybridization, denature the AR exon 2 probe for 10 min at 100°C and then put on ice. At the same time, wet a membrane with hybridization buffer for 5 min. Replace with new hybridization buffer and add the denatured DNA probe with membrane DNA side up. Incubate the blot on rotating platform at 68°C overnight.
10. After the blotting is completed, pour off the hybridization buffer with probe, add wash buffer #1, and incubate at room temperature for 5 min two times on rotating platform. Change into wash buffer #2 and again incubate at room temperature for 5 min two times on rotating platform. Remove the wash solution, rinse the membrane in 2X SSC buffer at room temperature, and blot excess liquid. The membrane is then wrapped in plastic wrap and the image obtained by autoradiograph on X-ray film.
11. After the screening, the ES cell clones with floxed AR are further amplified and re-electroporated using the conditions as described above to introduce pCMV-*Cre* vector into the cells. The transient expression of the *Cre* in the

cells results in three types of recombination because of the three loxP sites inside the inserted sequence (*see* **Note 12**).

12. The recombination types of ES cell clones are checked by Southern blot hybridization with the strategy described above (steps 6–10). The clone without recombination is 7 kb in size after the *Kpn*I digestion. The type 1 recombination is neomycin resistance gene (2 kb) knockout, resulting in 5 kb in size; type 2 recombination is knockout of the exon 2 including the artificially inserted *Kpn*I site, resulting in 11 kb in size; type 3 recombination is knockout of both the exon 2 and the neomycin resistance gene, resulting in 9 kb in size.
13. The ES cells with type 1 recombination are then injected into the inner cell mass of blastocysts, which are then implanted into the uteri of foster mothers for further development (*see* **Note 13**).
14. Four floxed AR germline chimeric male mice are then generated by microinjection of 129/jv recombinant embryonic stem cells into unsexed C57BL/6 blastocysts. To breed the pure floxed AR offsprings, each chimera male mouse is mated with six different female B6 mice. After producing 12 litters of offspring from each chimera male mouse, we successfully obtained two independent germ line transmissions from four of the chimera males.
15. These two independent floxed AR male offsprings have normal fertility, sperm count, serum hormone levels, and more than 2 years life span. They have been bred with β-actin *Cre* transgenic mice to produce the total ARKO male mice. There were no observed phenotype differences between these two floxed AR lines. We maintained two independent mouse lines. The line #1 floxed AR mouse was chosen for further breeding with different *Cre* transgenic mice.
16. To synchronize the mouse background, the line #1 floxed AR mouse are backcrossed with C57BL/6 mice at least seven generations to establish a C57BL/6 background.

3.3. Breeding with Tissue-Specific Cre *Mice*

1. To breed the tissue-specific ARKO male mice, the floxed AR/X female mice are mated with tissue-specific *Cre* expressed male mice (*see* **Note 14**).
2. The pregnant female mice are housed separately upon observation of a vaginal plug. Mice have an approximately 21 days gestation period following conception.
3. The mice are checked everyday in the late stage of pregnancy. Cages are labeled with the date of birth, the pups are weaned around 3–4 weeks of age. Following weaning, each

new pup is marked by ear punch and the tail snip (around 4 mm in length) is used for genotyping. (*see* **Note 15**).

4. To breed the tissue-specific ARKO female mice, the female mice with one allele of floxed AR transgene is used to mate with male mice carrying floxed AR and *Cre* transgene. If the tissue specific ARKO male mice are infertile, the female mice with *Cre* and one allele of floxed AR transgene should be mated with male mice carrying floxed AR transgene, as an alternative strategy.
5. The pregnant female mice are separated and checked for the pup delivery as described above. Each litter of pups should include both the female and the male ARKO mice.

3.4. Primer Design and Genotyping of AR Knockout Mice

1. Primer design: based on the sequence information obtained by sequencing the AR genomic DNA, one pair of primers are designed to distinguish the wild-type AR (wt AR), ARKO, and floxed AR in the mouse genome. The 5′ primer named "select" is located in the intron 1. The 3′ end primer is "2–9," which is located in intron 2. If the mice carry floxed AR, the PCR product size from this pair of primers is 638 bp. If the mice carry ARKO, the PCR product size from this pair of primers is 238 bp. If the mouse contains wt AR, the primers amplify a PCR product of 580 bp. The expression of *Cre* and internal control IL2 are confirmed by PCR during genotyping. The primers for *Cre* and IL2 genotyping follow The Jackson Laboratory's suggestions (*see* **Note 16**).
2. Each tail snip is immersed in 200 μl genomic DNA lysis buffer with proteinase K (50 μg/ml) and incubated overnight at 65°C. When the tail is completely digested, the sample is incubated at 85°C for 1 h to inactivate the proteinase K activity. Centrifuge the samples at 10,000 rpm for 1 min and transfer the clear supernatant into a new tube. Add additional 400 μl of distilled water to dilute before further PCR analysis.
3. Prepare the PCR reaction mixture: (1) floxed AR: 6.75 μl ddH_2O, 4 μl 5X PCR buffer (*GoTaq* green system, with DNA loading dye), 4 μl 25 mM $MgCl_2$, 1 μl 10 mM dNTP, 0.5 μl each of 10 mM primer, 0.35 μl of 5 U/μl Taq polymerase; the total volume is 20 μl. (2) *Cre* recombinase and IL-2 internal control: 3.97 μl ddH_2O, 2.4 μl 5X PCR buffer (*GoTaq* green system, with DNA loading dye), 0.96 μl 25 mM $MgCl_2$, 0.24 μl 10 mM dNTP, 0.6 μl each of 20 mM primer, 0.03 μl of 5 U/μl Taq polymerase; the total volume is 12 μl.
4. The optimized PCR conditions are (1) floxed AR: step #1, 94°C, 5 min (′); step #2, 94°C, 30 s (″), 58°C, 40″, 72°C,

1′ 20″, repeat step #2 for 35 cycles; step #3, 72°C, 7′, then hold at 10°C. (2) *Cre* recombinase and IL-2 internal control: step #1, 94°C, 3′; step #2, 94°C, 30″, 51.7°C, 1′, 72°C, 1′, repeat step #2 for 35 cycles; step #3, 72°C, 2′, then hold at 10°C.

5. The PCR products are then analyzed by electrophoresis. Put the gel into the electrophoresis tank filled with 0.5X TBE buffer, note that the gel must be fully covered by 0.5X TBE buffer. Load 10 μl of PCR products per well, and also load 0.5 μg of 1 kb plus DNA ladder (Invitrogen, Carlsbad, CA) with 6X loading dye as a standard to confirm the size of each band. Run at 100 V for 30 min.
6. After the electrophoresis is completed, remove the gel from the tank, observe the bands in a gel documentation system. In general ARKO mice, the knockout band with 238 bp in size will be observed. In tissue-specific ARKO mice, the floxed AR band with 644 bp in size and the knockout band may or may not be observed by PCR of genomic DNA from tails depending on the target tissue of *Cre* expression.

4. Notes

1. Ethidium bromide is a well-known carcinogen that can integrate into the double-stranded DNA; wear gloves and handle all equipment that may have been contaminated with caution. Reagents with lower toxicity, such as cyber green, can be used as alternatives.
2. A suitable tissue-specific *Cre* expression mouse model is very important for the experiment at the beginning. The expression specificity, intensity, and the time point that the *Cre* starts to express are all to be considered. If knockout at an early developmental stage can cause abnormal development in the target organ, it may interfere with the study of a disease model in the adult stage. In this case, the inducible *Cre* model, such as tetracycline, tamoxifen, or poly PI-PC inducible system, is an alternative approach. Many tissue-specific *Cre* expressing mice have been established and some of them are available in the research community. The investigators can find the available mice on the Mouse Genome Informatics (MGI) website (www.informatics.jax.org) for each line of available transgenic mice and corresponding principal investigators.
3. Always use the male mice with transgenic *Cre* expression. The female mice with tissue-specific *Cre* expression may

have leakage in the oocyte stage and therefore the mice become general knockout of targeted gene.

4. Carefully check that the reading frame after the floxed region is knocked out from the genome. The information about the sequence and mRNA splicing can be obtained from the web resources (for example, National Center of Biotechnology Information, www.ncbi.nlm.nih.gov). There are five floxed AR mice that have been generated, but not all of them can be used as ARKO model systems (see **Table 16.2** for summary of five floxed AR mouse models). In our case, the DNA binding domain includes exon 2 and exon 3, but only knockout of exon 2 can produce a non-in-frame transcript. Therefore only 14 additional amino acids can be found in the final protein product.
5. It should be noted that the floxed AR gene in male pups is always from the female breeder. Since there is only one X-chromosome in male mice, total loss of AR expression can be achieved by one floxed AR gene with *Cre* expression in male mice.
6. The bacteriophage genomic library or bacteria artificial chromosomes (BAC) are very useful techniques for generating of the floxed AR mice. With the advanced genomic sequence information in public resources, the researchers can also use PCR to get the template from 129/SEVE mouse genomic DNA for the construction.
7. The broth with bacteriophage is diluted at the optimized condition, so that the lytic infection of bacteria can be in a high density on each plate to reduce the sample volume for analysis.
8. Because the bacteria lytic plaques are in a high density at the first screening, more than one clone of bacteriophage will be harvested for the re-infection of *E. coli*. The re-infection step should be optimized to get the infected lytic bacteria in lower density, so that the single clone of bacteriophage can be obtained.
9. It is very important that the DNA sequence on right and left arms used for recombination is matched. Therefore the sequencing must be carefully confirmed. The sequence in the ES cell line must also be confirmed to make sure there are no differences in the right and left arm used for recombination.
10. The *Kpn*I site is chosen because there are two *Kpn*I sites found in the right and left arm sequences. The artificially inserted *Kpn*I site is then used for the identification of recombination type by Southern blot in the later steps. The

Table 16.2
Summary of five floxed AR mice models

PI and Institute	Floxed exon	Promoter of *Cre* for G-ARKO	Affected anatomical systems	Note
Chawnshang Chang (University of Rochester Medical Center, NY, USA)	Exon2	Beta-actin	Reproductive, endocrine/exocrine, homeostasis, tumorigenesis, behavior, endocrine/exocrine, hematopoietic, immune	PGK-NEO was removed from the floxed AR genome No detectable AR protein in all tissues examined by IHC (9)
Shigeaki Kato (University of Tokyo, Tokyo, Japan)	Exon1	CMV	Adipose, endocrine/exocrine, growth/size, renal/urinary, reproductive, cardiovascular, muscle	PGK-NEO retained in the floxed AR genome. CMV promoter expression is negative in several tissues including lung, liver, pancreas, and muscle (31). Therefore AR may not be generally knocked out in all the tissues (32)
Guido Verhoeven (University of Katholieke, Leuven, Belgium)	Exon2	PGK	Reproductive, growth/size, endocrine/exocrine, homeostasis, embryogenesis	PGK-NEO was removed from the floxed AR genome No detectable protein by IHC (33)
Jeffrey D Zajac (University of Melbourne, Melbourne, Australia)	Exon3	CMV	Reproductive, growth/size, homeostasis, endocrine/exocrine reproductive, homeostasis, digestive/alimentary, endocrine/exocrine	PGK-NEO retained in the floxed AR genome G-ARKO mice still express intact AR protein except second zing finger of DNA binding domain was deleted: more suitable to study the impact of loss of mice AR second zing finger domain, instead of loss of whole AR function in these ARKO mice (34)
Robert E Braun (University of Washington, Seattle, USA)	Exon1 (opposite direction of loxp)	Sycp1 and EIIa	Reproductive, endocrine/exocrine	PGK-NEO retained in the floxed AR genome *Cre*-mediated inverted exon1 instead of ARKO. Floxed AR mouse (without mating with *Cre*-mouse) has abnormal high serum T level (35)

This table was modified from the information obtained at Mouse Genome Informatics (http://www.informatics.jax.org/)

inserted site can be replaced by alternative experimental strategy to confirm the recombination results.

11. If the culture of ES cells is longer than 2 weeks, the ability to generate the mice after injection into the blastocysts might be lost. Therefore the sample used for Southern blot analysis should be at minimal amount for the detection. The probe labeled with isotope has higher sensitivity, which can help to reduce the required sample amount.
12. It is important to remove neomycin resistance gene since its existence in transgenic mice may cause unexpected side effects. However, it may not be easy to get the targeted clone due to high expression level of *Cre* recombinase controlled by CMV promoter. The FLP-FRT system is an alternative strategy to eliminate this concern.
13. This part is usually done by the core facility in the animal resource section of the investigation center. The detailed strategy may be different according to the standard operation protocols in each center.
14. Transferring the female mice into the male mice's cage can usually result in a higher breeding rate. Mice at the age around 12–24 weeks have the best breeding ability.
15. Carefully design and estimate the ratio by which you can get the knockout and control mice. For example, the heterozygous *Cre* expressing male mice and heterozygous floxed AR female mice can only have 1:8 ratio (1 out of 8 pups are the knockout genotype) for experiments.
16. Most of the tissue-specific *Cre* transgenic mice pups can be genotyped by following the same strategy. If distinguishing a *Cre* expression controlled by a special gene promoter, this can be achieved by designing the forward primer on the promoter and the reverse primer on the *Cre* gene. The internal control IL2 is required for the quality control of genomic DNA sample in case of the false negative results of genotyping.

Acknowledgments

The authors would like to thank Karen Wolf for manuscript preparation. This work was supported by RO1 CA127300 and the George H. Whipple Professorship Endowment.

References

1. Chang, C. S., Kokontis, J., and Liao, S. T. (1988) Molecular cloning of human and rat complementary DNA encoding androgen receptors, *Science* **240**, 324–326.
2. Heinlein, C. A., and Chang, C. (2002) Androgen receptor (AR) coregulators: an overview, *Endocr Rev* **23**, 175–200.
3. Donath, J., Michna, H., and Nishino, Y. (1997) The antiovulatory effect of the antiprogestin onapristone could be related to down-regulation of intraovarian progesterone (receptors), *J Steroid Biochem Mol Biol* **62**, 107–118.
4. Compston, J. E. (2001) Sex steroids and bone, *Physiol Rev* **81**, 419–447.
5. Liao, D. J., and Dickson, R. B. (2002) Roles of androgens in the development, growth, and carcinogenesis of the mammary gland, *J Steroid Biochem Mol Biol* **80**, 175–189.
6. Wang, S. W., Kim, B. S., Ding, K., Wang, H., Sun, D., Johnson, R. L., Klein, W. H., and Gan, L. (2001) Requirement for math5 in the development of retinal ganglion cells, *Genes Dev* **15**, 24–29.
7. Sasaki, M., Dahiya, R., Fujimoto, S., Ishikawa, M., and Oshimura, M. (2000) The expansion of the CAG repeat in exon 1 of the human androgen receptor gene is associated with uterine endometrial carcinoma, *Mol Carcinog* **27**, 237–244.
8. Holt, C. L., and May, G. S. (1993) A novel phage lambda replacement *Cre*-lox vector that has automatic subcloning capabilities, *Gene* **133**, 95–97.
9. Yeh, S., Tsai, M. Y., Xu, Q., Mu, X. M., Lardy, H., Huang, K. E., Lin, H., Yeh, S. D., Altuwaijri, S., Zhou, X., Xing, L., Boyce, B. F., Hung, M. C., Zhang, S., Gan, L., and Chang, C. (2002) Generation and characterization of androgen receptor knockout (ARKO) mice: an in vivo model for the study of androgen functions in selective tissues, *Proc Natl Acad Sci USA* **99**, 13498–13503.
10. Yeh, S., Hu, Y. C., Wang, P. H., Xie, C., Xu, Q., Tsai, M. Y., Dong, Z., Wang, R. S., Lee, T. H., and Chang, C. (2003) Abnormal mammary gland development and growth retardation in female mice and MCF7 breast cancer cells lacking androgen receptor, *J Exp Med* **198**, 1899–1908.
11. Hu, Y. C., Wang, P. H., Yeh, S., Wang, R. S., Xie, C., Xu, Q., Zhou, X., Chao, H. T., Tsai, M. Y., and Chang, C. (2004) Subfertility and defective folliculogenesis in female mice lacking androgen receptor, *Proc Natl Acad Sci USA* **101**, 11209–11214.
12. Chang, C., Chen, Y. T., Yeh, S. D., Xu, Q., Wang, R. S., Guillou, F., Lardy, H., and Yeh, S. (2004) Infertility with defective spermatogenesis and hypotestosteronemia in male mice lacking the androgen receptor in Sertoli cells, *Proc Natl Acad Sci USA* **101**, 6876–6881.
13. Tsai, M. Y., Yeh, S. D., Wang, R. S., Yeh, S., Zhang, C., Lin, H. Y., Tzeng, C. R., and Chang, C. (2006) Differential effects of spermatogenesis and fertility in mice lacking androgen receptor in individual testis cells, *Proc Natl Acad Sci USA* **103**, 18975–18980.
14. Wang, R. S., Yeh, S., Chen, L. M., Lin, H. Y., Zhang, C., Ni, J., Wu, C. C., di Sant'Agnese, P. A., deMesy-Bentley, K. L., Tzeng, C. R., and Chang, C. (2006) Androgen receptor in sertoli cell is essential for germ cell nursery and junctional complex formation in mouse testes, *Endocrinology* **147**, 5624–5633.
15. Zhang, C., Yeh, S., Chen, Y. T., Wu, C. C., Chuang, K. H., Lin, H. Y., Wang, R. S., Chang, Y. J., Mendis-Handagama, C., Hu, L., Lardy, H., and Chang, C. (2006) Oligozoospermia with normal fertility in male mice lacking the androgen receptor in testis peritubular myoid cells, *Proc Natl Acad Sci USA* **103**, 17718–17723.
16. Wu, C. T., Altuwaijri, S., Ricke, W. A., Huang, S. P., Yeh, S., Zhang, C., Niu, Y., Tsai, M. Y., and Chang, C. (2007) Increased prostate cell proliferation and loss of cell differentiation in mice lacking prostate epithelial androgen receptor, *Proc Natl Acad Sci USA* **104**, 12679–12684.
17. Lin, H. Y., Xu, Q., Yeh, S., Wang, R. S., Sparks, J. D., and Chang, C. (2005) Insulin and leptin resistance with hyperleptinemia in mice lacking androgen receptor, *Diabetes* **54**, 1717–1725.
18. Lin, H. Y., Yu, I. C., Wang, R. S., Chen, Y. T., Liu, N. C., Altuwaijri, S., Hsu, C. L., Ma, W. L., Jokinen, J., Sparks, J. D., Yeh, S., and Chang, C. (2008) Increased hepatic steatosis and insulin resistance in mice lacking hepatic androgen receptor, *Hepatology* **47**, 1924–1935.
19. Yu, I. C., Lin, H. Y., Liu, N. C., Wang, R. S., Sparks, J. D., Yeh, S., and Chang, C. (2008) Hyperleptinemia without obesity in male mice lacking androgen receptor in adipose tissue, *Endocrinology* **149**, 2361–2368.
20. Altuwaijri, S., Chuang, K. H., Lai, K. P., Lai, J. J., Lin, H. Y., Young, F. M., Bottaro, A.,

Tsai, M. Y., Zeng, W. P., Chang, H. C., Yeh, S., and Chang, C. (2009) Susceptibility to autoimmunity and B cell resistance to apoptosis in mice lacking androgen receptor in B cells, *Mol Endocrinol* **23**, 444–453.
21. Chuang, K. H., Altuwaijri, S., Li, G., Lai, J. J., Chu, C. Y., Lai, K. P., Lin, H. Y., Hsu, J. W., Keng, P., Wu, M. C., and Chang, C. (2009) Neutropenia with impaired host defense against microbial infection in mice lacking androgen receptor, *J Exp Med* **206**, 1181–1199.
22. Lai, J. J., Lai, K. P., Chuang, K. H., Chang, P., Yu, I. C., Lin, W. J., and Chang, C. (2009) Monocyte/macrophage androgen receptor suppresses cutaneous wound healing in mice by enhancing local TNF-alpha expression, *J Clin Invest* **119**, 3739–3751.
23. Kang, H. Y., Shyr, C. R., Huang, C. K., Tsai, M. Y., Orimo, H., Lin, P. C., Chang, C., and Huang, K. E. (2008) Altered TNSALP expression and phosphate regulation contribute to reduced mineralization in mice lacking androgen receptor, *Mol Cell Biol* **28**, 7354–7367.
24. Ma, W. L., Hsu, C. L., Wu, M. H., Wu, C. T., Wu, C. C., Lai, J. J., Jou, Y. S., Chen, C. W., Yeh, S., and Chang, C. (2008) Androgen receptor is a new potential therapeutic target for the treatment of hepatocellular carcinoma, *Gastroenterology* **135**, 947–955, e941–945.
25. Miyamoto, H., Yang, Z., Chen, Y. T., Ishiguro, H., Uemura, H., Kubota, Y., Nagashima, Y., Chang, Y. J., Hu, Y. C., Tsai, M. Y., Yeh, S., Messing, E. M., and Chang, C. (2007) Promotion of bladder cancer development and progression by androgen receptor signals, *J Natl Cancer Inst* **99**, 558–568.
26. Niu, Y., Altuwaijri, S., Lai, K. P., Wu, C. T., Ricke, W. A., Messing, E. M., Yao, J., Yeh, S., and Chang, C. (2008) Androgen receptor is a tumor suppressor and proliferator in prostate cancer, *Proc Natl Acad Sci USA* **105**, 12182–12187.
27. Niu, Y., Altuwaijri, S., Yeh, S., Lai, K. P., Yu, S., Chuang, K. H., Huang, S. P., Lardy, H., and Chang, C. (2008) Targeting the stromal androgen receptor in primary prostate tumors at earlier stages, *Proc Natl Acad Sci USA* **105**, 12188–12193.
28. Wu, M. H., Ma, W. L., Hsu, C. L., Chen, Y. L., Ou, J. H., Ryan, C. K., Hung, Y. C., Yeh, S., and Chang, C. Androgen receptor promotes hepatitis B virus-induced hepatocarcinogenesis through modulation of hepatitis B virus RNA transcription, *Sci Transl Med* **2**, 32ra35.
29. Wang, S. W., Kim, B. S., Ding, K., Wang, H., Sun, D., Johnson, R. L., Klein, W. H., and Gan, L. (2001) Requirement for math5 in the development of retinal ganglion cells, *Genes Dev* **15**, 24–29.
30. Kaczmarczyk, S. J., and Green, J. E. (2001) A single vector containing modified cre recombinase and LOX recombination sequences for inducible tissue-specific amplification of gene expression, *Nucleic Acids Res* **29**, E56–56.
31. Schmidt, E. V., Christoph, G., Zeller, R., and Leder, P. (1990) The cytomegalovirus enhancer: a pan-active control element in transgenic mice, *Mol Cell Biol* **10**, 4406–4411.
32. Matsumoto, T., Takeyama, K., Sato, T., and Kato, S. (2003) Androgen receptor functions from reverse genetic models, *J Steroid Biochem Mol Biol* **85**, 95–99.
33. De Gendt, K., Swinnen, J. V., Saunders, P. T., Schoonjans, L., Dewerchin, M., Devos, A., Tan, K., Atanassova, N., Claessens, F., Lecureuil, C., Heyns, W., Carmeliet, P., Guillou, F., Sharpe, R. M., and Verhoeven, G. (2004) A Sertoli cell-selective knockout of the androgen receptor causes spermatogenic arrest in meiosis, *Proc Natl Acad Sci U S A* **101**, 1327–1332.
34. Holdcraft, R. W., and Braun, R. E. (2004) Androgen receptor function is required in Sertoli cells for the terminal differentiation of haploid spermatids, *Development* **131**, 459–467.
35. Notini, A. J., Davey, R. A., McManus, J. F., Bate, K. L., and Zajac, J. D. (2005) Genomic actions of the androgen receptor are required for normal male sexual differentiation in a mouse model, *J Mol Endocrinol* **35**, 547–555.
36. Yu, S., Zhang, C., Lin, C. C., Niu, Y., Lai, K. P., Chang, H. C., Yeh, S. D., Chang, C., and Yeh, S. Altered prostate epithelial development and IGF-1 signal in mice lacking the androgen receptor in stromal smooth muscle cells, *Prostate* **71**, 517–524.

Chapter 17

Methodology to Investigate Androgen-Sensitive and Castration-Resistant Human Prostate Cancer Xenografts in Preclinical Setting

Holly M. Nguyen and Eva Corey

Abstract

Understanding the biology of prostate cancer and the roles of androgen receptor in prostate cancer progression is essential to the development of novel therapeutic strategies to effectively attack and eradicate this disease. Preclinical, in vivo, studies are critical to further evaluate potential clinical relevance of in vitro findings. Ideally, in vivo studies should employ models that mimic characteristics of prostate cancer from early diagnosis through the period of castration-resistant metastases. In this chapter we describe methodologies used to grow human prostate cancer xenografts in mice. In this setting, roles of androgen receptor signaling in prostate cancer progression and efficacy of novel treatment modalities, including those affecting androgen receptor signaling, can be investigated.

Key words: Prostate cancer, androgen receptor, xenograft, animal model, castration resistance, androgen-sensitive.

1. Introduction

Androgen receptor (AR) is a key modulator of growth and development of prostate and prostate cancer (PCa) progression (1–4). Androgen suppression has been the central theme in treatment of recurrent PCa since the 1940s. While initially effective, PCa progresses from androgen-sensitive (AS) tumors that respond favorably to androgen ablation to castration-resistant (CR) metastatic tumors that are invariably fatal. AR signaling is important not only in the AS state of the disease but also in CR PCal. Due to the medical significance of the transformation from AS to CR PCa, research into the mechanisms involved in this process is intensive.

F. Saatcioglu (ed.), *Androgen Action*, Methods in Molecular Biology 776,
DOI 10.1007/978-1-61779-243-4_17, © Springer Science+Business Media, LLC 2011

AR can stimulate both proliferation and differentiation, although the mechanism by which AR switches between these modes is unknown. A better understanding of AR signaling in AS and CR PCa, along with the development of more effective inhibitors of AR signaling may have a dramatic positive impact on PCa treatment strategies (2, 3, 5).

Preclinical studies are necessary in order to increase our understanding of the roles of AR in PCa progression and select the most promising agents for clinical testing. Ideally animal models should exhibit all characteristics of the human disease. However, PCa does not present itself clinically as a disease with limited, defined characteristics. It is notoriously heterogeneous in growth, biomarker expression, and responses to therapy. It is therefore critical that in vivo testing employs a variety of xenografts and cell lines that exhibit the wide range of characteristics observed in the human disease. In vivo models used should also portray the microenvironment at primary or secondary tumor sites. This is an extremely important issue because the environment alters the phenotype and behavior of tumor cells. For example, we have demonstrated that the efficacy of taxotere, a standard treatment for advanced PCa, is very different when taxotere is used on subcutaneous tumors as opposed to tumors growing in the bone (6). In other words, if one wants to investigate efficacy of new agents as treatment for bone metastases, in vivo models of experimental bone metastasis or growth of PCa cells in the bone should be used rather than subcutaneous tumors.

In summary, using relevant multiple models in preclinical testing is necessary to understand AR signaling in more detail, to determine the spectrum of therapeutic responses, including response to alterations in interactions among androgens, androgen receptor, and PCa cells in various cancer phenotypes, and to select the most promising agents for clinical testing.

2. Materials

2.1. Chemicals

1. Sterile 0.9% sodium chloride for injection
2. Ketaset (ketamine, 100 mg/mL)
3. Xyla-Ject (xylazine, 20 mg/mL)
4. 1× GIBCO Dulbecco's phosphate buffered saline, stored at 4°C
5. Gentamicin (40 mg/mL), injection, USP, stored at 20–25°C

6. Betadine Surgical Scrub, 7.5% povidone Iodine
7. Phenol Red-Free Matrigel (BD Biosciences, Franklin Lakes, NJ)
8. BrdU (5-bromo-2′-deoxy-uridine), 97%. 80 mg/kg solution is prepared in sterile saline just before injection
9. EDTA (ethylenediaminetetra-acetic acid), 10% solution prepared in sterile dH_2O
10. Clear Frozen Section Compound (OCT, VWR International, Batavia, IL)
11. Buprenex (buprenorphine 0.3 mg/mL)

2.2. Surgical Materials

1. 4/0 coated Visorb polyglyconic acid suture
2. 13 gauge cancer implant needle
3. Alcohol Prep Pads, 70% isopropyl alcohol
4. Deltaphase isothermal pad
5. Deltaphase operating board
6. Microdissecting forceps, 4″, serrated
7. Tungsten carbide iris scissors, 4.5″
8. Autoclip applier, 9 mm autoclips, and autoclip remover
9. Small vessel cauterizer
10. Sterile petri dish
11. Rib-Back carbon steel surgical blade, 22, sterile
12. Calibrated micropipet

2.3. Mice

Multiple strains of immunocompromized mice can be used for human tumor implantation/injection. In our studies we use C.B-17 SCID for subcutaneous, orthotopic, and subrenal implantation. SCID Beige mice are used for intra-tibial and cardiac injections, and Nu/Nu and IL-2 NOD SCIDs for maintenance of the colonies. Studies can be performed in nu/nu mice but, in our experience, use of C.B-17 SCID and SCID Beige mice gives better take rates and tighter data.

1. Nu/Nu (NU-*Foxn1nu)* (Charles River Laboratory, Wilmington, MA)
2. C.B-17 SCID (CB17/Icr-*Prkdcscid*/IcrCrl) C.B-17 SCID (Charles River Laboratory, Wilmington, MA)
3. SCID Beige (CB17.Cg-*Prkdcscid Lystbg*/Crl) (Charles River Laboratory, Wilmington, MA)
4. IL-2 NOD SCID (NOD.Cg-$Prkdc^{scid}$ $Il2rg^{tm1Wjl}$/SzJ) (Jackson Laboratory, Bar Harbor, ME)

3. Methods

3.1. General Procedures

All procedures are performed in certified biosafety cabinets, ASBL2/BSL2, and universal safety precautions are followed. All animal procedures are in compliance with the University of Washington's Institutional Animal Care and Use Committee and National Institutes of Health guidelines.

3.1.1. Anesthesia

1. Inject ketamine/xylazine solution (130 mg/8.8 mg/kg) intraperitoneally prior to surgery.
2. Wait approximately 3–5 min for mice to reach the surgical plane of anesthesia in which they will remain for approximately 15–20 min.

3.1.2. Preparation for Surgery

1. Sterilize all surgical instruments prior to surgeries.
2. Sterilize all instruments during surgical procedure between cages using a glass bead sterilizer.
3. Once the mouse reaches the surgical plane of anesthesia shave the surgical site, (if necessary – SCID mice), and sterilize the site with a betadine solution and 70% isopropyl alcohol by swabbing three times alternatively.
4. Place mice on a heated sterile stainless steel surgical plate in a biosafety cabinet.
5. Perform surgery (for details see below).

3.1.3. Tumor Sources

Multiple sources of tumors can be used for implantations/injections including tumor bits from tumors grown in mice, tumor bits harvested from patients or tumor cells grown in vitro. **Table 17.1** shows a list of the most common AR-expressing PCa cell lines (7–22) and **Table 17.2** shows some of the AR-expressing PCa xenografts used to study AR signaling (23–26; please note that not all of these have yet been published, as indicated).

3.1.3.1. Tumor Bits

1. Extract tumors under sterile conditions from the host and section into 20 mm slices in a sterile Petri dish using dissecting scissors or a sterile surgical blade.
2. Immediately place tissue sections in ~10 mL 1X DBPS and 20 mg of gentamicin for 5 min and then rinse with 1X DBPS and cut into 20 mg pieces (about 3–4 mm per cubed side).
3. These tissue pieces are ready for implantation and should be introduced to the host as soon as possible.

3.1.3.2. Tumor Cells Grown in Vitro

Prostate tumor cells are grown in vitro under appropriate tissue culture conditions. Media and supplements may vary depending on cell line and specific requirements for each study. Cells are

Table 17.1
Androgen-Receptor Expressing Prostate Cancer Cell Lines

Name	Origin	PSA	Response to Androgen Ablation	Reference
LNCaP	Lymph node metatasis	+	+	7
C4	LNCaP	+	–	8, 9
C4-2	C4	+	–	8, 9
C4-2B	C4-2	+	–	10
CWR22 Rvl	CWR22R	+	–	16
VCaP	Vertabrae metastasis	+	+	11
DuCaP	Dura metastasis	+	+	12
ARCaP	Ascite fluid	+	+	13
MDA PCa 2a	Bone metastasis	+	+	17
MDA PCa 2b	Bone metastasis	+	+	17
MDA PCa 2b HR	MDA Pca2b	+	–	18
TEN12	Primary tumor	+	+	19
TEN12F	TEN12	+	–	19
TEN12C	TEN12	+	–	19
BM18	Bone metastasis	+	+	20
LAPC-4	Lymph node metatasis	+	+	21
LACP-4squared	LAPC-4	+	–	22

harvested in a log phase, washed with HBSS, and injected. Cell numbers injected vary between xenografts and route of injection (see below).

3.1.4. Castration

Androgen ablation causes inhibition of prostate tumor growth and a drop in serum PSA levels at first but all patients on androgen ablation will eventually develop castration-resistant PCa. Our understanding of alterations in PCa phenotypes and tumor microenvironment is limited and better understanding of these conditions is critical to attack castration-resistant PCa. In vivo, responses of tumors to castration are used to investigate these conditions (24–28). Furthermore castration-resistant tumors grown in castrated mice can be used to examine efficacy of novel drugs in castration-resistant PCa (28, 29). Some of the castration-resistant PCa cells and xenografts that can be used for such investigations are listed in **Tables 17.1** and **17.2**.

1. Male mice to be used for subcutaneous and orthotopic growth of tumors are castrated once they reach 6–8 weeks of age. Mice to be used for intra-tibial injections are cas-

Table 17.2
Androgen Receptor-Expressing Prostate Cancer Xenografts

AR Expressing PCa Xenografts	Origin	PSA	Response to Androgen Ablation	Reference
CWR 22	Primary tumor	+	+	14
CWR 22R	CWR 22	+	–	15
LAPC-9	Bone metastasis	+	+	23
LuCaP 23.1	Lymph node metatasis	+	high	24
LuCaP 23.1 AI	LuCaP 23.1	+	non-responsive	–
LuCaP 23.8	Lymph node metatasis	+	high	24
LuCaP 23.12	Liver metastasis	+	high	24
LuCaP 35	Lymph node metatasis	+	high	25
LuCaP 35V	LuCaP 35	+	non-responsive	25
LuCaP 58	Lymph node metatasis	+	intermediate	–
LuCaP 70	Liver metastasis	+	nd	–
LuCaP 73	Primary tumor	+	intermediate	26
LuCaP 77	Femur metastasis	+	high	26
LuCaP 78	Lymph node metastasis	+	nd	–
LuCaP 81	Lymph node metastasis	+	non-responsive	–
LuCaP 86.2	Bladder metastasis	+	non-responsive	–
LuCaP 92	Lymph node metastasis	+	nd	–
LuCaP 96	Primary tumor	+	high	–
LuCaP 96 AI	LuCaP 96	+	non-responsive	–
LuCaP 105	Rib metastasis	+	intermediate	–
LuCaP 115	Lymph node metastasis	+	nd	–
LuCaP 136	Ascites	+	intermediate	–
LuCaP 141	TURP	+	high	–
LuCaP 147	Liver metastasis	+	non-responsive	–

trated once they reach 4 weeks of age. In older animals the bones are harder and easier to fracture during the injection.

2. Anesthetize animals and allow them to reach the surgical plane of anesthesia.
3. Swab the testicular region with betadine solution and 70% isopropyl alcohol three times alternatively.
4. Make a small incision on the scrotal sac using a sterile scalpel.
5. Apply slight pressure to the abdomen to move testicles to the scrotum.

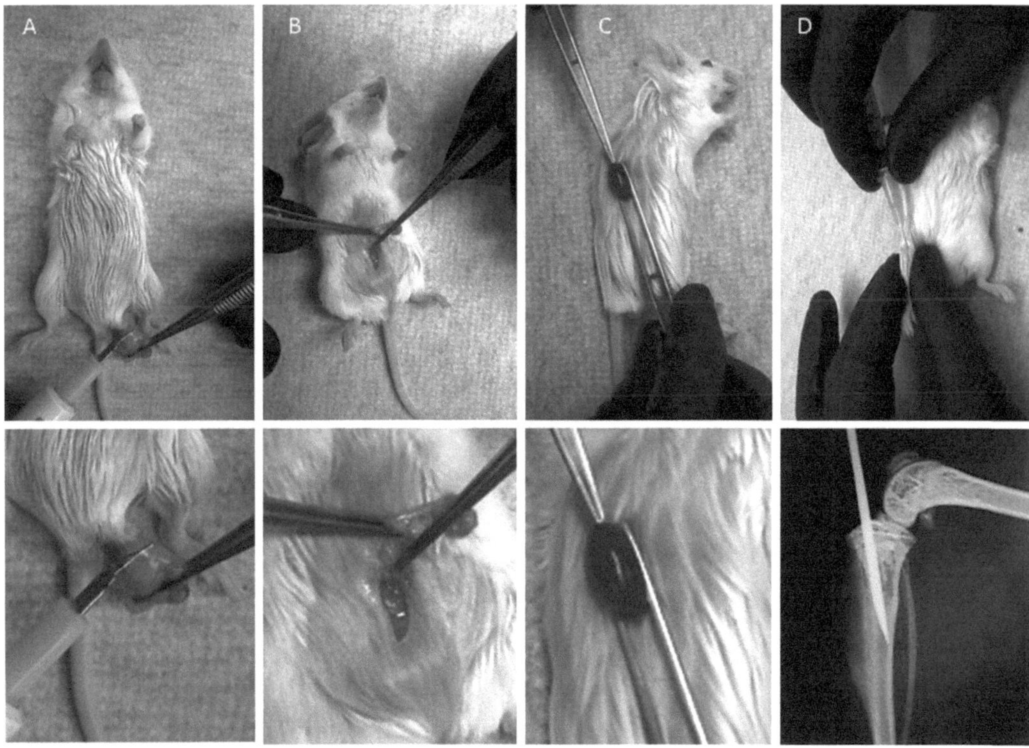

Fig. 17.1. Surgical procedures. (**a**) castration; (**b**) orthotopic implantation/injection; (**c**) subrenal capsule implantation; (**d**) intra-tibial injection.

6. Gently pull the testicle and attached fat pad through the scrotal incision.
7. Cauterize the vas deferens and associated blood vessels to remove the fat pad and testicle (**Fig. 17.1a**).
8. Repeat the procedure to remove the second testicle, using the same incision.
9. Once castration is complete, the scrotum is pulled out and away from the body and closed with a 9 mm autoclip.
10. Allow mice to recover on an isothermal heat pad until ambulatory.
11. Lift cutaneous layer between the shoulders, insert needle between the skin and the muscle and inject 0.05 mg/kg buprenorphine (100 μL volume) every 8–12 h for 48 h subcutaneously, and then as needed for pain management. Remove autoclips 7–10 days post surgery.
12. A 2-week period is allowed for the mouse to recover and the androgens to subside prior to introducing a castration-resistant tumor to the mouse.

3.1.5. Retro-orbital Bleed

Retro-orbital bleeds are performed once or twice weekly. In accordance with the IACUC's guidelines, no more than 1% of the animal's body weight should be collected within a 2 week period.

1. Place a narrow pipette or a micro-hematocrit tube in the medial or lateral canthus of the eye.
2. Using a slight rotation and slight pressure, apply the tube in the canthus to bleed from the retro-orbit of the mouse.
3. Once blood is acquired, sterile gauze is applied with gentle pressure to obtain hemostasis.

3.1.6. BrdU Injections

If interested in evaluation of effects on proliferation, mice are injected with 80 mg/kg BrdU via intraperitoneal injection. Mice bearing subcutaneous, orthotopic, and subrenal implants are injected 1 h prior to sacrifice. Mice with intra-tibial tumors are injected 4–8 h prior to sacrifice to allow for sufficient BrdU incorporation. Examples of BrdU staining used for evaluation of tumor growth can be found in (30).

3.1.7. Euthanasia

1. Anesthetize mice by intraperitoneal injection with 130 mg/8.8 mg/kg ketamine/xylazine solution.
2. Exsanguinate mice under the surgical plane of anesthesia via cardiac puncture then cervically dislocate.

3.1.8. Cardiac Puncture

1. Cardiac puncture is performed only after the mouse is anesthetized.
2. A 1 mL slip tip syringe fitted with a 26 gauge needle is used for this procedure and the mouse is placed in the supine position.
3. Insert the needle at the distal end of the rib cage and slowly move it upward toward the left ventricle of the heart while the syringe is very slightly drawn back to enable the observation of blood entering the hub.
4. Slowly pull the plunger once blood is observed while keeping the syringe and needle in place.

3.1.9. Cervical Dislocation

1. Cervical dislocation is a rapid and humane approach to euthanizing mice.
2. Secure the left forefinger and thumb around the base of the mouse's skull while the right hand is grasping the base of the tail.
3. In one smooth, secure, and quick motion both hands are pulled apart and the cervical vertebrae in the neck of the mouse are separated.

Animals are euthanized once tumor volumes (subcutaneous tumors that can be measured) reach 1000 mm^3, if animals lose

20% or more of body weight or if animals become otherwise compromised (hunched in posture, piloerected, rapid respiration, lethargic, etc.).

3.2. Tumor Implantation/Injection

Any of the described implantation/injection procedures can be performed in intact male mice as well as in castrated host. This choice depends on xenograft/cells used and experimental design.

3.2.1. Subcutaneous Implantation/Injection

The subcutaneous implantation/injection is the simplest to perform and allows easy monitoring of tumor take and growth rates. It also generates relatively large amounts of tumor tissue for analyses. Subcutaneous tumors have been used extensively for evaluation of multiple new agents, as well as to study the biology of PCa and AR signaling (25, 28, 31–33).

3.2.1.1. Implantation

1. Shave the area from hind leg to foreleg and prepare for surgery.
2. A trocar is a hollow stainless steel tube with a sharp tip which is combined with a blunt plunger that is used to implant solid tumor bits subcutaneously.
3. Draw a 20 mg tumor bit into a 13 gauge sterile trocar and insert the trocar subcutaneously closer to the hind leg.
4. Inject the tumor bit around the rib cage area and remove the trocar.
5. The site of the puncture heals quickly and no other postoperative care is necessary.

3.2.1.2. Injection

1. For subcutaneous injection (usually cells grown in vitro), the cutaneous layer on the right side of the mouse is lifted, and cells are injected subcutaneously.
2. Gently tap the cell suspension prior to drawing into syringe to minimize the variation of cell numbers injected. Keep all cell preparations on ice until all injections are complete. With cells that are not tumorigenic (e.g., LNCaP) or have a low take rate, a mixture of cells with Matrigel (1:1) is used. Matrigel is a mixture of gelatinous proteins that substitutes the extracellular environment present in many tissues, lending support to the cells to survive and start growing. Matrigel and Matrigel/cell mixtures need to be kept cold before the injection. At this temperature Matrigel is liquid, and when it warms up to 37°C the proteins self-assemble and solidify. Depending on cell type 1–10 million cells in ~100 μL volume are injected.
3. Draw 500 μL of the well mixed cell suspension of appropriate dilution into a 1 mL syringe fitted with a 26 gauge needle.

4. Inject 100 μL of the suspension subcutaneously on the right side of the mouse.

Subcutaneous tumor volumes are measured with digital calipers at least once weekly. Tumor volumes are then calculated using the ellipsoidal formula LxWxHx0.5236. Blood for serum prostate-specific antigen (PSA) levels are drawn once weekly via the retro-orbital sinus. Body weights are measured once weekly at approximately the same time of day.

The following example shows an experiment in which we have examined the response of subcutaneous LuCaP 23.1 tumors to castration to determine sensitivity of the xenograft to androgen ablation. In this experiment, tumor bits were implanted subcutaneously. When tumors reached 200 mm^3 the animals were castrated. Tumor growth and serum PSA levels were used to monitor the response to androgen ablation (**Fig. 17.2**). After castration, LuCaP 23.1 tumors shrank and serum PSA dropped. With the development of castration-resistant PCa the tumors started growing again and serum PSA levels increased. Castration of animals bearing LuCaP 23.1 increased their survival in comparison to intact animals. Additional examples of responses to castration, androgen ablation by other means and AR staining under these conditions can be found in the literature.

3.2.2. Orthotopic Implantation/Injection

Orthotopic growth of PCa tumors has the advantage of mimicking the tumor microenvironment of primary prostate cancer. Tumor cells are surrounded with mouse prostate stroma and fibroblast. The drawback is that the tumor volumes cannot be directly measured by calipers.

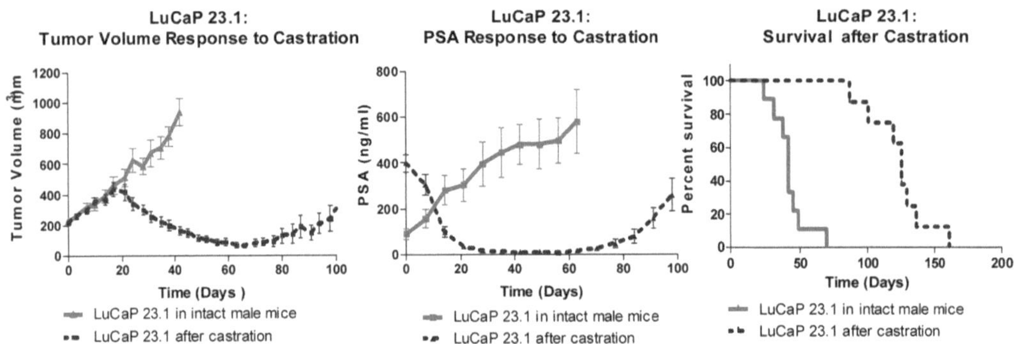

Fig. 17.2 Response of LuCaP 23.1 to androgen ablation. LuCaP23.1 tumor bits were implanted subcutaneously in SCID mice and when tumor reached 200 mm^3, the animals were randomized into two groups. (1) Control group – sham castration and (2) castration group. Tumor volumes were measured twice weekly and plotted as mean ± SEM. Blood was drawn weekly and serum PSA levels were measured. The results are plotted as mean ± SEM. LuCaP 23.1 is an androgen-sensitive PCa xenograft. Tumors responded to castration by decreases in volume and drop in serum PSA levels. With the development of castration-resistant PCa, tumor volume started to increase as well as PSA serum levels. Survival analysis show significant increase in survival of castrated animals bearing LuCaP 23.1 in comparison to intact animals bearing LuCaP 23.1.

1. Anesthetize C.B-17 SCID male mice.
2. Shave the abdominal region and sterilize for surgery.
3. Make a small incision on the ventral midline proximal to the penis in the skin and the underlying muscle (**Fig. 17.1b**).
4. Locate the coagulating gland on the ventral surface of the seminal and retract through the incision. For orthotopic implantations, a small hole is torn in the coagulating gland using sterile forceps and a 5 mg tumor bit is inserted into the gland. For orthotopic injections, a 1 mL syringe is fitted with a 26 gauge needle, loaded with 2×10^5 cells in 20 μL volume, and the coagulating gland is injected.
5. Once the vesicle is either implanted or injected with tumor, it is gently pushed back in the abdomen and the muscular layer is sutured with 4/0 coated visorb polyglyconic acid suture.
6. Clip the cutaneous layer with a 9 mm autoclip. Animals are allowed to recover on heat pads and are then injected subcutaneously with 0.05 mg/kg buprenorphine (100 μL volume) every 8–12 h for 48 h, and then as needed for pain management.
7. Remove autoclips 7–10 days later.
8. Blood for serum PSA levels can be collected once weekly via retro-orbital bleeds to monitor tumor growth. In addition to weekly PSA monitoring, weekly palpations of the abdomen are performed to ensure the tumor does not become too large. As mentioned above, tumor growth can also be monitored by in vivo imaging if cells used are labeled with GFP or luciferase.

3.2.3. Subrenal Capsule Implantation/Injection

Subrenal implantation is used because of the good blood supply in the kidney that supports tumor growth. Subrenal capsule implantation was the choice for growth of primary PCa samples (34) which are quite difficult to grow in animals:

1. Anesthetize C.B-17 SCID male mice and then shave and sterilize the right side.
2. Make a small incision using tungsten carbide scissors in the cutaneous layer just below the last rib once the mouse reaches the surgical plane of anesthesia (**Fig. 17.1c**.).
3. Cut the muscle and slip the kidney out. Keep the kidney moist with sterile 0.9% sodium chloride flushes.
4. Make a small opening in the renal capsule using sterile forceps and a 10 mg tumor bit is inserted into the pocket.
5. Gently push the kidney back into the incision and suture the muscle with a 4/0 coated visorb polyglyconic acid suture.
6. Clip the cutaneous layer with a 9 mm autoclip.

7. Allow recovery of the mice on heat pads and inject subcutaneously with 0.05 mg/kg buprenorphine (100 μL volume) every 8–12 h, and then as needed for pain management.
8. Remove autoclips 7–10 days later.
9. Blood for serum PSA levels are collected once weekly to monitor tumor growth in the subrenal capsule. Alternatively, tumor growth can be monitored by in vivo imaging if cells labeled with luciferase or GFP were used.
10. Excise and weigh tumors after sacrifice.

3.2.4. Intra-tibial Injection

About 60–90% of patients with advanced PCa suffer from bone metastasis (35–41). Unfortunately, unlike the human disease, PCa xenograft models rarely metastasize spontaneously to bone from the orthotopic site of primary tumor growth. Therefore, to study the phenotype of PCa cells in the bone environment and their responses to therapy, alternative models are used, including SCID-hu model (42), and intra-femoral and intra-tibial injections of tumor cells (6, 32, 43–50). We use intra-tibial injections as a model of PCa growth in the bone environment. This model does not mimic the processes of tumor cells spreading and adhesion in the bone, but it effectively recapitulates changes in tumor cells phenotype elicited by the bone environment and responses of tumor cells to treatments. We have used this model successfully to study behavior of PCa cells in the bone and alteration of PCa cells resulting from exposure to factors present in the bone as well as evaluate efficacy of new agents (6, 44–46, 51–53):

1. Use SCID Beige mice, 4–6 weeks of age for intra-tibial injections. It is important to use young mice to ensure that the tibiae are not yet fully calcified. Injection into tibiae of older mice often results in breakage of the tibiae after injection and tumor grows in the muscle instead of bone. Studies can be done in intact or castrated male mice.
2. Lay mice in the supine position on an isothermal heat pad and plate and the right leg is sterilized for surgery after the surgical plane of anesthesia is reached.
3. Inject 2×10^5 cells in 20 μL volume using a 1 mL syringe fitted with a 26-gauge needle. We do not recommend injecting larger volumes or larger number of cells since that often results in overfilling the tibia cavity and bone breakage.
4. Stabilize the right foot of the mouse with the left hand while the right hand moves the syringe and needle in a drill-like motion on the proximal end of the tibia (**Fig. 17.1d**).
5. Inject the cells into the tibial cavity once the needle is inserted approximately 3 mm into the tibia.
6. Allow mice to recover on heat pads and then inject them subcutaneously with 0.05 mg/kg buprenorphine (100 μL

volume) every 8–12 h for 48 h, and then as needed for pain management.

As intra-tibial tumor volumes cannot be measured directly using calipers, tumor growth is monitored indirectly. In vivo imaging can be employed if cells labeled with luciferase or GFP were used. Alternatively, serum PSA levels can be measured to evaluate tumor growth (if PSA producing cells were used). Effects of tumor cells in the bone are examined using radiographs and measurements of bone mineral density (BMD). For radiography and BMD measurements, mice are anesthetized with ketamine/ xylazine and are placed on their stomachs with all limbs laid outward. Once the total body BMD is collected, a region of interest is available to analyze between the tumored tibia and the nontumored tibia to see the osteolytic or osteoblastic responses within the bone.

In the following experiments, we evaluated growth of LuCaP 23.1 in the murine bone of intact male mice and growth of C4-2B, CR PCa cells, in the bone of castrated male mice, and the bone responses to these tumor cells (**Fig. 17.3**). Tumor progression in the bone is demonstrated by increases in serum PSA levels, and radiological appearance of the tumored tibiae (*see* **Notes 1** and **2**).

3.2.5. Cardiac Injection

As mentioned above there are no reliable models of spontaneous PCa bone metastasis in mice. Cardiac injection is another alternative to study some of the aspects of PCa cells spreading to the bone, adhesion in the bone, and their growth in the bone environment. This method is referred to as colonization rather than metastasis.

1. Anesthetize 4–6 week old male mice and allow them to reach the surgical plane of anesthesia.
2. Place mice in the supine position and shave and swab their chests with 70% ethanol.
3. Inject 2×10^5 cells in 100 μL volume using a 1 mL syringe fitted with a 26-gauge needle.
4. Leave a small bubble of air near the plunger of the syringe to allow blood from the left ventricle to enter the syringe once it is pierced.
5. Stabilize the mouse's chest with one hand while the other approaches the chest in a perpendicular manner with the needle and syringe.
6. Locate the left ventricle of the heart at the midpoint between the sternum and the xiphoid process, and slightly to the left.
7. Insert the needle at this site, approximately 6–8 mm in depth, until blood enters the hub of the needle.

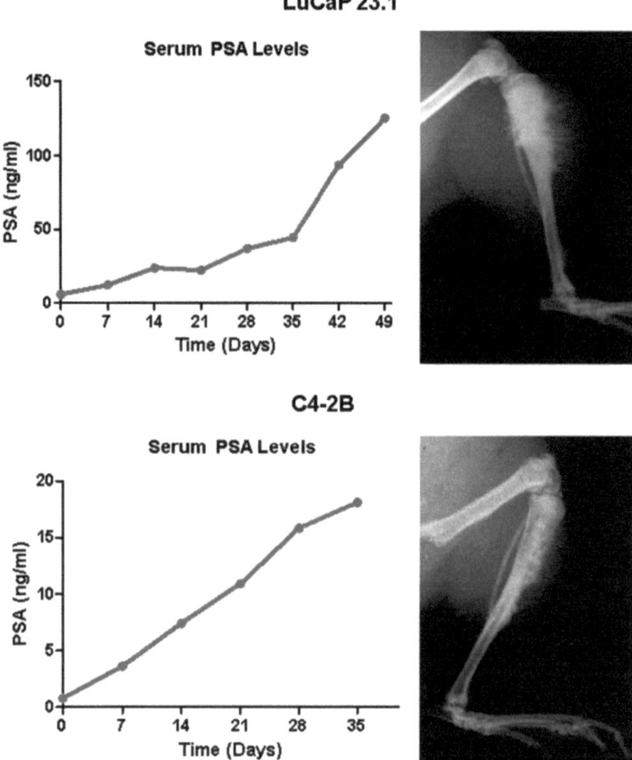

Fig. 17.3. Growth of LuCaP 23.1 in tibia of intact mice and C4-2B in tibia of castrated mice. Tumor cells were injected into tibiae as described in the text. Serum PSA levels were used to monitor tumor growth. The results are plotted as mean ± SEM. Increases in serum PSA levels indicate progression of tumor growth in the bone. Radiography shows bone response to the tumors. LuCaP 23.1 and C4-2B PCa cells growing in the bone elicited osteoblastic response.

8. Slowly inject the cells into the left ventricle. Once injected, quickly withdraw the needle and apply slight pressure on the heart for about 30 s.
9. Allow mice to recover on heat pads and inject subcutaneously with 0.05 mg/kg buprenorphine (100 μL volume) every 8–12 h for 48 h, and then as needed for pain management.
10. Tumor colonization of the bone and tumor growth can be monitored by serum PSA levels, by radiography or by in vivo imaging (*see* **Note 3**).

3.3. Tissue Processing

3.3.1. Freezing in OCT

1. Place tumor bits in a cryomold containing a clear freezing compound.
2. Place the tumor in the center of the mold and add more clear freezing compound to ensure that the tissue is completely covered.

3. Use tongs to slowly lower the cryomold into liquid nitrogen to set the freezing compound.
4. Once the tissue is properly frozen, store it in the –80°C freezer until sectioning and analysis are performed.

3.3.2. Processing for Tumors Paraffin Embedding

1. Place tumors to be paraffin embedded in 10% Neutral Buffered Formalin for 24 h.
2. After fixing, move tumors into 70% EtOH for paraffin embedding.

3.3.3. Processing Tibiae

1. Dissect out tibiae with tumor and contralateral tibiae and fix in 10% neutral buffered formalin for 24 h.
2. After fixing, place tibiae in 10% EDTA for decalcification of the bones. Replace 10% EDTA every other day until the bone feels soft to the touch (about 10–12 days).
3. Place decalcified tibiae in 70% ethanol for paraffin embedding.

3.3.4. Freezing Tissues

1. Cut tumors immediately after excision, then place in a sterile empty vial and snap freeze directly in liquid nitrogen for further analyses.
2. Cut tumors into pieces ranging from 50 to 300 mg, depending on the kind of future analyses.
3. Store tissues in –80°C.

4. Notes

1. Some researchers use intra-femoral implantation of tumor cells to study interactions between tumor and bone cells. This procedure requires surgical incision in a leg muscle, drilling a hole in the femur, plugging in the hole with wax after injection of tumor cells and suturing the muscle.
2. A procedure has been published wherein bone marrow is removed prior to injection of tumor cells into tibia or femur. We neither use nor do we recommend such a procedure. Presence of bone marrow is critical for biologically relevant interactions between tumor cells and the bone environment.
3. Unfortunately, at present, there are no AR-expressing PCa cells that colonize bone with rates that would allow study set-up. The percentage of colonization between various models is in a range of 0–20%. However, this procedure can be used to evaluate new lines for their potential to colonize bone.

Acknowledgments

We would like to thank the Richard M. Lucas Foundation and the Prostate Cancer Foundation for long-term support for generation and characterization of our prostate cancer xenograft lines. We would also like to thank Dr. Vessella, the Director of the GU Cancer Research Laboratory in the Department of Urology at the University of Washington, for his continuous support of our work. Our preclinical and biological studies have been supported over the years by NIH grants 5-P50-DK47656, 5-PO1-CA085859, 5-50-CA97186. Finally, we would like to acknowledge Abbott Laboratories for their generous support throughout the years by providing reagents for the quantification of PSA levels.

References

1. Culig Z, Klocker H, Bartsch G, Steiner H, Hobisch A. (2003) Androgen receptors in prostate cancer. *J. Urol.* **170**, 1363–1369.
2. Danielpour D. (2005) Functions and regulation of transforming growth factor-beta (TGF-beta) in the prostate. *Eur. J. Cancer* **41**, 846–857.
3. Quinn DI, Henshall SM, Sutherland RL. (2005) Molecular markers of prostate cancer outcome. *Eur. J. Cancer* **41**, 858–887.
4. Burd CJ, Morey LM, Knudsen KE. (2006) Androgen receptor corepressors and prostate cancer. *Endocr. Relat Cancer* **13**, 979–994.
5. Chatterjee B. (2003) The role of the androgen receptor in the development of prostatic hyperplasia and prostate cancer. *Mol. Cell Biochem.* **253**, 89–101.
6. Brubaker KD, Brown LG, Vessella RL, Corey E. (2006) Administration of zoledronic acid enhances the effects of docetaxel on growth of prostate cancer in the bone environment. *BMC Cancer* **6**, 15.
7. Horoszewicz JS, Leong SS, Kawinski E, Karr J, Rosenthal H, Chr TM, et al. (1983) LNCaP model of human prostatic carcinoma. *Cancer Res.* **43**, 1809–1818.
8. Thalmann GN, Anezinis PE, Chang S, Zhau HE, Kim EE, Hopwood VL, et al. (1994) Androgen-independent cancer progression and bone metastasis in the LNCaP model of human prostate cancer. *Cancer Res.* **54**, 2577–2581.
9. Thalmann GN, Sikes RA, Wu TT, Degeorges A, Chang SM, Ozen M, et al. (2000) LNCaP progression model of human prostate cancer: androgen-independence and osseous metastasis. *Prostate* **44**, 91–103.
10. Wu TT, Sikes RA, Cui Q, Thalmann GN, Kao C, Murphy CF, et al. (1998) Establishing human prostate cancer cell xenografts in bone: induction of osteoblastic reaction by prostate-specific antigen- producing tumors in athymic and SCID/bg mice using LNCaP and lineage-derived metastatic sublines. *Int. J. Cancer* **77**, 887–894.
11. Korenchuk S, Lehr JE, MClean L, Lee YG, Whitney S, Vessella R, et al. (2001) VCaP, a cell-based model system of human prostate cancer. *In Vivo* **15**, 163–168.
12. Lee YG, Korenchuk S, Lehr J, Whitney S, Vessela R, Pienta KJ. (2001) Establishment and characterization of a new human prostatic cancer cell line: DuCaP. *In Vivo* **15**, 157–162.
13. Terry S, Yang X, Chen MW, Vacherot F, Buttyan R. (2006) Multifaceted interaction between the androgen and Wnt signaling pathways and the implication for prostate cancer. *J. Cell Biochem.* **99**, 402–410.
14. Wainstein MA, He F, Robinson D, Kung HJ, Schwartz S, Giaconia JM, et al. (1994) CWR22: Androgen-dependent xenograft model derived from a primary human prostatic carcinoma. *Cancer Res.* **54**, 6049–6052.
15. Nagabhushan M, Miller CM, Pretlow TP, Giaconia JM, Edgehouse NL, Schwartz S, et al. (1996) CWR22: The first human prostate cancer xenograft with strongly androgen-dependent and relapsed strains

both in vivo and in soft agar. *Cancer Res.* **56**, 3042–3046.
16. Tepper CG, Boucher DL, Ryan PE, Ma AH, Xia L, Lee LF, et al. (2002) Characterization of a novel androgen receptor mutation in a relapsed CWR22 prostate cancer xenograft and cell line. *Cancer Res.* **62**, 6606–6614.
17. Navone NM, Olive M, Ozen M, Davis R, Troncoso P, Tu SM, et al. (1997) Establishment of two human prostate cancer cell lines derived from a single bone metastasis. *Clin. Cancer Res.* **3**, 2493–2500.
18. Hara T, Nakamura K, Araki H, Kusaka M, Yamaoka M. (2003) Enhanced androgen receptor signaling correlates with the androgen-refractory growth in a newly established MDA PCa 2b-hr human prostate cancer cell subline. *Cancer Res.* **63**, 5622–5628.
19. Harper ME, Goddard L, Smith C, Nicholson RI. (2004) Characterization of a transplantable hormone-responsive human prostatic cancer xenograft TEN12 and its androgen-resistant sublines. *Prostate* **58**, 13–22.
20. McCulloch DR, Opeskin K, Thompson EW, Williams ED. (2005) BM18: A novel androgen-dependent human prostate cancer xenograft model derived from a bone metastasis. *Prostate* **65**, 35–43.
21. Klein KA, Reiter RE, Redula J, Moradi H, Zhu XL, Brothman AR, et al. (1997) Progression of metastatic human prostate cancer to androgen independence in immunodeficient SCID mice. *Nat. Med.* **3**, 402–408.
22. Davies MR, Lee YP, Lee C, Zhang X, Afar DE, Lieberman JR. (2003) Use of a SCID mouse model to select for a more aggressive strain of prostate cancer. *Anticancer Res.* **23**, 2245–2252.
23. Craft N, Chhor C, Tran C, Belldegrun A, deKernion J, Witte ON, et al. (1999) Evidence for clonal outgrowth of androgen-independent prostate cancer cells from androgen-dependent tumors through a two-step process. *Cancer Res.* **59**, 5030–5036.
24. Ellis WJ, Vessella RL, Buhler KR, Bladou F, True LD, Bigler SA, et al. (1996) Characterization of a novel androgen-sensitive, prostate-specific antigen-producing prostatic carcinoma xenograft: LuCaP 23. *Clin. Cancer Res.* **2**, 1039–1048.
25. Corey E, Quinn JE, Buhler KR, Nelson PS, Macoska JA, True LD, et al. (2003) LuCaP 35: A New Model of Prostate Cancer Progression to Androgen Independence. *Prostate* **55**, 239–246.
26. Corey E, Vessella RL. Xenograft models of human prostate cancer. In: Chung LWK, Isaacs WB, Simons JW, editors, Prostate Cancer: biology, genetics and the new therapeutics. Totowa, NJ: Humana Press; 2007, Second Edition, Chapter 1, pp. 3–32.
27. Snoek R, Cheng H, Margiotti K, Wafa LA, Wong CA, Wong EC, et al. (2009) In vivo knockdown of the androgen receptor results in growth inhibition and regression of well-established, castration-resistant prostate tumors. *Clin. Cancer Res.* **15**, 39–47.
28. Coleman IM, Kiefer JA, Brown LG, Pitts TE, Nelson PS, Brubaker KD, et al. (2006) Inhibition of androgen-independent prostate cancer by estrogenic compounds is associated with increased expression of immune-related genes. *Neoplasia* **8**, 862–878.
29. Wu JD, Odman A, Higgins LM, Haugk K, Vessella R, Ludwig DL, et al. (2005) In vivo effects of the human type i insulin-like growth factor receptor antibody A12 on androgen-dependent and androgen-independent xenograft human prostate tumors. *Clin. Cancer Res.* **11**, 3065–3074.
30. Kiefer JA, Vessella RL, Quinn JE, Odman AM, Zhang J, Keller ET, et al. (2004) The effect of osteoprotegerin administration on the intra-tibial growth of the osteoblastic LuCaP 23.1 prostate cancer xenograft. *Clin. Exp. Metastasis* **21**, 381–387.
31. Trauger R, Corey E, Bell D, White S, Garsd A, Stickney D, et al. (2009) Inhibition of androstenediol-dependent LNCaP tumour growth by 17alpha-ethynyl-5alpha-androstane-3alpha, 17beta-diol (HE3235). *Br. J. Cancer* **100**, 1068–1072.
32. Corey E, Brown LG, Quinn JE, Poot M, Roudier M.P, Higano CS, et al. (2003) Zoledronic acid exhibits inhibitory effects on osteoblastic and osteolytic metastases of prostate cancer. *Clin. Cancer Res* **9**, 295–306.
33. Corey E, Quinn JE, Emond MJ, Buhler KR, Brown LG, Vessella RL. (2002) Inhibition of androgen-independent growth of prostate cancer xenografts by 17 beta-estradiol. *Clin. Cancer Res* **8**, 1003–1007.
34. Wang Y, Xue H, Cutz JC, Bayani J, Mawji NR, Chen WG, et al. (2005) An orthotopic metastatic prostate cancer model in SCID mice via grafting of a transplantable human prostate tumor line. *Lab Invest* **85**, 1392–1404.
35. Rubin MA, Putzi M, Mucci N, Smith DC, Wojno K, Korenchuk S, et al. (2000) Rapid ("warm") autopsy study for procurement of metastatic prostate cancer. *Clin Cancer Res 2000.* Mar **6**(3), 1038–1045.
36. Bubendorf L, Schopfer A, Wagner U, Sauter G, Moch H, Willi N, et al. (2000) Metastatic patterns of prostate cancer: an

autopsy study of 1,589 patients. *Hum. Pathol.* **31**, 578–583.

37. Harada M, Iida M, Yamaguchi M, Shida K. (1992) Analysis of bone metastasis of prostatic adenocarcinoma in 137 autopsy cases. *Adv. Exp. Med. Biol.* **324**, P173–182.
38. Saitoh H, Hida M, Shimbo T, Nakamura K, Yamagata J, Satoh T. (1984) Metastatic patterns of prostatic cancer. Correlation between sites and number of organs involved. *Cancer* **54**, 3078–3084.
39. Roudier MP, Vesselle H, True LD, Higano CS, Ott SM, King SH, et al. (2003) Bone histology at autopsy and matched bone scintigraphy findings in patients with hormone refractory prostate cancer: the effect of bisphosphonate therapy on bone scintigraphy results. *Clin. Exp. Metastasis* **20**, 171–180.
40. Roudier MP, Corey E, True LD, Higano CS, Ott SM, Vessella RL. Histological, immunological and histomorphometrical characterization of prostate cancer bone metastases. In: Keller ET, Chung LWK, editors. The biology of skeletel metastases. Boston, MA: Kluwer; 2004, pp. 311–339.
41. Roudier MP, True LD, Higano CS, Vesselle H, Ellis W.J., Lange PH, et al. (2003) Phenotypic Heterogeneity of Androgen-Independent Prostate Cancer Bone Metastases. *Hum. Pathol.* **34**, 646–653.
42. Nemeth JA, Harb JF, Barroso UJ, He Z, Grignon DJ, Cher ML. (1999) Severe combined immunodeficient-hu model of human prostate cancer metastasis to human bone. *Cancer Res.* **59**, 1987–1993.
43. Brown LG, Vessella RL, Corey E. (2000) Effects of zoledronc acid on prostate cancer cells. *J Bone Miner. Res.* **15**, S446.
44. Corey E, Quinn JE, Bladou F, Brown LG, Roudier MP, Brown JM, et al. (2002) Establishment and characterization of osseous prostate cancer models: intra-tibial injection of human prostate cancer cells. *Prostate* **52**, 20–33.
45. Koreckij TD, Trauger RJ, Montgomery RB, Pitts TE, Coleman I, Nguyen H, et al. (2009) HE3235 inhibits growth of castration-resistant prostate cancer. *Neoplasia.* **11**, 1216–1225.
46. Morgan TM, Pitts TE, Gross TS, Poliachik SL, Vessella RL, Corey E. (2008) RAD001 (Everolimus) inhibits growth of prostate cancer in the bone and the inhibitory effects are increased by combination with docetaxel and zoledronic acid. *Prostate* **68**, 861–871.
47. Pfitzenmaier J, Quinn JE, Odman AM, Zhang J, Keller ET, Vessella RL, et al. (2003) Characterization of C4-2 prostate cancer bone metastases and their response to castration. *J. Bone Miner. Res.* **18**, 1882–1888.
48. Kundra V, Ng CS, Ma J, Bankson JA, Price RE, Cody DD, et al. (2007) In vivo imaging of prostate cancer involving bone in a mouse model. *Prostate* **67**, 50–60.
49. Lu Y, Cai Z, Xiao G, Keller ET, Mizokami A, Yao Z, et al. (2007) Monocyte chemotactic protein-1 mediates prostate cancer-induced bone resorption. *Cancer Res.* **67**, 3646–3653.
50. Zhang J, Dai J, Qi Y, Lin DL, Smith P, Strayhorn C, et al. (2001) Osteoprotegerin inhibits prostate cancer-induced osteoclastogenesis and prevents prostate tumor growth in the bone. *J Clin. Invest.* **107**, 1235–1244.
51. Koreckij T, Nguyen H, Brown LG, Yu EY, Vessella RL, Corey E. (2009) Dasatinib inhibits the growth of prostate cancer in bone and provides additional protection from osteolysis. *Br. J. Cancer* **101**, 263–268.
52. Morrissey C, Brown LG, Pitts TE, Vessella RL, Corey E. (2010) Bone morphogenetic protein 7 is expressed in prostate cancer metastases and its effects on prostate tumor cells depend on cell phenotype and the tumor microenvironment. *Neoplasia* **12**, 192–205.
53. Morrissey C, Kostenuik PJ, Brown LG, Vessella RL, Corey E. (2007) Host-derived RANKL is responsible for osteolysis in a C4-2 human prostate cancer xenograft model of experimental bone metastases. *BMC Cancer* 7, 148.

Chapter 18

Automated Microscopy and Image Analysis for Androgen Receptor Function

Sean M. Hartig, Justin Y. Newberg, Michael J. Bolt, Adam T. Szafran, Marco Marcelli, and Michael A. Mancini

Abstract

Systems-level approaches have emerged that rely on analytical, microscopy-based technology for the discovery of novel drug targets and the mechanisms driving AR signaling, transcriptional activity, and ligand independence. Single cell behavior can be quantified by high-throughput microscopy methods through analysis of endogenous protein levels and localization or creation of biosensor cell lines that can simultaneously detect both acute and latent responses to known and unknown androgenic stimuli. The cell imaging and analytical protocols can be automated to discover agonist/antagonist response windows for nuclear translocation, reporter gene activity, nuclear export, and subnuclear transcription events, facilitating access to a multiplex model system that is inherently unavailable through classic biochemical approaches. In this chapter, we highlight the key steps needed for developing, conducting, and analyzing high-throughput screens to identify effectors of AR signaling.

Key words: Fluorescence microscopy, high content analysis, nuclear receptor, automated cytometry, androgen, endocrine disruptor, coregulator.

1. Introduction

The androgen receptor (AR) is a member of the nuclear receptor (NR) superfamily and regulates gene transcription to promote male sexual development and also cell proliferation during prostate cancer progression. Emerging functional data in several organ systems have extended the importance of AR across several endocrine targets including skeletal muscle, bone, and adipose tissue (1). As a type I NR, AR is predominantly cytoplasmic in the absence of agonists, principally testosterone (T) and

F. Saatcioglu (ed.), *Androgen Action*, Methods in Molecular Biology 776,
DOI 10.1007/978-1-61779-243-4_18, © Springer Science+Business Media, LLC 2011

5α-dihydrotestosterone (DHT). Following ligand binding, AR sheds its interaction with heat shock proteins and translocates to the nucleus, undergoing concomitant association with coregulators and binding to promoters/enhancers of AR-regulated genes, as reviewed recently (2). Although less understood than genomic AR action, non-genomic AR effects are controlled in both ligand-dependent and independent manners. Depending on context and stimulus, AR can contribute to gene regulation or cell adaptation through rapid, non-genomic mechanisms, whereby AR interacts with signaling elements (PI3K, Src, and MAPK) or acts as scaffolds to other transcription factors (STAT family) (3).

Technical advances during the last decade have made it feasible to use microscopy as a primary tool for probing several aspects of cellular function at increasing throughput (4–10). Although still significantly slower that traditional single-point readout high-throughput assays, high-throughput microcopy (HTM) approaches can integrate multiple coexisting variables within a cellular framework that is inherently unavailable in standard biochemical approaches (ELISA, Western blotting, luciferase, ChiP). Single cell-based imaging assays allow simultaneous analysis of multiple features using fluorescent labels, which is often necessary to accurately describe underlying biological processes that are hidden in whole population-based measurements. Due to the data-rich nature of HTM applications, the approach has been broadly dubbed high content analysis (HCA, *see* **Note 1**).

HCA provides unique opportunities to learn important mechanistic information about AR biology. The classic steps of AR action can be quantified by image-based methods enabling researchers to correlate AR protein levels and subcellular localization with transcriptional activity. Our lab (11–16) and others (17) have used microscopy-based techniques to elucidate the actions of standard agonists and antagonists, characterize endocrine disruptors, and identify non-competitive inhibitors of AR-regulated transcription. In recent years, we have established automated microscopy and HCA as primary approaches to understand AR biology. Routinely, we are able to simultaneously quantify three central mechanistic steps that contribute to AR activation: (1) nuclear translocation via the percent localization of signal in the nucleus (e.g., the nuclear:cellular AR ratio); (2) formation of AR-rich subnuclear "speckles" that correlate with AR transcriptional activity (18); and (3) transcription via fluorescent reporter gene activity. These assays of androgenic response have Z′ scores of 0.4–0.9 in several cell types AR(−) HeLa cells that transiently or stably express GFP-AR fusion proteins and an ARR_2PB-dsRED2skl reporter (13–15) and AR(+ or −) prostate cancer cell lines. Moreover, these Z′ scores have been reproduced in primary cultures generated from patients affected by various conditions of AR malfunction (i.e., genital skin fibroblasts of normal individuals and patients affected by androgen insensitivity syndromes), where the endogenous AR is immunolabeled with antibodies (14–16).

This chapter summarizes experimental AR HCA designs used by our group to quantify nuclear receptor functions at an integrated, systems level. Importantly, these protocols are also applicable to smaller scale experiments where automated, high-throughput microscopes are unavailable (e.g., using cells plated onto glass coverslips). The chapter closes with a discussion of typical bioinformatic approaches utilizing our central image analysis toolbox and data reduction techniques that can be used to characterize AR ligands in the context of complex biological responses.

2. Materials

2.1. Cell Culture and Stable Cell Line Creation

1. Phenol red-free DMEM supplemented with L-glutamine, sodium pyruvate, penicillin/streptomycin, and 5% fetal bovine serum (FBS). Experimental and standard culture media conditions are shown in **Table 18.1**.
2. Stripped-dialyzed FBS (*see* **Note 2**).
3. Antibiotic selection markers: hygromycin B and puromycin.
4. Trypsin/ethylenediamine tetraacetic acid (EDTA).
5. Transfection reagent of choice.
6. GFP-AR and ARR_2PB-dsRED2skl constructs.
7. Androgens and anti-androgens.

Table 18.1
Growth and experimental media components for GFP-AR/ARR_2PB-dsRED2skl cell culture

	Regular culture	Experiments
DMEM/phenol red free	X	X
Glutamine	X	X
Pen/Strep	X	
FBS (5%)	X	
SD-FBS (5%)		X
Puromycin (1.2 μg/ml)	X	
Hygromycin (200 μg/ml)	X	
OHF (0.1 nM)	X	

2.2. Plate Handling

1. 96- or 384-well glass bottom plates (*see* **Note 3**).
2. 1 mg/ml poly-D-lysine prepared in water, sterile-filter 0.22 μm or 0.45 μm (*see* **Note 4**).
3. Phosphate-buffered saline containing Ca^{2+} and Mg^{2+} (PBS++), pH = 7.4. Concentrations of constituents are as follows: 138 mM NaCl, 2.67 mM KCl, 8.1 mM Na_2HPO_4, and 1.47 mM KH_2PO_4. pH must be reduced (pH = 7.0) before adding $CaCl_2$ and $MgCl_2$ but can be re-adjusted to pH = 7.4 after dissolution. Concentrations for $CaCl_2$ and $MgCl_2$ are 0.1 g/l.
4. Tris-buffered saline (20 mM Tris–HCl, 150 mM NaCl) with 0.1% Tween 20 (TBS-T), pH = 7.6.
5. PEM: 80 mM K-PIPES, pH = 6.8; 5 mM EGTA, pH = 7.0; and 2 mM $MgCl_2$.
6. Formaldehyde: EM-grade, 16% stock solutions (sold as paraformaldehyde by Electron Microscopy Sciences, Hatfield, PA). For fixation, prepare a 4% solution in PEM. Opened ampoules should be tightly sealed with Parafilm® and stored at 4°C in the dark for only a few days. Always dispose of used formaldehyde in appropriate waste accumulation units.
7. Triton X-100: Prepare 10% (w/v) aliquots in PEM. Keep covered in foil or in amber-colored Eppendorf tubes. Store parent stocks at –20°C. The working solution of Triton X-100 for plates is 0.5%, diluted in PEM buffer. After thawing 10% aliquots, do not refreeze. Triton X-100 dilutions are unstable and should be stored at 4°C and used quickly. Use of high-quality (peroxide and carbonyl free) and fresh Triton X-100 is fundamentally important if HCA experiments utilize immunofluorescence protocols.
8. 4′,6-diamidino-2-phenylindole (DAPI) can be prepared at 1 mg/ml concentration in PEM, aliquoted in amber-colored Eppendorf tubes, and stored at 4°C. Protect from light. Reasonable working solutions are 1 μg/ml in PEM.
9. Quench: 100 mM NH_4Cl or 1 mg/ml $NaBH_4$ prepared in PEM can be used to quench residual formaldehyde from fixation. This step reduces autofluorescence by minimizing non-specific antibody binding to free aldehydes.
10. CellMask fluorescent dyes (Invitrogen, Carlsbad, CA): These different color non-specific protein compounds are reconstituted to a final stock concentration of 10 mg/ml in DMSO. Working concentrations should be determined empirically for each experiment and cell type (*see* **Note 5**).
11. TiterTek Multidrop 384 (TiterTek, Huntsville, AL) or equivalent cell plating instrument.

12. Fluid handling robotics, e.g., BioMek NXS8 or S96 (Beckman Coulter, Mountain View, CA) coupled with plate washer module (e.g., ELx405; BioTek, Winooski, VT).

2.3. Image Acquisition, Background Subtraction, and Analysis

1. High-throughput microscope or analogous imaging platform with automated stage and focus (*see* **Note 6**).
2. Three Dell workstations with the following minimum specifications: 500 GB hard drives, dual-core 2 GHz processors, and 3 GB RAM.
3. Windows XP operating system SP2 with the following software (and their required toolboxes): Pipeline Pilot 7.5 (Accelrys, San Diego, CA) with the Imaging Toolbox, Python 2.6 with numpy, scipy, and matplotlib, and R 2.10 with heatmap.plus (*see* **Note 7**).
4. Workstations should be networked so computing jobs can be parallelized across machines (using an option in Pipeline Pilot).

3. Methods

Assays for HCA begin with routine standard microscopy to establish the response of both positive and negative controls of robust stable cell lines expressing GFP-AR and AR-responsive fluororeporter constructs. With our GFP-AR/ARR_2PB-dsRED2skl cell line, in vehicle controls we look for primarily cytoplasmic localization, diffuse intranuclear organization, and low background reporter expression (dsRED2skl). Conversely, with standard AR agonists (DHT, R1881, and mibolerone), positive control responses are identified by predominantly nuclear, "hyperspeckled" GFP-AR and dsRED2skl-labeled peroxisomes in the cytoplasm. Once the assay is robust and highly reproducible using a coverslip-based approach, we scale the experiment up to 96- and 384-well plates. In a two-step process, we detail the creation of a double-stable cell line expressing GFP-AR and an AR-dependent reporter construct with a dsRED2skl reporter readout. Identical protocols may be used to create stable receptor/reporter cell lines in different cell backgrounds (e.g., PC3 and U2OS).

3.1. Construction of GFP-AR/ARR_2PB-dsRED2skl Biosensor

1. Cotransfect GFP-AR and a linearized hygromycin marker into HeLa cells using standard chemical transfection methods. After 24 h, re-plate in media (DMEM-F12, 5% FBS) supplemented with hygromycin.
2. Expand surviving colonies and analyze for GFP-AR distribution and expression by wide-field fluorescence microscopy.

Single cell clone positive or desired clones by limiting dilution or cell sorting (FACS).

3. Infect low-expressing GFP-AR HeLa cell clone(s) with a virus encoding a derivative of the widely used androgen-responsive pARR$_2$PB transcriptional reporter construct (19). Our reporter encodes a dsRED2 protein (Clontech) fused at the C-terminus with a peroxisome targeting sequence (SKL, serine, lysine, and leucine) that serves to localize and concentrate the fluorescence signal. Select for antibiotic-resistant cultures with both puromycin (1.5 μg/ml) and hygromycin (200 μg/ml) for 2 weeks.
4. Following DHT treatment overnight, dual sort single cell clones by FACS (*see* **Note 8**) by GFP-AR and ARR$_2$PB-dsRED2-skl responsiveness (e.g., low GFP/high dsRED2). Maintain single cell clones in phenol red-free DMEM with FBS and antibiotics. Ultimately, when the double-stable cell line (GFP-AR HeLa ARR$_2$PB-dsRED2skl) is exposed to AR agonists (R1881, DHT, and mibolerone), AR rapidly (~60 min) translocates to the nucleus and initiates a "hyper-speckled" pattern and reporter gene transcription (13, 18). Maximal levels of dsRED2skl accumulate in cytoplasmic peroxisomes by 24–48 h (**Fig. 18.1**). The quality and reproducibility of the assay should be assessed before scale-up to multi-well formats (*see* **Note 9**).

3.2. Surface Preparation for Plates

1. Coat with a 1 mg/ml, sterile-filtered solution of poly-D-lysine by adding 25 μl to each well, sealing the plate edges with Parafilm® and incubating at room temperature (RT) for 4 h with gentle rocking or overnight at 4°C.
2. Remove the PDL. To avoid toxicity, wash plates extensively (more than six times) with PBS (50 μl) to remove residual PDL. After the last wash, cells can be plated. Well plates can be sealed and stored for a few days at 4°C in 20–30 μl PBS.
3. Alternatively, coat overnight at 37°C with a 10–20% FBS solution prepared in PBS. After incubation, remove excess serum. Plates are ready to be seeded with cells. We detect no basal hormone response from residual FBS in the GFP-AR/ARR$_2$PB-dsRED2skl cell line. For all substrate coatings, it is critical that all solutions remain sterile.

3.3. Cell Treatment in Multi-well Plate Formats

Below, we have summarized the steps used for detection of compound-specific effects on AR function and transcriptional activity. A schematic of this process is depicted in **Fig. 18.2**. Microfluidic robotics are used for all cell processing to ensure high efficiency, accuracy, and throughput.

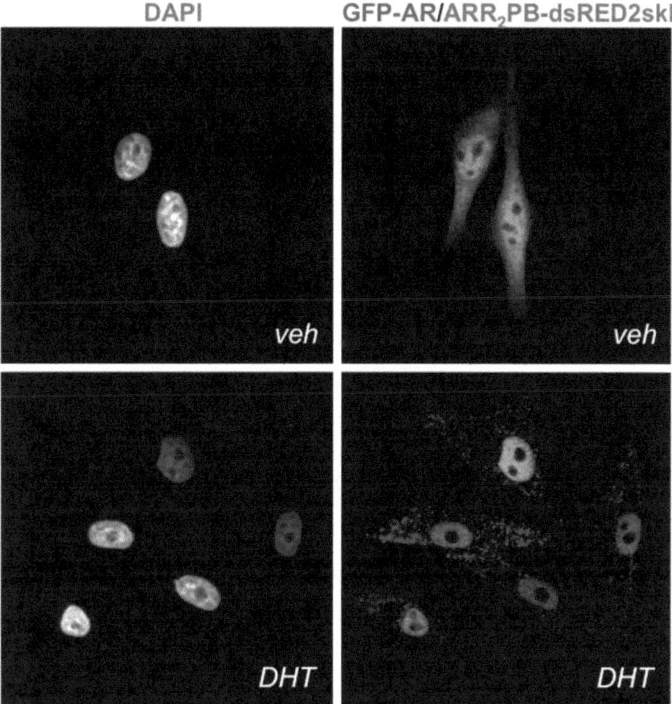

Fig. 18.1. A stable HeLa AR biosensor cell line that expresses both GFP-AR and the androgen-responsive fluoro-reporter construct ARR_2PB-DsRED2skl. Upon exposure to 10 nM DHT for 48 h AR translocates to the nucleus exhibiting visible alterations in subnuclear structure and dsRED2-labeled peroxisomes indicative of an AR agonist response.

1. Before experiments, grow cells in the appropriate growth media and serum to 90–95% confluency. Change media 1 day prior to subculturing the cells for experimentation.
2. Trypsinize the cells with trypsin/EDTA and wash cultures three times with phenol red-free DMEM containing 5% SD-FBS.
3. Determine cell density. We use 20,000 cells/well (100 μl volume) for 96 wells and 4,000 cell/well (30 μl) in 384-well plates. It is critical to determine the appropriate cell number that results in a subconfluent cell density; confluent cultures are incompatible with HCA for multiple reasons, including heightened probability of washing away during pipetting, altered immunofluorescence results, and greatly reduced ability to perform accurate automated cell segmentation for image analysis. If available, cells can be seeded using the TiterTek Multidrop (TiterTek) or comparable equipment. Otherwise, multi-channel pipettors can be used, albeit with less throughput and accuracy.
4. Allow cells to attach and incubate in experimental media for 48 h.

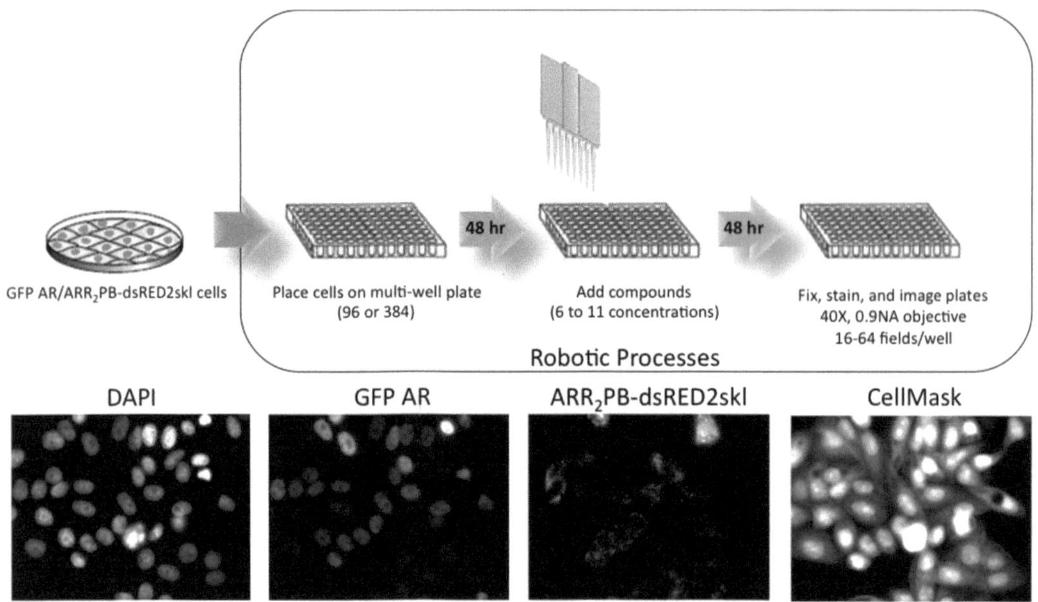

Fig. 18.2. Experimental workflow to study AR activation. (**a**) HeLa cells with stable expression of GFP-AR and ARR_2PB-dsRED2skl are maintained in sufficient culture size for the desired experiment. 48 h before ligand exposure, cells are trypsinized and seeded to either 96- or 384-well plates. After 48 h of cell attachment and synchronization in hormone-free media, cells are treated for 24–48 h with compounds of interest (6–11 concentrations) and fixed and stained using robotic protocols (Biomek FXS8). Automated imaging is carried out at 40× using a dry 0.9 NA objective with the Beckman IC 100 HTM. Images are collected in four colors, DAPI (*blue*), GFP-AR (*green*), reporter (*red*), and cell mask (*Far red*).

5. In multi-well format, expose cells to 10–11 compounds with 8–11 doses. An example plate layout for 384 wells is shown in **Fig. 18.3**. Columns 21, 22, and 23 are used for agonist (10 nM R1881), antagonist (1 μM OHF), and vehicle controls. The other 20 rows across the top are used for 10 different compounds. In this example, we exposed GFP-AR/ARR_2PB-dsRED2skl cells to 10 common AR agonists, antagonists, and endocrine disruptors. A maximum concentration is prepared for row 1 and dilutions are performed using the BioMek S8 (Beckman Coulter, Brea, CA), all at 2× concentrations; we also use the BioMek S8 to transfer compounds (30 μl) to each well.
6. Following compound transfer, incubate cells for 24–48 h to allow for simultaneous detection of AR translocation and reporter expression (*see* **Note 10**).

3.4. Plate Processing and Image Acquisition by HTM

Subsequent to compound treatment, plates are transferred to a microfluidic robot for plate processing. The buffers and solutions needed for the robotic protocols are described above. All solutions are filter-sterilized and stored at 4°C until needed. PBS and TBS-T buffers can be pipetted from basins; also, upside down

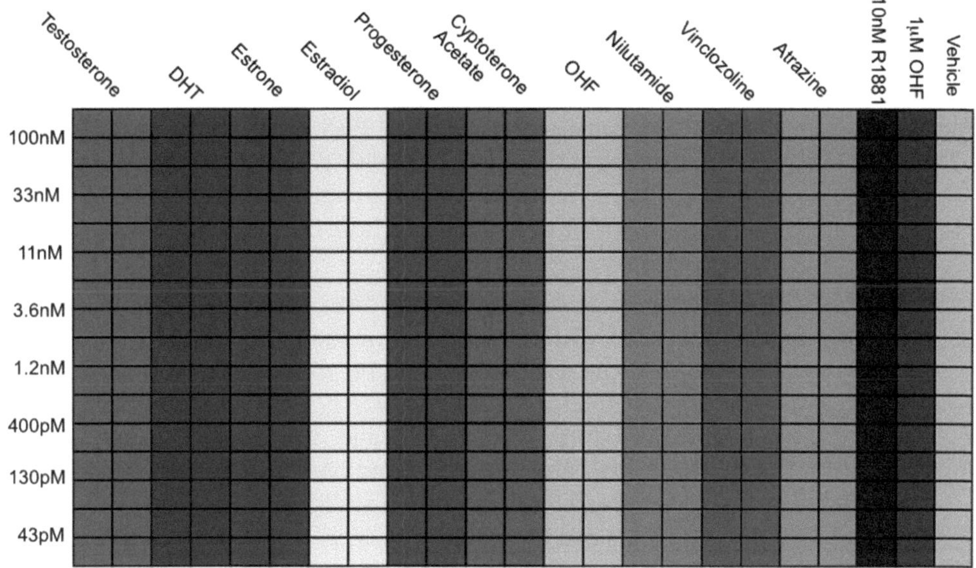

Fig. 18.3. Example plate layout for HCA of AR agonists, antagonists, and endocrine disruptors. Concentrations decrease from rows A–H, while the ligands vary across columns 1–20. Each experimental treatment is performed in quadruplicate.

tip box lids can suffice. Other reagents such as formaldehyde, quench, Triton X-100, and DAPI are set up in modular basins that fit the microfluidic robot. For plate processing protocols, we regularly make 2 l PEM, connected directly to the plate washer. After buffers are prepared and organized in appropriate basins, the following steps are used to maximize signal integrity and minimize background noise:

1. Microfluidic robots and plate washer can be programmed to achieve final aspirate heights that leave an approximate residual volume of 5–7 μl. This is the lowest the aspiration can go without disrupting the cell monolayer. Aspiration of each well is performed in the back right corner to minimize disruption of the cell monolayers by fluid transfers.
2. Transfer plates to robot pods, where all aspects of the protocol will be controlled by software. All volumes added to wells are 25 μl.
3. Wash two times with cold PBS++, add 4% PFA in PEM (initially cold to minimize cell autolysis and then allow to warm to RT). Incubate 20 min at RT.
4. Wash three times with PEM, add 100 mM NH_4Cl (quench). Incubate 15 min at RT. A shorter incubation time can be used with an alternate quenching agent, $NaBH_4$ (5 min); however, NaBH4 is a naturally bubbling agent that can complicate pipetting. Both solutions should be prepared fresh.
5. Wash three times with PEM, add 0.5% Triton X-100, and incubate 5 min RT.

6. Wash three times with PEM and once with TBS-T.
7. Stain cells with DAPI and CellMask FarRed 1:200,000 or CellMask Blue in PEM. Incubate 45 min at RT (*see* **Note 5**).
8. Aspirate wells, add 60 μl PBS++/0.01% sodium azide. Seal plates with adhesive or with parafilm to reduce evaporation. Plates are now ready to image. To hold plates more than a day prior to imaging, include 0.4% formaldehyde as a preservative.
9. Acquire images using HTM or microscope with automated stage with either laser- or image-based auto-focus. Images for the GFP-AR/ARR$_2$PB-dsRED2skl HeLa cell line are collected with a 40 × 0.9 NA S-Fluor objective (or 40 × 0.95 NA) (Nikon, Melville, NY). 8-bit images, written as bitmap files, are generated with 2 × 2 binning (672 × 512 pixels, 0.344 × 0.344 μm^2/pixel). For 384-well plates, 16 fields can be acquired (4 × 4), while 96-well formats permit the collection of 8×8 fields.

3.5. Image Analysis

1. Export data to image analysis server/database (*see* **Note 7**).
2. Remove background fluorescence from the images by subtraction of the minimum pixel intensity (MIP) plus noise (the square root of the MIP) from each image. This also corrects for variable background across different wells and helps to ensure measurements across images are comparable (*see* **Note 11**).
3. Segment nuclei to permit both whole cell segmentation and nucleus measurement extraction. Apply a white top-hat transform to the DAPI images to extract peaks in the image corresponding to the nuclei. Threshold the image using a global, K-means threshold. Apply binary morphology operations to smooth out the resulting objects, giving tight nucleus regions.
4. Segment cells to extract single cell-level features from the whole cell. Invert CellMask image and use seeded watershed manipulations, with seeds defined by the nucleus regions. This splits the image into single-cell regions. The original CellMask image is thresholded to foreground from background, and this binary image is masked by the watershed result to produce tightly segmented cells. Binary morphology operations are performed to smooth the cell regions; the cytoplasm is segmented by subtracting the nucleus from cell regions.
5. After segmentation, cells touching the edge of the image field are removed. A sample segmentation schematic is shown in **Fig. 18.4**.

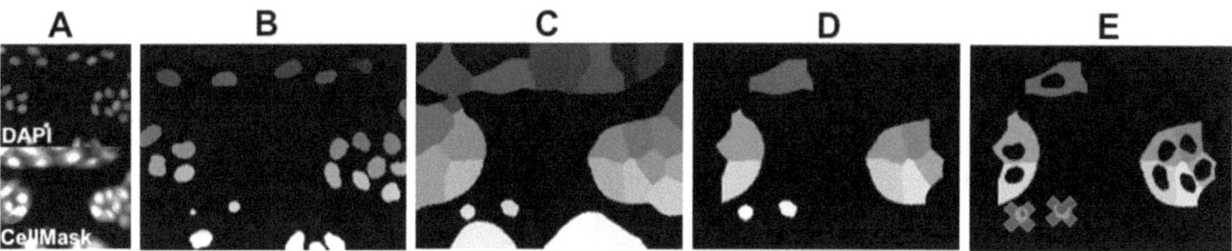

Fig. 18.4. Image segmentation using Pipeline Pilot. (**a**) Raw images are grouped into an image reader. (**b**) Nuclear and (**c**) whole cell masks are created to extract cell-by-cell measurements (**d**) with removal of edge touching regions. (**e**) Cytoplasmic masks are generated by subtraction of the nuclear mask from the whole cell masks. Filtering (*gray strikethroughs*) is used to discard dead cells or mis-segmented objects.

3.6. Feature Extraction Using the Pipeline Pilot Advanced Imaging Collection

Extract androgen receptor (AR) object and intensity features using the AR images in the image region shape statistics component (IRSC) and image region intensity statistics component (IRISC).

1. Extract AR in nucleus-object intensity and object colocalization features using AR and nucleus-regions images in the IRISC and the object colocalization component (OCC).
2. Extract AR in cytoplasm-object intensity and object colocalization features using the AR and cytoplasm-regions images in the IRISC and OCC.
3. Extract the IRISC and OCC features for the reporter in the whole cell using the reporter and cell-regions images.
4. For filtering purposes, extract IRSC features from the nucleus- and cell-regions images.
5. Remove cells that have misshapen nuclei (too large, too small, and not ellipsoidal enough) or that are expressing AR outside of the physiological range (*see* **Note 12**).
6. From the features extracted, derive translocation, hyperspeckling, and reporter activity features. The nuclear/cytoplasmic ratio, representative of AR translocation, is determined by measuring the percent of total AR signal (sum of pixel intensities) within the nucleus mask. Nuclear "hyperspeckling" is the statistical variance in AR pixel intensity within the nucleus mask. Accumulation of the AR-sensitive ARR2PB-dsRED2skl fluorescent reporter protein is measured as the sum of pixel intensities under the cytoplasmic mask. These measurements are highlighted in **Fig. 18.5a** for all 10 compounds using a heat map representation (generated in Python).

3.7. Data Reduction

In order to define treatment "fingerprints," we perform hierarchical clustering using responses at the treatment level (single cell-based measurements of features are averaged at the well

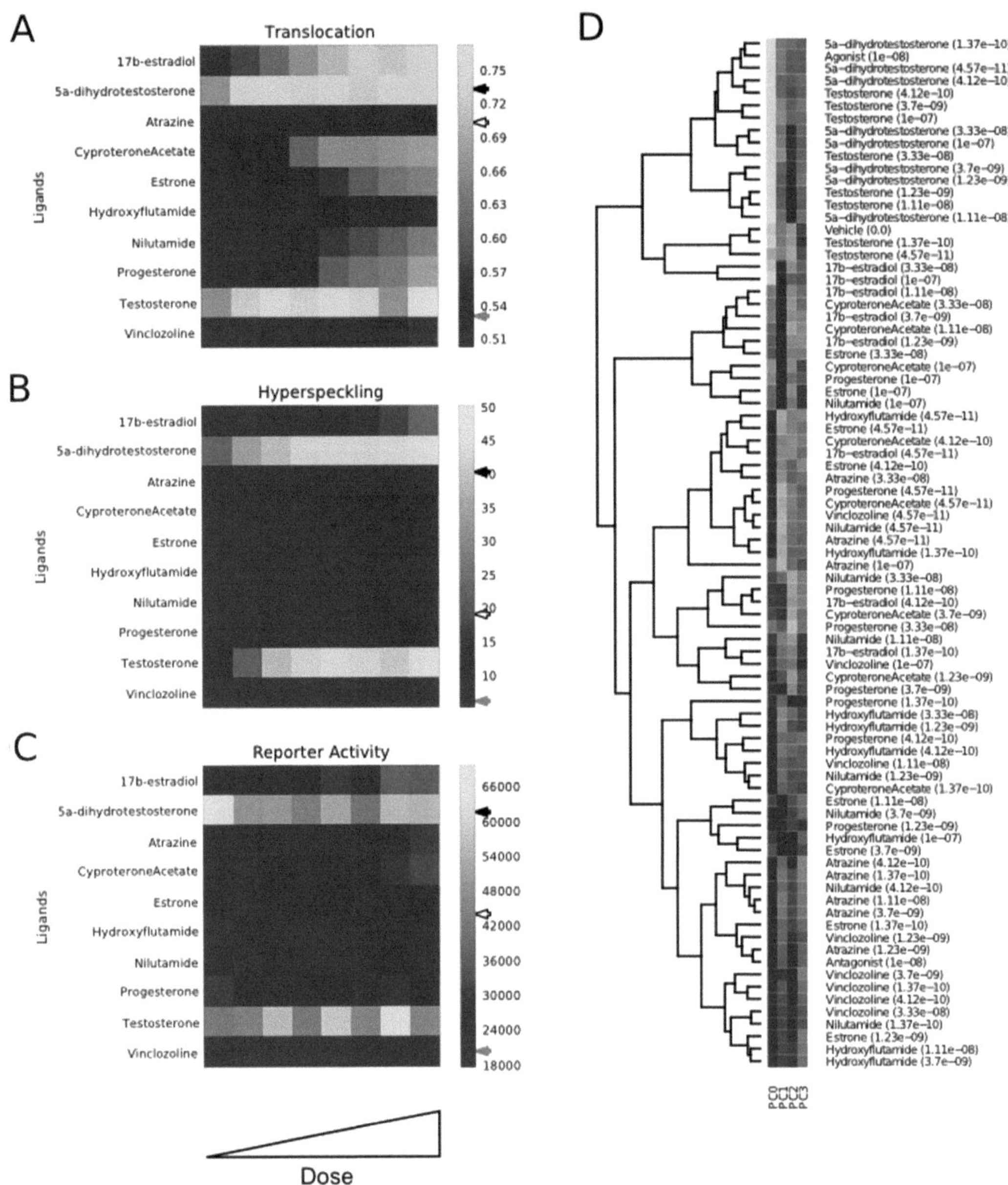

Fig. 18.5. Example data generated from exposure of the AR biosensor cell line to a group of AR effectors. Dose response data are depicted for the three central measurements: (**a**) nuclear AR/cellular AR, (**b**) nuclear hyperspeckling, and (**c**) ARR_2PB-dsREDskl activity. Note that agonist, antagonist, and vehicle treatments are denoted on the color bars by *black*, *gray*, and *white arrows*, respectively. (**d**) Hierarchical cluster analysis and associated heat map responses of image-derived features over the dose regime of the 10 compounds in the experimental design. Four PCA features (PC0, PC1, PC2, and PC3) capture 80% of the data set variance.

level across treatments). Sorting through hundreds of features can be challenging; furthermore, not all features are informative in defining a unique treatment type. In order to produce a compact fingerprint, principal component analysis (PCA; (20)) can be used to reduce measurements to a smaller set of features useful for grouping compound treatments. Features derived from cytological measurements and object/shape colocalization functions are highly effective in distinguishing between subcellular patterns as has been established for location proteomics studies (21). Increasingly these features are being integrated into assays for mechanism- and/or screening-based applications. Considerable effort has been put into developing and using phenotype-based assays for siRNA or compound library screens (22). Open source-based feature processing and selection was performed in Python 2.6.4, while clustering was done in R 2.10, using the heatmap.plus package. Our group has wrapped these methods in Pipeline Pilot to produce graphics-based workflows that can be utilized and modified by end users. Other recent examples (5–7, 10, 23) of this type of analytical approach have been used in a similar manner to maintain data-rich features, but provide a simpler interpretation of a handful of condensed variables as opposed to hundreds of esoteric measurements. A basic approach for data reduction of HCA data sets is as follows:

1. Calculate the average (median) feature values across each field (using the cell-level features). Then, calculate the average feature values across each well (using the field-level features). Finally calculate the average feature values across each treatment (using the well-level data). This step can reduce the data set to a more manageable number of samples (dealing with dozens to thousands of treatments instead of hundred of thousands to tens of millions of cells).
2. Standardize features by subtracting from them their mean and then dividing them by their standard deviation.
3. Normalize treatment samples by dividing them by their magnitude.
4. Perform PCA to reduce measurements to a smaller set of features that are useful for grouping compound treatments. Set the PCA to return features that capture 75% of the variance in the data set. Increasing the variance parameter in PCA causes it to return more features.
5. Using the PCA features, perform hierarchical clustering on processed features. Cluster using the Euclidean distance metric (**Fig. 18.5b**).

4. Notes

1. HTM and HCA are increasingly being adopted as standardized approaches to understand the complex biology of cell signaling, not just nuclear receptors and gene regulation. Further ease of use and speed-associated developments in auto-focus, image analysis, illumination, and data reduction/visualization will continue to progress HCA. As other, more advanced technologies, like FRET, FLIM, and RNA FISH, become integrated into HCA approaches, the untapped biology of the genome will be more readily made accessible. Despite the progress of HCA, the use of complex data reduction approaches to understand AR cytology, and other proteins for that matter, remains in its infancy. A merger of structure–activity relationships with multiplex measurement capability, available only by HCA, should provide detailed mechanistic insights to guide pharmacological leads that modulate nuclear receptor activity and other difficult drug discovery bottlenecks.
2. For all AR response experiments, cells are plated and treated in phenol red-free DMEM with 5% charcoal-stripped, dialyzed FBS (SD-FBS). SD-FBS can be prepared from dialyzed FBS with dextran/charcoal stripping of hormones.
3. When using HCA as a means to quantify cellular response, the choice of well plate is critical. We have found that Greiner SensoPlate Plus 96 and 384 low skirt glass plates provide both excellent image quality (low background in FITC/GFP and dsRED2/Texas Red/mCherry) and appear free of intrinsic estrogenic activity from plate plastic, sealant, or adhesive derivatives. The issue of multi-well plate-derived estrogenic activities from (presumably) plasticizers (24) like bisphenol A is a much unappreciated problem in general cell culture, and we have utilized our sensitive estrogen receptor HCA model (11, 25) to empirically eliminate "estrogenic" plates from all of our steroid receptor studies. For strong antibody or XFP signals, Aurora 384 optical bottom plastic plates are a markedly less-expensive alternative, do not possess estrogenic activity, and do not require pre-experiment coating; low antibody or XFP signals, however, can be obscured by an intrinsic high background signal specifically in the dsRED2/Texas Red/mCherry wavelengths.
4. Many cell types, including HeLa or HeLa stables, do not adhere well to glass surfaces. Hence, we have used both

poly-D-lysine (PDL) and FBS coatings to facilitate optimal cell adhesion. For experiments involving several 96- or 384-well plates, coatings can be added using the TiterTek Multidrop or equivalent cell plating systems. In brief, coating protocols involve overnight treatment with PDL or FBS prior to cell seeding; additional details on acid etching and coating of coverslips can be found in a previous methods chapter (12).

5. We predominantly use CellMask Blue or CellMask FarRed to allow segmentation of cell bodies. It is essential to determine empirical working concentrations for each experiment and cell type. We find that CellMask Blue generally works best at 0.1 μg/ml (in PEM) while CellMask FarRed at 5 ng/ml (in PEM). It is important to indicate that for the FarRed CellMask variant, excessive incubation (time or concentration) can result in bleeding of the dyes into dsRED2/Texas Red/mCherry fluorescence filter sets. CellMask Blue has the advantage of opening up the Cy5 channel for antibodies while using the UV channel specifically for detecting the cell body features. When staining is optimal, lower exposures can be used to detect only the nucleus outline from DAPI while longer exposures slightly saturate the DAPI signal while allowing the cell edge detection with CellMask Blue. It is important to note that combinations of CellMask Blue and DAPI may be detrimental for detecting both cell border/size properties and cell cycle position as CellMask dyes also label the nucleus.
6. Our lab has predominantly used the Cell Lab IC 100 Image Cytometer (IC 100) platform (Beckman Coulter) for high-resolution HTM. The imaging platform consists of (1) Nikon Eclipse TE2000-U inverted microscope (Nikon, Melville, NY) with a Chroma 82000 triple band filter set (Chroma, Brattleboro, VT); (2) Hamamatsu ORCA-ER digital CCD camera (Hamamatsu, Bridgewater, NJ); and (3) Photonics COHU progressive scan camera (Photonics, Oxford, MA). Due to auto-focus feedback loops, the IC 100 performs accurate and fast image-based focus, minimizing image acquisition times and providing more accurate focus compared to laser reflection methods.
7. We use the Pipeline Pilot data management, workflow, and analysis software platform (Accelrys, San Diego, CA; www.accelrys.com). Other platforms available that perform similar functions include CyteSeer ((26, 27); Vala Sciences, San Diego, CA; www.valasciences.com) and CellProfiler ((28, 29); www.cellprofiler.org; Broad Institute, MIT), in addition to software that comes with equipment manufacturers. We choose to use Pipeline Pilot because it provides

us with a single software package that is highly customizable for all image and data analysis methods that further allows us to report data internally or to third party software. As many industrial, biotech, and research environments have adopted Pipeline Pilot, protocols can be easily transferred and exchanged. However, research groups engaging in HCA generally find it necessary for HTM users to devote sufficient time learning the functionalities to apply PLP or similar software adequately and efficiently.

8. Upon single cell sorting the GFP-AR/ARR_2-dsRED2skl biosensor cell line, 81 single cells clones were isolated, with 2 clones exhibiting marked androgen responsiveness; the clone that responded maximally to DHT was used for all additional studies and maintained in phenol red-free DMEM/5% FBS growth media with 1.5 μg/ml puromycin and 200 μg/ml hygromycin. We additionally added 0.1 μM *o*-hydroxyflutamide (OHF) to cultures during routine maintenance to minimize basal dsRED2skl expression, and we removed OHF 24 h prior to experimental setup.

9. To determine the quality of the assay, we use the *Z*-factor calculation (30) where σ represents the standard deviation of both positive (10 nM DHT or R1881) and negative (vehicle) controls and μ represents the mean of the populations (equation [1]). Z' values of 1 are theoretically "perfect" and values between 0.2 and 0.6 are typical for cell-based assays.

$$Z' = 1 - \frac{(3\sigma_{+\text{control}} + 3\sigma_{-\text{control}})}{|\mu_{+\text{control}} - \mu_{-\text{control}}|} \qquad [1]$$

10. To detect unknown antagonists, titrate compounds against 0.5–10 nM DHT or R1881 as described previously (15). As anti-androgen concentration is increased against constant agonist, an inhibition of transcriptional activity is detected. Depending on the antagonist, reduced transcriptional competence may accompany effects on AR subnuclear organization and localization. This approach is particularly useful in terms of not only defining a transcriptional antagonist but also, simultaneously, reporting on possible mechanisms in play. Here, the time and cost-efficiency of an HCA approach is readily apparent.

11. We have additionally created alternative background subtraction methods that remove background signal from all images using plate- and channel-specific correction images. An average of >500 randomly selected images for each channel are amplified with a correction factor based on the

median pixel intensity of experimental images. The combination of correction factor and average image (from randomly selected images) allows for a pixel-by-pixel subtraction that compensates any consistent, uneven background artifacts in the image set. Object detection algorithms are highly dependent on the specific fluorescent marker morphologies and signal-to-noise properties of the HCA assay. Reliable identification methods and background subtraction methods are key for accurate quantification of image data sets.

12. Data streams, containing cell feature data, can be filtered to remove spurious sources of error by removing cell aggregates, mitotic cells, cellular debris, or events with expression levels that are outside the physiological range. The applied filters are based upon cell morphological measurements derived from cell and nucleus segmentation: nucleus area, nucleus circularity, nucleus/cell size ratio, and DNA content (derived from DAPI). Additional filters are used to remove cells with non-physiological expression of GFP-AR expression, using levels found in LnCaP as a reference (14, 16). Generally, these filters remove 10–15% of the cells in our GFP-AR HeLa cell line with individual limits determined by analyzing the Gaussian distribution of pixel intensities (e.g., 5% outliers); in transient GFP-AR transfection assays, up to ~70% of the cells are generally discarded (14–16).

Acknowledgments

This work was funded by NIH 5R01DK055622, the Hankamer Foundation, DOD Prostate Cancer Research Program (DAMD W81XWH-10-1-0390) and pilot grant, and equipment support from the John S. Dunn Gulf Coast Consortium for Chemical Genomics (MA Mancini). Additional funding was provided by NIH 1F32DK85979 (SM Hartig), 5T32HD007165 (BW O'Malley), and 5K12DK083014 (DJ Lamb). Imaging resources were supported by SCCPR U54 HD-007495 (BW O'Malley), P30 DK-56338 (MK Estes), P30 CA-125123 (CK Osborne), and the Dan L. Duncan Cancer Center of Baylor College of Medicine. The authors thank the members of the Mancini Lab for thoughtful discussion and J.H. Price (Vala Sciences) and Tim Moran (Accelrys) for longstanding support in automated cytometry.

References

1. Zitzmann, M. (2009) Testosterone deficiency, insulin resistance and the metabolic syndrome, *Nat. Rev. Endocrinol.* **5**, 673–681.
2. Echeverria, P. C., and Picard, D. (2010) Molecular chaperones, essential partners of steroid hormone receptors for activity and mobility, *Biochim. Biophys. Acta Mol. Cell Res.* **1803**, 641–649.
3. Rahman, F., and Christian, H. C. (2007) Non-classical actions of testosterone: an update, *Trends Endocrinol. Metab.* **18**, 371–378.
4. Bakal, C., Aach, J., Church, G., and Perrimon, N. (2007) Quantitative morphological signatures define local signaling networks regulating cell morphology, *Science* **316**, 1753–1756.
5. Loo, L. H., Lin, H. J., Steininger, R. J., Wang, Y. Q., Wu, L. F., and Altschuler, S. J. (2009) An approach for extensibly profiling the molecular states of cellular subpopulations, *Nat. Methods* **6**, 759–U718.
6. Loo, L. H., Wu, L. F., and Altschuler, S. J. (2007) Image-based multivariate profiling of drug responses from single cells, *Nat. Methods* **4**, 445–453.
7. Perlman, Z. E., Slack, M. D., Feng, Y., Mitchison, T. J., Wu, L. F., and Altschuler, S. J. (2004) Multidimensional drug profiling by automated microscopy, *Science* **306**, 1194–1198.
8. Ramadan, N., Flockhart, I., Booker, M., Perrimon, N., and Mathey-Prevot, B. (2007) Design and implementation of high-throughput RNAi screens in cultured Drosophila cells, *Nat. Protoc.* **2**, 2245–2264.
9. Tanaka, M., Bateman, R., Rauh, D., Vaisberg, E., Ramachandani, S., Zhang, C., Hansen, K. C., Burlingame, A. L., Trautman, J. K., Shokat, K. M., and Adams, C. L. (2005) An unbiased cell morphology-based screen for new, biologically active small molecules, *PLoS. Biol.* **3**, 764–776.
10. Young, D. W., Bender, A., Hoyt, J., McWhinnie, E., Chirn, G. W., Tao, C. Y., Tallarico, J. A., Labow, M., Jenkins, J. L., Mitchison, T. J., and Feng, Y. (2008) Integrating high-content screening and ligand-target prediction to identify mechanism of action, *Nat. Chem. Biol.* **4**, 59–68.
11. Berno, V., Amazit, L., Hinojos, C. A., Zhong, J., Mancini, M. G., Sharp, Z. D., and Mancini, M. A. (2008) Activation of estrogen receptor-alpha by E2 or EGF induces temporally distinct patterns of large-scale chromatin modification and mRNA transcription, *PLoS One* **3**, e2286.
12. Berno, V., Hinojos, C. A., Amazit, L., Szafran, A. T., and Mancini, M. A. (2006) High-resolution, high-throughput microscopy analyses of nuclear receptor and coregulator function, in *Measuring Biological Responses With Automated Microscopy* (Methods in Enzymology, Inglese, J., ed), pp 188–210, Academic, Waltham, MA.
13. Marcelli, M., Stenoien, D. L., Szafran, A. T., Simeoni, S., Agoulnik, I. U., Weigel, N. L., Moran, T., Mikic, I., Price, J. H., and Mancini, M. A. (2006) Quantifying effects of ligands on androgen receptor nuclear translocation, intranuclear dynamics, and solubility, *J. Cell. Biochem.* **98**, 770–788.
14. Szafran, A. T., Hartig, S., Sun, H., Szwarc, M., Chen, Y., Mediwala, S., Bell, J., McPhaul, M. J., Mancini, M. A., and Marcelli, M. (2009) Androgen receptor mutations associated with androgen insensitivity syndrome: A high content analysis approach leading to personalized medicine, *PLoS One* **4**, e8179.
15. Szafran, A. T., Szwarc, M., Marcelli, M., and Mancini, M. A. (2008) Androgen receptor functional analyses by high throughput imaging: Determination of ligand, cell cycle, and mutation-specific effects, *PLoS One* **3**, e6205.
16. Szafran, A. T., Szwarc, M., Marcelli, M., and Mancini, M. A. (2008) High throughput multiplex image analyses for androgen receptor function, *Proc Intl Symp Biomed Imaging* **5**, 320–324.
17. Jones, J. O., Bolton, E. C., Huang, Y., Feau, C., Guy, R. K., Yamamoto, K. R., Hann, B., and Diamond, M. I. (2009) Non-competitive androgen receptor inhibition in vitro and in vivo, *Proc. Natl. Acad. Sci USA* **106**, 7233–7238.
18. van Royen, M. E., Cunha, S. M., Brink, M. C., Mattern, K. A., Nigg, A. L., Dubbink, H. J., Verschure, P. J., Trapman, J., and Houtsmuller, A. B. (2007) Compartmentalization of androgen receptor protein-protein interactions in living cells, *J. Cell Biol.* **177**, 63–72.
19. Zhang, J. F., Thomas, T. Z., Kasper, S., and Matusik, R. J. (2000) A small composite probasin promoter confers high levels of prostate-specific gene expression through regulation by androgens and glucocorticoids *in vitro* and *in vivo*, *Endocrinology* **141**, 4698–4710.
20. Wall, M. E., Rechtsteiner, A., and Rocha, L. M. (2003) Singular value decomposition and principal component analysis, in

A Practical Approach to Microarray Data Analysis (Berrar, D. P., Dubitzky, W., Granzow, M., and Norwell, M. A., eds), pp 91–109, Kluwer, Dordrecht.

21. Glory, E., and Murphy, R. F. (2007) Automated subcellular location determination and high-throughput microscopy, *Dev. Cell* **12**, 7–16.
22. Neumann, B., Walter, T., Heriche, J. K., Bulkescher, J., Erfle, H., Conrad, C., Rogers, P., Poser, I., Held, M., Liebel, U., Cetin, C., Sieckmann, F., Pau, G., Kabbe, R., Wunsche, A., Satagopam, V., Schmitz, M. H. A., Chapuis, C., Gerlich, D. W., Schneider, R., Eils, R., Huber, W., Peters, J. M., Hyman, A. A., Durbin, R., Pepperkok, R., and Ellenberg, J. (2010) Phenotypic profiling of the human genome by time-lapse microscopy reveals cell division genes, *Nature* **464**, 721–727.
23. Newberg, J., and Murphy, R. F. (2008) A framework for the automated analysis of subcellular patterns in human protein atlas images, *J. Proteome Res.* 7, 2300–2308.
24. McDonald, G. R., Hudson, A. L., Dunn, S. M. J., You, H. T., Baker, G. B., Whittal, R. M., Martin, J. W., Jha, A., Edmondson, D. E., and Holt, A. (2008) Bioactive contaminants leach from disposable laboratory plasticware, *Science* **322**, 917.
25. Sharp, Z. D., Mancini, M. G., Hinojos, C. A., Dai, F., Berno, V., Szafran, A. T., Smith, K. P., Lele, T. T., Ingber, D. E., and Mancini, M. A. (2006) Estrogen-receptor-alpha exchange and chromatin dynamics are ligand- and domain-dependent, *J. Cell Sci.* **119**, 4101–4116.
26. McDonough, P. M., Agustin, R. M., Ingermanson, R. S., Loy, P. A., Buehrer, B. M., Nicoll, J. B., Prigozhina, N. L., Mikic, I., and Price, J. H. (2009) Quantification of lipid droplets and associated proteins in cellular models of obesity via high-content/high-throughput microscopy and automated image analysis, *Assay Drug Dev. Technol.* 7, 440–460.
27. Prigozhina, N. L., Zhong, L., Hunter, E. A., Mikic, I., Callaway, S., Roop, D. R., Mancini, M. A., Zacharias, D. A., Price, J. H., and McDonough, P. M. (2007) Plasma membrane assays and three-compartment image cytometry for high content screening, *Assay Drug Dev. Technol.* **5**, 29–48.
28. Carpenter, A. E., Jones, T. R., Lamprecht, M. R., Clarke, C., Kang, I. H., Friman, O., Guertin, D. A., Chang, J. H., Lindquist, R. A., Moffat, J., Golland, P., and Sabatini, D. M. (2006) CellProfiler: image analysis software for identifying and quantifying cell phenotypes, *Genome Biol.* 7, 30.
29. Lamprecht, M. R., Sabatini, D. M., and Carpenter, A. E. (2007) CellProfiler(TM): free, versatile software for automated biological image analysis, *Biotechniques* **42**, 71–75.
30. Zhang, J. H., Chung, T. D., and Oldenburg, K. H. (1999) A simple statistical parameter for use in evaluation and validation of high throughput screening assays, *J. Biomol. Screen.* **4**, 67–73.

Section VI

Cross-talk of Androgen and Other Signalling Pathways

Chapter 19

Androgen Regulation of ETS Gene Fusion Transcripts in Prostate Cancer

Delila Gasi and Jan Trapman

Abstract

Fusion between androgen-regulated *TMPRSS2* and ETS transcription factor gene *ERG* is the most frequent genetic alteration that occurs in 40–70% of prostate cancers. Not only *ERG* but also other ETS transcription factor genes are involved in gene fusions. *ETV1*, *ETV4*, and *ETV5* have all several fusion partners. One common feature shared by the majority of these partners is androgen-regulated expression. Despite its high frequency, the biological and molecular effects of ETS gene fusion in prostate cancer development and progression are unknown. In this chapter quantitative polymerase chain reaction (Q-PCR) is used for detection and further studying the incidence and properties of these fusion transcripts. The focus is on the expression of *TMPRSS2–ERG* transcripts in clinical prostate samples. Androgen regulation of *TMPRSS2* is measured in commonly used LNCaP prostate cancer cells grown with and without the synthetic androgen R1881. Furthermore, combining Q-PCR with 5′ RLM-RACE and sequencing are described for the identification of novel ETS fusion partners.

Key words: Prostate cancer, TMPRSS2–ERG fusion, ETS transcription factors, androgen regulation, quantitative PCR.

1. Introduction

Gene fusions are common genetic alterations in prostate cancer. The fusion between *TMPRSS2* and the ETS transcription factor gene *ERG* has been reported in 40–70% of prostate cancers (1–4). *TMPRSS2* is an androgen-regulated gene that is preferentially expressed in the prostate. *ERG* is a known oncogene and due to the fusion with *TMPRSS2* it is androgen-regulated overexpressed in the prostate. The high frequency of *TMPRSS2–ERG* fusion is probably due to the close proximity of the two

F. Saatcioglu (ed.), *Androgen Action*, Methods in Molecular Biology 776,
DOI 10.1007/978-1-61779-243-4_19, © Springer Science+Business Media, LLC 2011

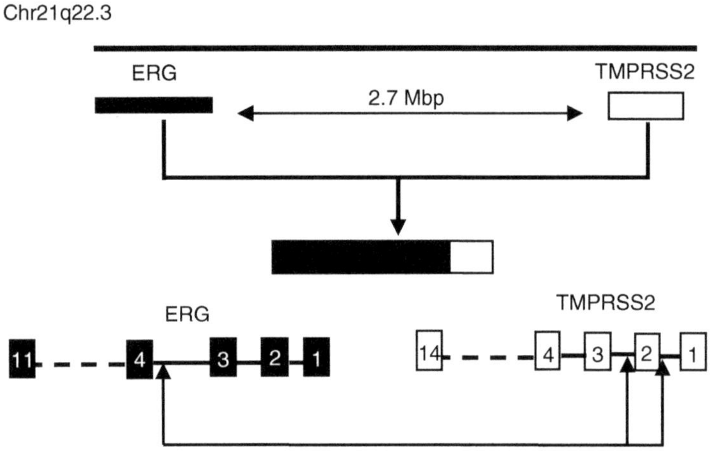

Fig. 19.1. Schematic representation of the fusion between *TMPRSS2* and *ERG*. Both genes are located in the same orientation on chromosome 21, separated by 2.7 Mbp. The *black arrows* indicate the most frequently occurring fusion events, which are either in intron 1 or 2 of *TMPRSS2* or in intron 3 *of ERG*.

genes on chromosome 21 (*see* **Fig. 19.1**). Most breakpoints in *TMPRSS2* are in intron 1 or 2 and most breakpoints in *ERG* are in intron 3; however, occasionally other fusions have been found. Moreover, alternative splicing and differential promoter usage have been described. It is assumed that the major ERG protein that is translated from *TMPRSS2–ERG* fusion transcripts is a slightly N-terminal truncated version of the protein. In addition to *TMPRSS2*, two other *ERG* fusion partners, *SLC45A3* and *NDRG1*, have been described (*see* **Table 19.1**). Fusions with these genes occur at much lower frequencies. Both fusion partners share with *TMPRSS2* the property of androgen-regulated expression (5, 6).

Fusion of ETS factor genes *ETV1*, *ETV4*, and *ETV5* occurs at low frequencies in prostate cancer (7–9). *ETV1* has at least 10 fusion partners, but again, almost all of its partners are androgen regulated and most are also prostate-specific expressed (*see* **Table 19.1**). Different from *ERG*, *ETV1* can be overexpressed also as the full-length gene.

ETS transcription factors are important mediators of many biological processes including proliferation, metastasis, angiogenesis, and growth, all also important in tumor development (10). The precise role that the fusion transcripts play in prostate cancer is still unclear. To understand the molecular mechanisms that determine the specific gene fusions, it is important to elucidate their common regulatory expression mechanisms and to distinguish in which stages of prostate cancer they occur.

Because the gene fusions are unique, they are promising functional biomarkers for diagnosis of prostate cancer (11). However,

Table 19.1
Fusion partners of ETS genes in prostate cancer

5′ fusion partner	Prostate specific	Androgen regulated	3′ fusion partner	Present (%)
TMPRSS2 (chr 21q)	+	+	ERG (chr 21q)	40–70
SLC45A3 (chr 1q)	+	+		<1
NDRG1 (chr 8)	–	+		~1
TMPRSS2 (chr 21q)	+	+	ETV1 (chr 7p)	<1
FOXP1 (chr 3q)	ND	ND		<1
EST14 (chr 14q)	+	+		<2
HERVK17 (chr 17p)	+	+		<2
SLC45A3 (chr 1q)	+	+		<1
HERV-K_22q11.23	+	+		<1
C15orf21 (chr 15q)	+	+ (down)		<1
HNRPA2V1 (chr 7p)	–	–		<1
ACSL3 (chr 2q)	–	+		<1
CANT1 (chr 17)	+	+		<1

it is unclear whether they are also prognostic markers. So far, conflicting data have been published on this subject (12–14).

Several methods to detect gene fusions in clinical samples can be discriminated. For known fusion events, these include as classical methods interphase fluorescence in situ hybridization (FISH) and specific Q-PCR. More recently, genome-wide methods like genomic or transcript paired-end sequencing have been described as global approaches to find novel fusion events (15). Affymetrix exon microarrays can also function as a first step to identify genes overexpressed by gene fusion (16). Here we describe Q-PCR to identify and to monitor *TMPRSS2–ERG* expression in prostate cancer. Moreover, we describe 5′ RACE-PCR to discover novel fusion partners of ETS genes.

2. Materials

2.1. Cell Culture and Clinical Prostate Cancer Samples

1. LNCaP prostate cancer cells (ATCC).
2. Dulbecco's modified Eagle's medium (DMEM) (BioWhittaker/Lonza) supplemented with 5% fetal bovine serum (FBS), 100 U penicillin, and 100 U streptomycin (PS) stored at 4–8°C.

3. Trypsin (170.000 U/L)/EDTA (200 mg/L) (BioWhittaker/Lonza) stored at 4–8°C.
4. Phosphate-buffered saline (PBS) (BioWhittaker/Lonza) stored at room temperature (RT).
5. Synthetic androgen R1881 (NEN Chemicals) 10^{-6} M stock solution in 100% EtOH stored at –20°C in the dark (*see* **Note 1**).
6. 100% EtOH (Sigma Aldrich) stored at RT.
7. Clinical prostate cancer samples, fresh-frozen, and stored in liquid nitrogen.

2.2. RNA Isolation and cDNA Preparation

1. RNeasy Mini kit (Qiagen) for RNA extraction from LNCaP cells.
2. QIAshredder columns (Qiagen) for homogenization of the sample.
3. RNA-Bee solution (Bio-Connect) for RNA isolation from clinical prostate samples.
4. 100% and 70% EtOH (Sigma Aldrich) for RNA isolation from in vitro cell cultures and from clinical samples. 100% EtOH is diluted to 70% with distilled and autoclaved water (dH_2O).
5. Chloroform (Fluka) for RNA isolation with RNA-Bee solution.
6. 2-Propanol (Fluka) for RNA precipitation with the RNA-Bee method.
7. RNase-free DNase (Qiagen) to eliminate genomic DNA.
8. Spectrophotometer NanoDrop, ND-1000 with software v3.3.1 to measure RNA concentration at 260 nm.
9. SuperScript™ II Reverse Transcriptase kit (Invitrogen) for cDNA preparation. The kit contains 0.1M DTT, 5X first-strand buffer (250 mM Tris–HCl, pH 8.3; 375 mM KCl; 15 mM $MgCl_2$) and Moloney murine leukemia virus reverse transcriptase (M-MLV RT 200 U/μl), stored at –20°C.
10. 10 mM dNTP's 1:1:1:1 (Roche) for cDNA synthesis.
11. T12 primer (Invitrogen) for cDNA synthesis.
12. RiboLock RNase Inhibitor (Fermentas) for cDNA synthesis.
13. DNA Thermal cycler, Perkin Elmer 480.

2.3. Quantitative PCR

1. Power SYBR-green master mix (Applied Biosystems) for Q-PCR and for product detection.
2. Gene-specific and control gene primers (Invitrogen) for Q-PCR (manually designed).

3. MicroAmp fast optical 96-well reaction plate with barcode (0.1 ml) (Applied Biosystems).
4. MicroAmp optical adhesive film, PCR compatible, DNA/RNA/RNase-free (Applied Biosystems) to cover the plate in order to avoid evaporation of the samples.
5. 7900HT Fast Real-Time PCR system with SDS 2.3 software (Applied Biosystems).

2.4. 5′ RLM-RACE

1. GeneRacer Kit for full-length RNA ligase-mediated rapid amplification of 5′ and 3′ cDNA ends (RML-RACE) (Invitrogen).
2. 95%, 70% EtOH and dry ice.
3. Gene-specific primers to perform 5′ RLM-RACE, both for reverse transcribing mRNA and for the different PCR steps.
4. Biometra T1 Thermocycler (Westburg) for PCR.
5. Instead of Platinum Taq DNA polymerase high fidelity as suggested in the kit, QiaTaq DNA polymerase (Qiagen) can be used in the first PCR reaction "amplifying cDNA ends."
6. GoTaq flexi DNA polymerase (Promega) in case of nested PCR.

2.5. Sequencing

1. Bigdye terminator (BDT) v3.1 cycle sequencing kit (Applied Biosystems).
2. Biometra, T1 Thermocycler (Westburg).
3. 3130xl genetic analyzer and software for data collection v3.0 (Applied Biosystems).
4. MicroAmp™ optical 96-well reaction plates for sequence reactions (Applied Biosystems).
5. Gene-specific primers (3.2 pmol).
6. Sodium acetate (3M stock solution; pH 4.6) for DNA precipitation.
7. 95% and 70% EtOH for DNA precipitation and washing.
8. Formamide (Applied Biosystems) to dissolve the precipitated DNA.
9. DNAMAN and ChromasLite software for sequence analysis.

3. Methods

mRNA expression of *ERG* or another ETS factor gene can be measured by Q-PCR in clinical prostate cancer samples to identify overexpression of the gene. In this chapter product detection with SYBR-green dye is described. A fusion event is indicated by low expression of the first exons of the gene compared to high expression of the last exons of the same gene. This is because the gene fusion results in overexpression of the essential protein coding part of the ETS oncogene, regulated by the fusion partner, *TMPRSS2* or other.

If overexpression of (part of) an ETS mRNA is detected, the sample can be tested for fusion transcripts with previously described fusion partners by standard RT-PCR, using a forward primer in the ETS part and a reverse primer in the candidate fusion partner. If none of the known fusion partners is detected, this can be an indication of a novel fusion partner.

Taking the advantage of the fact that the 3′ end of the fusion transcript is a known ETS mRNA, it is possible to identify the unknown fusion partners by 5′ RLM-RACE, followed by sequencing of an amplified product. If a novel fusion partner is detected its frequency of expression can be studied by Q-PCR in a clinical sample cohort with fusion-specific primers. **Figure 19.2a**

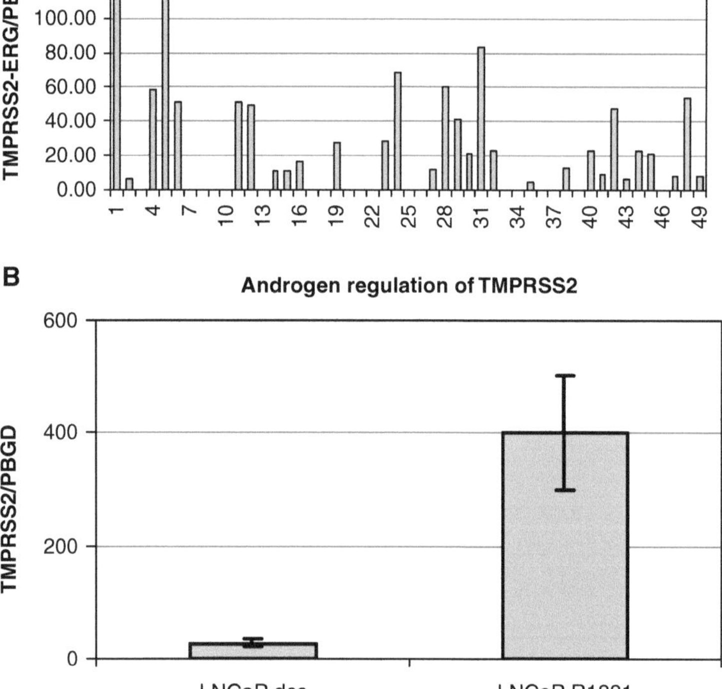

Fig. 19.2. (**a**) Expression of *TMPRSS2–ERG* fusion transcripts in 50 clinical prostate cancer samples measured by Q-PCR relative to the housekeeping gene *PBGD*. (**b**) Expression of *TMPRSS2* in LNCaP cells incubated with or without the synthetic androgen R1881 in DCC-supplemented medium for 24 h. The mRNA levels are depicted relative to *PBGD*.

shows an example of the analysis of *TMPRSS2–ERG* fusion transcript expression in prostate cancers.

Q-PCR can also be used to study changes in expression of a fusion gene in in vitro propagated cell lines under varying experimental conditions. This is important for studying the properties of a novel found fusion partner. In this chapter we show as an example the upregulation of *TMPRSS2* mRNA expression by incubating LNCaP prostate cancer cells in the presence of R1881 (**Fig. 19.2b**). Note that LNCaP cells do not harbor the *TMPRSS2–ERG* gene fusion.

3.1. Cell Culture and Clinical Prostate Cancer Samples

1. To study androgen regulation of *TMPRSS2*, LNCaP cells are grown in two T25 cell culture flasks to 50% confluence in Dulbecco's modified Eagle's medium (DMEM) supplemented with 5% FBS, and PS in a humidified incubator at 5% CO_2 and at 37°C .
2. After reaching 50% confluence the medium is replaced by DMEM supplemented with 5% dextran-coated charcoal (DCC; dextran 0.1%, charcoal 1%) treated FBS (i.e., steroid depleted).
3. 4 μl R1881 (10^{-6} M in EtOH) is added to one of the two T25 flasks in 4 ml medium, giving a final R1881 concentration of 10^{-9} M (*see* **Note 1**). To the second, control flask 4 μl 100% EtOH is added (*see* **Note 2**). Next, the cell cultures are incubated for 24 h (see above).
4. Medium is removed and cells are washed with PBS and harvested by trypsinization. Trypsinized cells are transferred to 15 ml tubes and pelleted by centrifugation for 5 min at 1000×*g*. RNA is isolated from the pellet as described in **Section 3.2**, Step 1–4.
5. Clinical prostate cancer samples contain at least 50% tumor cells, as determined by a pathologist. To this end 5 μm tissue sections are cut and stained by routine hematoxylin/eosine (HE) staining. RNA is isolated from 20 consecutive 100 μm sections. Finally again a 5 μm section is cut and HE stained. RNA is isolated from the pooled tissue sections as described in **Section 3.2**, Steps 5–11.

3.2. RNA Isolation and cDNA Preparation

1. For RNA isolation from LNCaP cells the RNeasy Mini kit (Qiagen) is used and the RNeasy Mini Handbook followed under the section "Purification of total RNA from Animal Cells" (*see* **Note 3**).
2. The pellet contains approximately 10^7 cells and the recommended buffer volumes for this number of cells are used.
3. QIAshredder tubes are used for homogenization of the lysed cells as the RNA yield is generally higher than

by pipetting or passing the sample through a needle with a syringe.

4. A DNase treatment step is performed to eliminate possible DNA contamination.
5. For RNA isolation of fresh-frozen clinical prostate cancer samples RNA-Bee (*see* **Note 4**) (Bio-Connect) is used according to the protocol provided by the supplier.
6. The tissue is homogenized in a Polytron homogenizer in RNA-Bee solution with 1 ml solution per 50 mg tissue.
7. 0.2 ml chloroform (*see* **Note 5**) is added to every 1 ml sample (homogenized tissue in RNA-Bee solution), vigorously shaken for 30 s and kept on ice for 5 min.
8. The solution is then centrifuged at 12,000×*g* for 15 min at 4°C. This separates the RNA to an aqueous top phase, which is transferred to a new tube.
9. RNA is precipitated by adding one sample volume 2-propanol and incubating at RT for 15 min.
10. Next, centrifuge to pellet the RNA at 12,000×*g* for 5 min at 4°C. Wash the pellet with 1 ml 75% EtOH, centrifuge at 7,500×*g* for 5 min at 4°C.
11. The pellet is air-dried and finally dissolved in 50 μl distilled and autoclaved water (dH_2O) by pipetting.
12. cDNA is prepared from 2 μg total RNA.
13. dH_2O is added to the RNA to a volume of 18 μl. 2 μl 100 ng/μl T12 primer is added and the sample is denaturated at 70°C for 10 min in the thermal cycler.
14. The sample is cooled on ice for 2 min and spinned down by a short centrifugation to collect the fluid at the bottom of the tube. Following reagents are added on ice: 8 μl 5× first-strand buffer, 4 μl 0.1 M DTT, 2 μl dNTP's, 0.5 μl RiboLock RNase Inhibitor, and 2 μl M-MLV RT (400 U).
15. The mixture is then incubated for 1 h at 37°C followed by 10 min at 95°C in a DNA thermal cycler. After the incubation samples are cooled on ice for at least 2 min, centrifuged shortly to collect fluid, and then stored at –20°C.

3.3. Quantitative PCR

1. To facilitate pipetting and the analysis of a larger scale Q-PCR experiment, a scheme of a 96-well plate where samples are pipetted is made before the start of the experiment. In addition to the samples that will be analyzed, a serial dilution of (mixture of) standard samples is added to each plate for preparation of a standard line. Other controls are

a minus RT sample and a genomic DNA sample. Expression of each gene is calculated relative to the expression of a housekeeping gene. So, all samples are tested also for the expression of this gene.

2. A standard line is composed of five serial fourfold dilutions of a mix of cDNA samples. 5 μl of four different samples expressing the gene of interest is mixed and diluted 10 times to 200 μl, this is standard 1. Values for standard line are set to 25600 for standard 1, 6400 for standard 2, and fourfold decreased till 100 for standard 5. These preset values in the standard line are later used to quantify the samples.
3. A primer mix is prepared from two 100 μM primer stock solutions by adding 10 μl of each primer to 580 μl dH_2O. For primer design, *see* **Notes 6–8**.
4. It is advisable to run a melting curve program with each primer set and the standard line. This gives peaks at the melting temperature of the samples for each different set of primers. Primers that anneal to each other (primer-dimers) can be detected because they will give a peak at a lower temperature than the product of interest. Normally in this experiment the standard line is run together with a water control and a DNA control as to make sure that the primers form a proper standard line (*see* **Note 9**), do not amplify DNA and do not form primer-dimers.
5. The housekeeping gene porphobilinogen deaminase (PBGD) is chosen as an internal control. Thus, PCR reactions are carried out with all cDNA samples and the standard line using PBGD primers (*see* **Note 10**). The PCR values of all samples are divided by their corresponding PBGD value to adjust for small differences in cDNA levels present between the samples.
6. cDNA is diluted 20 times and 5 μl diluted cDNA is added per well in the 96-well plate.
7. 5 μl of the five serially diluted standard samples is added to the plate for every primer pair.
8. 5 μl of the primer mix prepared in Step 3 of **Section 3.3** is added to each well.
9. 12.5 μl Power SYBR-green master mix (Applied Biosystems) and 2.5 μl dH_2O is added to each well. cDNA, primer mix, and SYBR-green PCR mix together give a total volume of 25 μl (*see* **Note 11**).
10. The 7900HT Fast Real-Time PCR system from Applied Biosystems with accompanying software SDS 2.3 is used for amplification. The exact settings are variable depending

on the primer annealing temperature (*see* **Note 12**). A PCR program can be designed as follows:
Step 1. 5 min at 95°C

Step 2. 15 s at 95°C

Step 3. 30 s at 60°C

Step 4. 30 s at 72°C, Steps 2–4 is repeated 39 cycles.

11. Analysis is performed using SDS 2.3 software. First the standard line and control points are defined. A threshold that will define the Ct values is determined (*see* **Note 13**). The quantification of the sample is automatically calculated by the program during analysis by comparing the Ct values (*see* **Note 14**) with those of the standard line and its pre-set values.

3.4. 5′-RACE

1. GeneRacer Kit, for full-length, RNA ligase-mediated rapid amplification of 5′- and 3′-cDNA ends (RML-RACE) (Invitrogen), is used and the manual provided by the supplier is followed with minor modifications.
2. In the first step, "dephosphorylating RNA," 5 μg total RNA is used.
3. For the reverse transcription step a gene-specific primer is used (2 pmol).
4. The first PCR step "Amplifying cDNA ends" is done according to the protocol, but using QiaTaq DNA Polymerase and 10× PCR buffer (Qiagen), dH_2O is added to a final volume of 50 μl.
5. The amplified products are separated over a 1% agarose gel by electrophoresis. Clear bands can be isolated for sequencing (*see* **Section 3.5**). Often the product of this step appears as a smear and a second, nested PCR, is needed to increase the specificity and the amount of product.
6. A nested PCR is performed as described in the protocol but using GoTaq flexi DNA polymerase (Promega) and 5× GoTaq flexi buffer. The buffer volume is then increased two times and dH_2O added to a final volume of 50 μl (*see* **Note 15**).

3.5. Sequencing

1. 5′-RACE-PCR products for sequencing are isolated from the agarose gel using QIAquick Gel Extraction kit 250 (Qiagen).
2. Check the yield and quality of DNA after isolation by measuring the absorbance at 260 nm and analyzing part of the sample on an agarose gel. The DNA is added to the reaction mix described in Step 3 of **Section 3.5** as final step.

3. Add on ice 1 μl BDT and 4 μl 5× sequencing buffer provided by the Bigdye terminator v3.1 cycle sequencing kit (Applied Biosystems). 1 μl of the gene-specific sequence primer (3.2 pmol) is added and dH_2O is added to a final volume of 20 μl.
4. The Biometra T1 Thermocycler (Westburg) is programmed for the sequencing reaction as follows:
   Step 1. 1 min at 96°C

   Step 2. 10 s at 96°C

   Step 3. 5 s at 50°C

   Step 4. 4 min at 60°C, steps 2–4 is repeated 24 cycles

   The samples are kept at 4–8°C until continuing with precipitation.
5. In 1.5 ml Eppendorf tubes, 13 μl dH_2O, 3 μl 3 M NaAc (pH 4.6), and 64 μl 95% EtOH is added to the 20 μl sequence reaction and product is precipitated at RT for 15 min up to max 24 h.
6. The orientation of the tube is marked (since pellet might be hard to visualize) and tubes are spinned at maximum speed in an Eppendorf centrifuge for 20 min at RT. Discard the supernatant and wash the pellets with 250 μl ice-cold 70% EtOH, place the tubes in the same orientation as before, and spin for 10 min at maximum speed in the Eppendorf centrifuge. Aspirate the supernatant carefully and leave the tubes to dry a few min at RT (*see* **Note 16**). Resuspend the pellets in 20 μl formamide.
7. Add the samples to a MicroAmp™ optical 96-well reaction plate (Applied Biosystems) and analyze in an Applied Biosystems 3130xl genetic analyzer.
8. The sequence can be analyzed using different available software programs. Here ChromasLite and DNAMAN are used (*see* **Note 17**).

4. Notes

1. R1881 is light sensitive and is stored in the dark at –20°C.
2. Do not use more than 0.3% EtOH in the cell culture, because this might be toxic to the cells.
3. It is highly important to keep all tubes and samples on ice and work as efficiently as possible to preserve RNA from degrading. During RNA isolation and handling it is very

important to work RNase-free, wear gloves, and use autoclaved/sterilized materials and solutions.

4. When working with RNA-Bee, use gloves and eye protection. The RNA-Bee solution contains guanidine thiocyanate and phenol that is an irritant and causes skin burns.
5. Chloroform can dissolve disposables. Most grades polythene and polypropylene are resistant but most other disposables are not (e.g., PET, perspex, PVC, polystyrene). However, it is good practice to test all disposables before use in RNA isolation.
6. There are commercial programs for design of primers. These programs take into consideration most important parameters. It is also possible to design primers yourself. The two primers of a PCR are complementary to one of each DNA strands of the amplified fragment. Long primers not only increase the specificity but also increase annealing time. Most appropriate PCR primers have a length of 18–20 nucleotides. The GC content of the primers should be 40–60% because it affects the annealing temperature and specificity of the primers. Interaction between the two primers should also be avoided (forming of primer-dimers).
7. If the product yield is low and no primer-dimer is seen in the optimization step, the cause can still be in primer design. Primers might form secondary structures. How prone primers are to form secondary structures is difficult to predict, but there is software that can predict and calculate the ΔG, G standing for the Gibbs free energy. The ΔG represents the energy needed to break a secondary structure; the acceptable values are between 2 and 6 kcal/mol dependent on the type of secondary structure. The GC content, size, and structure of the amplified fragment are also important aspects of PCR efficacy.
8. Primer sequences used are as follows:

   TMPRSS2 exon 1 forward: GAGCTAAGCAGGAG-GCGGA

   TMPRSS2 exon 3 reverse: AGGGGTTTTCCGGTTG-GTATC

   ERG exon 4 reverse: CATCAGGAGAGTTCCTTGAG

   PBGD forward: CATGTCTGGTAACGGCAATG

   PBGD reverse: GTACGAGGCTTTCAATGTTG
9. Check the slope of the standard line and the R^2 value. The slope should not deviate much from –3.0 and R^2 should be close to 1.0.
10. Although a housekeeping gene by definition is supposed to be expressed at equal levels and by all cells this might not

always be completely correct. Therefore, more than one housekeeping gene should be tested for every tissue or cell line studied to distinguish that the expression levels do not vary significantly over the sample cohort.

11. Mix the PCR reaction by pipetting and spin the plate before performing the Q-PCR, this not only does collect the liquid to the bottom of the well but also helps to decrease air bubbles in the sample.
12. The design of the Q-PCR will be dependent on the design of primers, the equipment, the software, and the manufacturer's advice.
13. Threshold is the allowed background level. It is important to use the same threshold if several plates/runs are needed for an experiment. This allows comparing the results. In the experiments described in this chapter the threshold is always manually set to 0.02.
14. Ct value is a value defining at which of the amplification cycles the fluorescence signal crosses the pre-set threshold value.
15. If there are several fragments (bands) from the (nested) PCR reaction, sequence all bands. The bands can be specific and represent different transcripts present in the sample, e.g., different splice-forms or alternative first exons.
16. Do not over-dry the sample because this can cause dye blobs and disturb reading of the sequence.
17. ChromasLite is free software available on Internet. DNAMAN can be purchased from several different providers.

References

1. Tomlins SA, Rhodes DR, Perner S, et al. (2005) Recurrent fusion of TMPRSS2 and ETS transcription factor genes in prostate cancer. *Science* **310**:644–8.
2. Hermans KG, van Marion R, van Dekken H, Jenster G, van Weerden WM, Trapman J. (2006) TMPRSS2:ERG fusion by translocation or interstitial deletion is highly relevant in androgen-dependent prostate cancer, but is bypassed in late-stage androgen receptor-negative prostate cancer. *Cancer Res* **66**:10658–63.
3. Perner S, Demichelis F, Beroukhim R, et al. (2006) TMPRSS2-ERG fusion-associated deletions provide insight into the heterogeneity of prostate cancer. *Cancer Res* **66**:8337–41.
4. Wang J, Cai Y, Ren C, Ittmann M. (2006) Expression of variant TMPRSS2/ERG fusion messenger RNAs is associated with aggressive prostate cancer. *Cancer Res* **66**:8347–51.
5. Pflueger D, Rickman DS, Sboner A, et al. (2009) N-myc downstream regulated gene 1 (NDRG1) is fused to ERG in prostate cancer. *Neoplasia* **11**:804–11.
6. Han B, Mehra R, Dhanasekaran SM, et al. (2008) A fluorescence in situ hybridization screen for E26 transformation-specific aberrations: identification of DDX5-ETV4 fusion protein in prostate cancer. *Cancer Res* **68**:7629–37.
7. Tomlins SA, Mehra R, Rhodes DR, et al. (2006) TMPRSS2:ETV4 gene fusions define a third molecular subtype of prostate cancer. *Cancer Res* **66**:3396–400.
8. Helgeson BE, Tomlins SA, Shah N, et al. (2008) Characterization of TMPRSS2:ETV5

and SLC45A3:ETV5 gene fusions in prostate cancer. *Cancer Res* **68**:73–80.
9. Tomlins SA, Laxman B, Dhanasekaran SM, et al. (2007) Distinct classes of chromosomal rearrangements create oncogenic ETS gene fusions in prostate cancer. *Nature* **448**: 595–9.
10. Seth A, Watson DK. (2005) ETS transcription factors and their emerging roles in human cancer. *Eur J Cancer* **41**:2462–78.
11. Turner DP, Watson DK. (2008) ETS transcription factors: oncogenes and tumor suppressor genes as therapeutic targets for prostate cancer. *Expert Rev Anticancer Ther* **8**:33–42.
12. Lapointe J, Kim YH, Miller MA, et al. (2007) A variant TMPRSS2 isoformand ERG fusion product in prostate cancer with implication for molecular diagnosis. *Mod Pathol* **20**: 497–73.
13. Saramaki OR, Harjula AE, Martikainen PM, et al. (2008) TMPRSS2:ERG fusion identifies a subgroup of prostate cancers with a favourable prognosis. *Clin Cancer Res* **14**:3395–400.
14. Demichelis F, Fall K, Perner S, et al. (2007) TMPRSS2:ERG gene fusion associated with lethal prostate cancer in a watchful waiting cohort. *Oncogene* **26**:4596–9.
15. Maher CA, Palanisamy N, Brenner JC, et al. (2009) Chimeric transcript discovery by paired-end transcriptome sequencing. *Proc Natl Acad Sci USA* **106**: 12353–8.
16. Jhavar S, Reid A, Clark J, et al. (2008) Detection of *TMPRSS2-ERG* translocations in human prostate cancer by expression profiling using GeneChip human exon 1.0 ST Arrays. *J Mol Diagn* **10**:50–7.

Chapter 20

Regulation of Apoptosis by Androgens in Prostate Cancer Cells

Yke Jildouw Arnoldussen, Ling Wang, and Fahri Saatcioglu

Abstract

The balance between proliferation and cell death is often disrupted in cancer leading to tumor growth. In prostate cancer, these events are regulated, at least in part, through androgen signaling. Prostate cancer is dependent on androgens for growth in the initial stages where apoptosis is simultaneously inhibited. Androgen signaling remains important in later stages of prostate cancer as well. Here, we provide methods to study apoptosis in prostate cancer cells and its regulation by androgens. In prostate cancer cells grown in vitro, apoptosis can be induced by different stimuli, such as the endoplasmic reticulum Ca^{2+} ATPase inhibitor Thapsigargin (TG) through the intrinsic apoptosis pathway, or tumor necrosis factor-related apoptosis-inducing ligand (TRAIL) plus the inhibition of PI3K, through the extrinsic signaling pathway; both of these apoptotic events can be blocked by androgens. Here, we provide protocols to assess apoptosis triggered by TG or TRAIL plus PI3K inhibitor LY294002, in prostate cancer cells in vitro using nuclear fragmentation and TUNEL assays aided by fluorescence microscopy or flow cytometry.

Key words: Androgens, apoptosis, prostate cancer, TUNEL, fluorescence microscopy, flow cytometry.

1. Introduction

Programmed cell death, or apoptosis, is characterized by genomic DNA cleavage giving rise to double-stranded DNA fragments and single-strand breaks (1, 2). In prostate cancer, androgens and androgen signaling through the androgen receptor (AR) have an important role both in the normal prostate and during prostate cancer development and progression (3, 4). It is well established that androgens tightly regulate the events of cell

F. Saatcioglu (ed.), *Androgen Action*, Methods in Molecular Biology 776,
DOI 10.1007/978-1-61779-243-4_20, © Springer Science+Business Media, LLC 2011

division and cell death in prostate cancer cells (3, 5). Previous studies have indicated that androgens protect the androgen-responsive prostate cancer cell line LNCaP from apoptosis induced by the apoptosis inducers thapsigargin (TG), 12-*O*-tetradecanoyl-13-phorbol-acetate (TPA), and ultraviolet irradiation (UV) (5). All these agents require the activation of c-Jun NH2 terminal kinase (JNK) to mediate their apoptotic effects (6), which is inhibited by androgens (5). Furthermore, TNF-related apoptosis-inducing ligand (TRAIL) induces apoptosis by JNK activation in prostate cancer cells by activation of the extrinsic apoptosis signaling pathway (7). These findings implicate the mitogen-activated protein kinase (MAPK) cascades and their regulation for control of apoptosis in prostate cancer cells (5, 8); however, other signaling pathways have also been implicated in this regard (9).

Here we provide methods to investigate the regulation of apoptosis in prostate cancer cells in vitro by androgens. Methods include visualization of nuclear fragmentation, the terminal deoxynucleotidyl transferase dUTP nick end labeling (TUNEL) assay (*see* **Note 1**), and flow cytometry to assess the degree of apoptosis in response to apoptosis inducers, for example, TG and TRAIL, in the presence or absence of androgens (8, 10). DNA fragmentation as an indicator of apoptosis is visualized by 4,6-diamidino-2-phenylindole (DAPI) staining of nuclei (*see* **Note 1**).

2. Materials

2.1. Cell Culture

2.1.1. Nuclear Fragmentation and TUNEL Assay by Fluorescence Microscopy

1. LNCaP prostate cancer cells (ATCC) are maintained in RPMI-1640 supplemented with 10% fetal calf serum (FBS). RPMI-1640 supplemented with 2 and 0.5% charcoal-stripped (CS) FBS are used in TUNEL experiments. All media are supplemented with 50 μg/ml streptomycin and 2 mM L-glutamine.
2. CS-FBS is prepared by treating it with activated charcoal (Sigma-Aldrich) and then sterile filtering (*see* **Note 2**).
3. Trypsin/EDTA.
4. Synthetic androgen methyltrienolone (R1881) (DuPont NEN Research Products) is dissolved at 3.52 mM in 100% ethanol and stored at –20°C; 1000X stock solutions are prepared by dilution at 10^{-4} M in 100% ethanol.
5. Thapsigargin (TG) (Sigma-Aldrich) dissolved at 100 μM as a 1000X stock solution in dimethyl sulfoxide (DMSO) and stored at –20°C.

6. Unless stated otherwise, all solutions are prepared in ultrapure water from the Milli-Q system (Millipore) that is sterilized by autoclaving (20 min, 121°C).
7. Six-well plates

2.1.2. For TUNEL Assay by Flow Cytometry

8. Same as Steps 1 and 2.
9. LY294002 (Invitrogen) dissolved at 10 mM in DMSO and stored at –20°C.
10. SuperKiller TRAIL (Enzo Life Science) dissolved at 0.5 μg/μl in water and stored at –80°C (*see* **Note 3**).
11. 60 × 15 mm cell culture dishes.
12. 15 ml conical centrifuge tubes used for cell harvesting and washing.

2.2. Nuclear Fragmentation and TUNEL Assay by Fluorescence Microscopy

1. Microscope cover slips (22×22 mm).
2. Phosphate buffered saline (PBS) prepared as a 10X stock with 137 mM NaCl, 100 mM Na_2HPO_4, and 2 mM NaH_2PO_4 (adjust to pH 7.0 with 37% HCl solution). Autoclave and store at 4°C. Prepare working solution by diluting with water to 1X.
3. Paraformaldehyde (PFA) (Electron Microscopy Sciences). Prepare a 4% solution in PBS adding 10 mM sodium hydroxide (NaOH), heat at 65°C while stirring to dissolve, cool down to room temperature, and adjust to pH 7.4 with 37% HCl solution.
4. Permeabilization solution: 0.1% Triton X-100 in PBS
5. Parafilm and a box with lid.
6. In situ cell death detection kit (Roche Diagnostics) (*see* **Note 4**).
7. DnaseI and DnaseI buffer.
8. 4,6-Diamidino-2-phenylindole (DAPI) (Sigma-Aldrich) dissolved 1/1000 in PBS.
9. Mounting medium: Mowiol solution
   a. Mix 2.4 g poly(vinyl alcohol) (Mowiol) with 6 g glycerol in a 50 ml tube.
   b. Add 6 ml sterile water and rotate at room temperature for 2 h.
   c. Add 12 ml 200 mM Tris-HCl solution (pH 8.5) and incubate at 50°C until the Mowiol is dissolved.
   d. Clarify by centrifugation at 1900 RCF for 20 min at room temperature.

e. Add 1,4-diazabicyclo[2.2.2]octane (DABCO) from a 20% (w/v) stock (dissolved in sterile water) to a final concentration of 0.2% (w/v).

f. Aliquot and store at –20°C.

2.3. TUNEL Assay by Flow Cytometry

1. Same as for Steps 2–4, 6, and 7 in **Section 2.2**.
2. 5 ml round bottom 12 × 15 mm tubes.
3. Becton Dickinson FACS Calibur flow cytometer (or equivalent).
4. Cell Quest software (or equivalent).

3. Methods

3.1. Nuclear Fragmentation and TUNEL Assay by Fluorescence Microscopy

For the experiments where microscopy is used to detect nuclear fragmentation and TUNEL staining, the cells are grown on coverslips. When working with cells growing on coverslips and cells that are undergoing apoptosis, it is important to remember that cells can easily detach (*see* **Note 5**).

3.1.1. Cell Culture

1. Place coverslips in a 50 ml tube containing 70% EtOH and make sure all of them are submerged.
2. Before passaging the cells, use a tweezer to place one coverslip in each well of a six-well plate, one well for each experimental condition. It is highly desirable to perform the experiments in duplicate or triplicate. Remember to include three extra wells with coverslips for positive and negative controls. Wash the coverslips in the wells twice with sterile 1X PBS.
3. Passage the cells by washing once with PBS and then trypsinize for 1 min. Add a few milliliters of media to stop trypsin and collect the cells in a 50 ml tube (*see* **Note 6**).
4. Count cells using a hemocytometer and plate out 1.4×10^5 cells/well in 2 ml RPMI-1640 containing 10% FBS. Leave cells overnight in a humidified 5% CO_2 and 95% air incubator at 37°C.
5. When the cells are completely attached and evenly spread (1–2 days), change to medium containing 2% CS-FBS.
6. Leave cells for 48 h and then change to 0.5% CS-FBS for approximately 10 h (*see* **Note 7**).
7. Treat with TG for 36 and 48 h to induce apoptosis. Make sure that all the treatments end at the same time so that the TUNEL assay is performed on all cells simultaneously.

3.1.2. Nuclear Fragmentation and TUNEL Assay

1. Prepare 1X PBS, 4% PFA, and permeabilization solution.
2. When the treatments are done, bring the six-well plates to the bench in the laboratory.
3. Wash cells once with 1X PBS. Be careful as the cells detach easily (*see* **Note 5**).
4. In a fume hood add 2 ml of 4% PFA to each well. Leave at room temperature for 20 min.
5. Carefully aspirate off the PFA solution and wash immediately with 1X PBS. Do not let the cells dry.
6. Remove PBS and add 2 ml permeabilization solution to each well and leave on a horizontal shaker for 5 min at approximately 50 rpm. Avoid vigorous shaking which will lead to detaching of cells.
7. Wash 3 × 5 min with PBS while shaking.
8. Take the coverslip for the positive control with a pair of tweezers after the first washing step and incubate with 50 μl of DNaseI (3 U/ml in DNaseI buffer) at room temperature for 10 min. Do this by placing the DNaseI solution onto a piece of parafilm (it will form a drop) and then placing the coverslip on top of the solution with the cells facing down. Avoid bubbles; however, if bubbles do form, gently press the coverslip down to remove them (*see* **Note 8**).
9. During the last wash in Step 7, thaw the TUNEL reagents, remove 100 μl from vial 2 for negative controls, and add the enzyme solution (vial 1) to vial 2 to obtain 500 μl TUNEL reaction mixture (*see* **Note 9**). Mix well and leave on ice.
10. Prepare a humid chamber by laying some wet tissue paper close to the walls of the box and place a piece of parafilm in the middle of the box. The box should be light impermeable; use aluminum foil to cover the box if needed.
11. For each coverslip, place 50 μl of the TUNEL reagent on the parafilm. Take the coverslip out of the well with a pair of tweezers, carefully dry the edges with a tissue paper and lay it with the cells down on the TUNEL reagent. Avoid bubbles. Include the positive control and also the two negative controls that are only incubated with the labeling solution (vial 2) without the enzyme (vial 1).
12. Close the box and incubate at 37°C for exactly 1 h (*see* **Note 10**).
13. Pipette few drops of PBS under the coverslips to detach them more easily from the parafilm, pick them up with tweezers, and place back in the six-well plate with the cells facing up. From now on, the coverslips should be exposed

to light minimally, only during manipulation; therefore, place a cover on top or wrap in aluminum foil.

14. Wash in PBS 3 × 5 min while shaking as in Step 7. Light exposure for the brief periods during the changing of PBS is fine; however, avoid intense light.
15. During the washing in Step 14, prepare the DAPI solution in PBS and make sure it is evenly resuspended. After removing the last PBS wash in Step 14, add 2 ml of DAPI solution to each well. Shake gently for 5 min at room temperature to evenly stain the DNA.
16. Wash in PBS 3 × 5 min with shaking (*see* **Note 11**).
17. Mounting: Mark the corner of the glass slide with the treatment name/code and add 30 μl of Mowiol in the middle of the slide (*see* **Note 12**). Take up the coverslip with tweezers and carefully dry the edges with a soft tissue paper to remove excess PBS. Gently lay it down onto the Mowiol drop with cells facing down. Avoid bubbles (*see* **Note 13**).
18. Keep the slides away from light and leave at room temperature for 2–4 h to dry. Then keep the slides at 4°C until examination under the microscope.
19. View the slides with a fluorescence microscope. Depending on the type of TUNEL reagent used, apoptotic cells will either emit green or red light. DAPI fluorescence is blue and can be viewed with the UV light.
20. First check if the positive and negative controls are as they should be. Cells undergoing apoptosis will have fragmented nuclei as can be seen by DAPI staining and will also be stained with the TUNEL reagent (**Fig. 20.1**). Then count a large number of cells (at least few hundred), both dead and alive, in at least three different areas of the coverslip for each treatment and calculate the percentage of dead cells. Remember to capture representative pictures (*see* **Note 14**).

3.2. TUNEL Assay Using Flow Cytometry

The flow cytometry-based TUNEL method is significantly more time efficient and avoids the possible errors/bias in the counting step. It is also more quantitative and has been widely used to monitor apoptosis in cells in suspension. With some minor modifications, it can be applied to detect apoptosis of adherent cells, such as LNCaP cells.

3.2.1. Cell Culture

1. Cells (4 × 10^5 per 60 mm dish) are cultured in medium containing 2% CS-FBS for 2 days and then in medium containing 0.5% CS-FBS for 1 day.

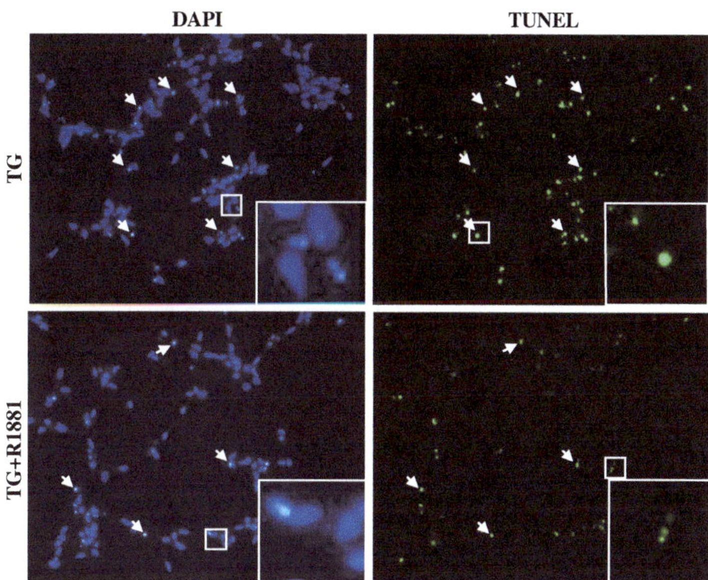

Fig. 20.1. Assessment of apoptosis by the TUNEL assay. LNCaP cells were grown on coverslips, starved and treated with R1881 (10^{-7} M) or vehicle for 40 h as described in **Section 3**. Then TG (100 nM) was added for an additional 36 h to induce apoptosis. *Arrows* point to some of the apoptotic cells and *insets* are magnifications of areas in the *small squares*.

2. Treat cells with vehicle (DMSO) or LY294002 (20 μM) plus TRAIL (50 ng/ml) in medium containing 0.5% CS-FBS for 8 h (*see* **Notes 15** and **16**).
3. Collect the medium containing floating cells in 15 ml conical centrifuge tubes.
4. Wash the cells adhering to the dish with 1X PBS, trypsinize gently, and collect cells in the corresponding tube from Step 3 (*see* **Note 17**). Centrifuge at 150×g for 10 min and remove the supernatant.
5. Wash cells twice with cold PBS. Centrifuge at 150×g for 10 min at 4°C and remove the supernatant. Resuspend cells in 0.5 ml PBS.

3.2.2. TUNEL Assay

1. Prepare 4% PFA in PBS and the permeabilization solution.
2. Fix the cells by addition of 0.5 ml 4% PFA in PBS (final concentration 2% PFA). Incubate for 60 min on a horizontal shaker at 20 rpm at room temperature.
3. Collect cells by centrifugation at 350×g for 10 min. Carefully aspirate the fixing solution without touching the cell pellets.

4. Wash cells once with 2 ml PBS by turning the tube upside down for three times. Centrifuge at 350×g for 10 min and carefully remove the supernatant.
5. Resuspend cells gently in 0.5 ml permeabilization solution for 5 min on ice (*see* **Note 18**).
6. Collect cells by centrifugation at 400×g for 10 min (*see* **Note 19**) and carefully remove PBS.
7. Wash cells gently once with 1 ml PBS as in Step 4. Centrifuge as in Step 6 and remove PBS carefully.
8. Incubate the positive control cells with 50 μl of DNase I (3 U/ml in DNase I buffer) at room temperature for 10 min.
9. Add 50 μl of label solution (vial 2) to each of the two negative controls. Mix the remaining 450 μl label solution in vial 2 with 50 μl enzyme solution in vial 1. Resuspend each sample from Steps 7 and 8 in 50 μl of this reaction mixture.
10. All the samples including the positive control and the negative controls are incubated at 37°C for 1 h with gentle agitation every 10 min. Protect the samples from light.
11. Wash once in 1 ml cold PBS as in Step 4. Centrifuge as in Step 6 and carefully remove PBS.
12. Resuspend the cells in 0.5 ml PBS and transfer to a round bottom 12 × 15 mm tube for FACS analysis.
13. Assess samples in a Becton Dickinson FACS Calibur flow cytometer (or equivalent) with Cell Quest software for data acquirement and analysis (**Fig. 20.2**). Set up the parameters based on the negative control. The dead cells showing red staining will be monitored with phycoerythrin emission signal detector (usually FL2). A total of at least 10,000 single cells should be analyzed per sample.

4. Notes

1. The TUNEL reagent contains terminal deoxynucleotidyl transferase (TdT) which attaches fluorescently labeled nucleotides to the free 3′-OH ends of the broken DNA. As a result, quantification of cells undergoing apoptosis is possible through the use of fluorescence microscopy or flow cytometry. In parallel to TUNEL staining, the cell nuclei and subsequent fragmentation can be visualized by fluorescence microscopy through 4,6-diamidino-2-phenylindole

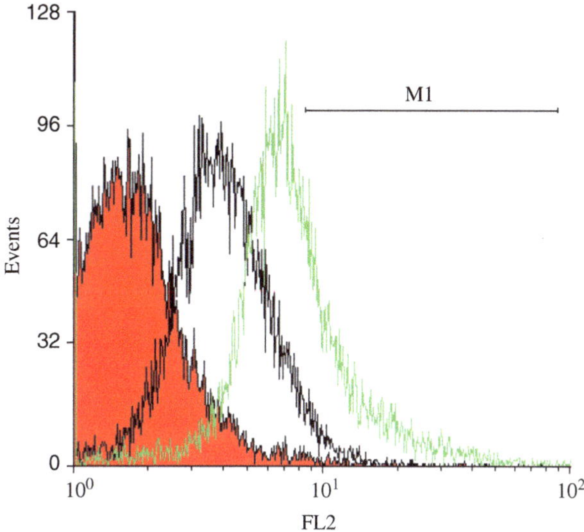

Fig. 20.2. Flow cytometry analysis of LY294002 plus TRAIL-induced apoptosis in LNCaP cells by TUNEL. *Red peak*: cells that are incubated with only label solution in the absence of enzyme solution (negative control). *Black line*: cells treated with vehicle DMSO in medium containing 0.5% CS-FBS for 8 h. *Green line*: cells treated with LY294002 (20 μM) plus TRAIL (50 ng/ml) in medium containing 0.5% CS-FBS for 8 h. Cells analyzed under marker *M1* are apoptotic (TUNEL positive). This assay and conditions can be applied to assess the effect of androgen on LY+TRAIL-induced apoptosis in LNCaP cells or other androgen-responsive cell lines.

(DAPI) staining. Cells undergoing apoptosis will have fragmented nuclei and this is easily visible as small blue dots in close proximity to each other.

2. 3 g of active charcoal powder is added to 45 ml of FBS in a 50 ml sterile centrifuge tube and incubated on a rotator for 18 h at 4°C. The tube is centrifuged at 3500×g for 20 min at 4°C and the liquid is transferred to a new tube containing a new batch of 3 g active charcoal powder. The tube is incubated for 2 h at 4°C on a rotator and centrifuged at 3500×g for 20 min at 4°C. All liquid is filtered first through a sterile filter with a pore size of 0.45 μm and then one with a pore size of 0.2 μm. Charcoal-stripped serum aliquots are stored at –20°C until use.
3. LY294002 and TRAIL should be aliquoted into smaller portions to avoid repeated cycles of freezing and thawing.
4. Different companies provide in situ cell death detection kits (TUNEL assay kits). Carefully follow the manufacturers' protocol and recommendations as these may be different from the in situ cell death detection kit by Roche Diagnostics described here.

5. In order to keep the cells from detaching when they grow on coverslips, add medium or PBS to the wells by touching the wall of the well with the tip of the pipette. Even though precautions are taken and media is slowly added, some cells may detach from the edges of the coverslips and are observed as flakes floating in the media, sometimes still being attached to the coverslip; this is normal.
6. Cells should not be trypsinized for too long as this may result in increased cell debris. When the cells are collected in the 50 ml tube, they may be resuspended by pipetting up and down three times to break apart cell clumps.
7. In our experiments we have repeatedly experienced that the TUNEL reagents only work on LNCaP cells that have been starved. The reagents do not work on cells that are grown in full medium. To our knowledge there are no studies using TUNEL on LNCaP cells in full medium. The reason for this problem is not known, but might be due to impairment of the cells' susceptibility to taking up the reagents when grown in full medium.
8. To avoid air bubbles and secure even spread of the solution, take up the coverslip with tweezers and touch the side of the coverslip to one side of the DnaseI solution and slowly let it down onto the drop with the cells facing down.
9. 500 μl of TUNEL reaction mixture is enough for 10 coverslips when using 50 μl/coverslip. However, if 10–15 coverslips are used, 30–50 μl of TUNEL reaction mixture may be used per coverslip.
10. Extended incubation with the TUNEL reagent at 37°C will dry out the slides and give false-positive TUNEL staining. All the cells, both those undergoing apoptosis and those that are not, will be stained. This is especially relevant for cells at the edges of the coverslip where they can dry out easier.
11. When the cells are not adequately washed after TUNEL staining there will be a high degree of background. Although this background is easy to distinguish from cells undergoing apoptosis, it will decrease the quality of the pictures that are taken.
12. The Mowiol solution is very viscous; therefore, cutting the tip of the pipette (yellow tip) is useful.
13. To avoid air bubbles and secure even spreading of the solution, take up the coverslip with a pair of tweezers and touch the side of the coverslip to one side of the Mowiol drop and slowly let it down onto the drop with the cells

facing down. Any trapped bubbles may easily be observed by laying a piece of black paper underneath the glass slide. If there are any bubbles, gently squeeze them out.

14. A total of approximately 1000–1200 cells from at least 3 different areas should be counted per coverslip. We have never experienced problems with the negative controls; however, DNA cleavage in the positive control may sometimes not work efficiently. When working with strong apoptosis inducers, such as TG, staining of apoptotic cells indicate that the TUNEL reagent is working properly.
15. Each treatment should be checked in triplicate.
16. Set up the optimum conditions for apoptosis induction (e.g., the concentration of the compounds and the period of the treatment). Do not let the apoptosis progress to a very late stage. If this occurs, there will be a large amount of cell debris which may block visualization of the apoptotic bodies resulting in erroneous data.
17. It is important to collect both floating and adherent cells for analysis.
18. Make sure to work gently with the cells at each step. Especially after permeabilization of the cells, each wash and resuspension should be very gentle. Otherwise, many cells could detach and be lost and there may not be enough cells to perform the FACS analysis.
19. A higher speed than normal facilitates centrifugation of the permeabilized cells.

Acknowledgments

This work was supported by grants from the Norwegian Cancer Society and the Norwegian Research Council.

References

1. Krysko, D. V., Vanden Berghe, T., D'Herde, K., and Vandenabeele, P. (2008) Apoptosis and necrosis: detection, discrimination and phagocytosis, *Methods* **44**, 205–221.
2. Orrenius, S., Nicotera, P., and Zhivotovsky, B. (2010) Cell death mechanisms and their implications in toxicology, *Toxicol Sci* **119**, 3–19.
3. Heinlein, C. A., and Chang, C. (2004) Androgen receptor in prostate cancer, *Endocrine Rev* **25**, 276–308.
4. Kaarbo, M., Klokk, T. I., and Saatcioglu, F. (2007) Androgen signaling and its interactions with other signaling pathways in prostate cancer, *Bioessays* **29**, 1227–1238.
5. Lorenzo, P. I., and Saatcioglu, F. (2008) Inhibition of apoptosis in prostate cancer cells by androgens is mediated through downregulation of c-Jun N-terminal kinase activation, *Neoplasia* **10**, 418–428.
6. Engedal, N., Korkmaz, C. G., and Saatcioglu, F. (2002) C-Jun N-terminal

kinase is required for phorbol ester- and thapsigargin-induced apoptosis in the androgen responsive prostate cancer cell line LNCaP, *Oncogene* **21**, 1017–1027.

7. Sah, N. K., Munshi, A., Kurland, J. F., McDonnell, T. J., Su, B., and Meyn, R. E. (2003) Translation inhibitors sensitize prostate cancer cells to apoptosis induced by tumor necrosis factor-related apoptosis-inducing ligand (TRAIL) by activating c-Jun N-terminal kinase, *J Biol Chem* **278**, 20593–20602.
8. Arnoldussen, Y. J., Lorenzo, P. I., Pretorius, M. E., Waehre, H., Risberg, B., Maelandsmo, G. M., Danielsen, H. E., and Saatcioglu, F. (2008) The mitogen-activated protein kinase phosphatase vaccinia H1-related protein inhibits apoptosis in prostate cancer cells and is overexpressed in prostate cancer, *Cancer Res* **68**, 9255–9264.
9. Lorenzo, P. I., Arnoldussen, Y. J., and Saatcioglu, F. (2007) Molecular mechanisms of apoptosis in prostate cancer, *Crit Rev Oncogen* **13**, 1–38.
10. Wang, L., Jin, Y., Arnoldussen, Y. J., Jonson, I., Qu, S., Maelandsmo, G. M., Kristian, A., Risberg, B., Waehre, H., Danielsen, H. E., and Saatcioglu, F. (2010) STAMP1 is both a proliferative and an antiapoptotic factor in prostate cancer, *Cancer Res* **70**, 5818–5828.

Chapter 21

Analysis of Androgen Receptor Rapid Actions in Cellular Signaling Pathways: Receptor/Src Association

Antimo Migliaccio, Gabriella Castoria, and Ferdinando Auricchio

Abstract

Much evidence indicates that, with few exceptions, non-genomic actions of steroids are mediated by receptors universally known as nuclear receptors. Steroid receptors do not exhibit intrinsic tyrosine kinase activity. Nevertheless, they stimulate different signaling pathways in cytoplasm of target cells, including those dependent on Src, a cytoplasmic tyrosine kinase. Steroid-induced Src activation regulates cell cycle progression, survival, migration, and associated processes, such as cell growth and differentiation. Androgen stimulation of human prostate cancer-derived LNCaP cells triggers cell cycle progression and proliferation. The key event in this process is the association of androgen receptor (AR) with Src. This association triggers activation of the Src/Ras/Erk pathway and finally impacts cell cycle. Androgen stimulation of fibroblasts also induces AR/Src association, which triggers DNA synthesis. Prevention of this association by a receptor-derived peptide competing for AR interaction with Src specifically inhibits the androgen receptor-dependent proliferative effect in vitro and in vivo.

Key words: Non-genomic action, androgen receptor, Src, protein/protein association, prostate cancer.

1. Introduction

Androgen induces cell cycle progression and cell proliferation in human prostate cancer cells (1–3). These effects are triggered by androgen-induced stimulation of the Src/Raf/Erk-2 pathway. This stimulation depends on the association of a proline-rich sequence of hAR with the SH3 regulatory domain of Src. Since in LNCaP cells extranuclear AR is in complex with ERbeta, association of Src with AR occurs in parallel with association of ERbeta with the SH2-Src domain. Simultaneous association of the two

F. Saatcioglu (ed.), *Androgen Action*, Methods in Molecular Biology 776,
DOI 10.1007/978-1-61779-243-4_21, © Springer Science+Business Media, LLC 2011

receptors with Src abolishes the intra-molecular inhibitory interactions of the two regulatory Src domains, thus triggering a strong stimulation of Src activity by androgens (2). Prevention of AR/Src association by an AR-derived small peptide inhibits DNA synthesis triggered by androgen in LNCaP cells and drastically reduces growth of mouse LNCaP xenografts (4). These findings highlight the role of AR/Src association in regulating cell cycle progression and tumor growth.

Androgen stimulation also induces AR/Src association with a proliferative effect in NIH3T3 fibroblasts (5). Here again, the AR-derived peptide reduces androgen-induced DNA synthesis. On the basis of these findings, it is evident that AR/Src association is an initial event in the rapid action of AR leading to the growth of both epithelial and mesenchymal cells in vitro and in vivo. Commercial availability of antibodies specifically recognizing AR and Src in biochemical approaches provides assay protocols for the study of this association in combined immunoprecipitation and Western blotting techniques. Detection of AR/Src complex upon androgen stimulation of target cells allows confirmation that Src-dependent extranuclear androgen signaling has been activated. It might also help in identifying other partners involved in this initial event in androgen action and in screening novel inhibitors of androgen-stimulated Src-dependent signaling.

2. Materials

2.1. Cell Culture, Lysis, and Immunoprecipitation

1. Cell medium A: RPMI-1640 medium with phenol red supplemented with 10% fetal bovine serum (FBS, EU approved origin South America), L-glutamine (2 mM), penicillin (100 U/ml), streptomycin (100 U/ml), gentamycin (50 μg/ml), sodium pyruvate (1 mM), non-essential amino acids (MEMNEAA 100×). All these reagents are from Gibco.

   Cell medium B: phenol red-free RPMI-1640 (Gibco), containing human insulin (Humulin I, 20 U/l, Lilly, Italy) and 10% charcoal-stripped bovine serum (CSS; *see* **Note 1**).

   Charcoal-stripped bovine serum: 0.25% activated charcoal (C5260; Sigma-Aldrich, St. Louis, U.S.A.), 0.005% dextran (D-1390; Sigma-Aldrich), and 0.01 M Tris-HCl pH 8.0 (Sigma-Aldrich).
2. Solution of 0.05% trypsin-EDTA (Gibco) stored at −20°C.
3. R1881 (Perkin Elmer) is dissolved in ethanol at 10 mM, stored at −20°C in aliquots of 100 μl in glass vials and used at 10 nM (final concentration).

4. Casodex (Zeneca) is dissolved in ethanol at 10 mM, stored at –20°C in aliquots of 100 μl in glass vials and used at 10 μM (final concentration).
5. Dulbecco's phosphate buffered saline (PBS; Sigma-Aldrich): 9.55 g in 1 l of deionized water.
6. Lysis buffer: 50 mM Tris-HCl pH 7.4, 1 mM EDTA, 1 mM EGTA, 1% Triton X-100 (1%), 150 mM NaCl (S9888; Sigma-Aldrich), 5 mM $MgCl_2$ (1 M, M1028; Sigma-Aldrich), protease inhibitor cocktail tablets (Complete Mini 11836153001; Roche; dissolved in distilled water according to Manufacturer's instructions), 1 mM sodium orthovanadate (Na_3VO_4, S-6508; Sigma-Aldrich; dissolved in distilled water), and 1 mM phenyl-methyl-sulfonyl-fluoride (PMSF, P7626; Sigma-Aldrich; dissolved in ethanol). All these inhibitors are freshly prepared and dissolved for each experiment of cell lysis. Avoid the use of dithiothreitol (DTT) in combination with PMSF.
7. Polyethylene cell scraper (Costar, Corning Incorporated).
8. Bio-Rad protein assay (500-0006; Bio-Rad, Hercules).
9. Mouse monoclonal anti-v-Src antibody (clone 327 at 0.1 mg/ml; Calbiochem).
10. IgG from mouse serum purified immunoglobulin reagent grade (Sigma-Aldrich).
11. Protein A/G PLUS-Agarose Immunoprecipitation reagent (sc-2003; Santa Cruz Biotechnology).
12. Sample buffer, Laemmli 2× concentrate (S3401; Sigma-Aldrich).

2.2. SDS-Polyacrylamide Gel Electrophoresis (SDS-PAGE)

1. Separating buffer: 1.5 M Tris-HCl pH 8.8 (Bio-Rad) and 10% sodium dodecyl sulfate (SDS; Bio-Rad).
2. Stacking buffer: 0.5 M Tris-HCl pH 6.8 (Bio-Rad) and 10% sodium dodecyl sulfate (SDS; Bio-Rad).
3. 30% acrylamide-bis solution (37, 5:1) (SERVA, Germany) and *N,N,N,N′*-tetramethyl-ethylenediamine (TEMED; Bio-Rad).
4. 10% ammonium persulfate (APS) solution prepared by dissolving 1 g of ammonium persulfate powder (Bio-Rad) in 10 ml of distilled water (store at –20°C in single use aliquots, 200 μl).
5. Running buffer: 192 mM glycine powder (Bio-Rad), 25 mM Tris base powder (Promega, Promega Corporation, Madison, USA), and 0.1% SDS powder (Bio-Rad).
6. Full-Range Rainbow Molecular weight markers (Amersham, GE Healthcare).

2.3. Western Blotting for AR

1. Transfer buffer: 20% (v/v) methanol (Sigma-Aldrich), 192 mM glycine powder (Bio-Rad), 25 mM Tris base powder (Promega), and 0.1% SDS powder (Bio-Rad).
2. Protran nitrocellulose transfer membrane: 0.45 μm pore size (Whatman, Germany), gel blotting filter paper (Schleicher & Shuell, Germany).
3. Tris-buffered saline with 0.1% Tween (TBS-T 0.1%): 150 mM NaCl (Sigma-Aldrich), 20 mM Tris-HCl pH 8.0 (Sigma-Aldrich), and 0.1% Tween 20 (Bio-Rad).
4. Blocking buffers:
   BB1: 5% Non-fat dry milk (Bio-Rad) in TBS-T 0.1% (pH 7.5–8.0) (*see* **Note 1**).
   BB2: 3% Non-fat dry milk (Bio-Rad) in TBS-T 0.1% (pH 7.5–8.0) (*see* **Note 1**).
5. Primary antibody: rabbit polyclonal anti-AR N-20 antibody (sc-816; Santa Cruz Biotechnology).
6. Secondary antibody: donkey horseradish peroxidase-conjugated anti-rabbit IgG (Amersham, GE Healthcare).
7. ECL Western blotting substrate (Pierce, Thermo Scientific); write-on transparency film (AF 4300; 3M, Italy); chemiluminescence film (Amersham, GE Healthcare).

2.4. Stripping and Re-probing for Src

1. Stripping buffer: 62.5 mM Tris-HCl pH 6.8 (Bio-Rad), 2% sodium dodecyl sulfate (SDS; Bio-Rad), and 100 mM β-mercaptoethanol (Bio-Rad).
2. Wash buffer: tap water and TBS-T.
3. Secondary antibody: sheep horseradish peroxidase-conjugated anti-mouse IgG (Amersham, GE Healthcare).

3. Methods

3.1. Cell Treatment with Androgen and AR/Src Complex Immunoprecipitation

Subconfluent (80%) fast-growing LNCaP cells (American Type Cell Culture, USA) at low passage (8–12 passage) are cultured in the cell medium A, then detached by trypsin/EDTA, and re-plated at 40–50% confluence on 100 mm Petri dishes (BD Falcon). After 12 h, cell medium A is substituted with cell medium B requiring DCSS (*see* **Note 2**).

Dextran-coated charcoal-stripped serum is used to remove steroid hormones. Treatment with charcoal removes hydrophobic, low molecular weight compounds such as steroid hormones. Some growth factors and cytokines are also removed by this treatment resulting in suppression of cell differentiation and proliferation.

Although charcoal/dextran-stripped serum is now commercially supplied, we carefully prepare the medium as indicated below.

For the treatment of 200 ml of FBS (fetal bovine serum):

1. Suspend 0.04 g of dextran T70 (Sigma-Aldrich) and add 8.0 ml of 1 M Tris-HCl pH 8.0 in 800 ml distilled deionized water. Stir gently at room temperature for about 15 min and add 2.0 g of charcoal. Thereafter, continue to stir gently for 60 min.
2. Aliquot the charcoal/dextran suspension in sixteen 50 ml disposable centrifuge tubes (50 ml in each tube) and centrifuge at room temperature at 2,000 rpm for 25 min.
3. Remove the supernatant.
4. Use only eight tubes (keep the other eight for the next step).
5. Add to each tube 25 ml of FBS, resuspend the pellet and incubate the tubes in a water bath with agitation at 57°C for 30 min.
6. Centrifuge the tubes at room temperature at 4,000 rpm for 30 min.
7. Set the water bath temperature to 37°C.
8. Collect the supernatant in a sterile bottle and throw away the eight tubes.
9. Use the eight new tubes and aliquot in each tube 25 ml of the supernatant collected above.
10. Resuspend the pellet and keep it under agitation in a water bath at 37°C for 30 min.
11. Centrifuge the tubes at room temperature at 4,000 rpm for 30 min.
12. Collect the supernatant in a sterile bottle and repeat the entire treatment twice.
13. Filter the stripped serum through a 0.22 μm filter, aliquot, and store in 50 ml aliquots at –20°C.
14. Analyze by DELFIA assay (Perkin Elmer) the levels of 17-beta estradiol, progesterone, and testosterone before and after the dextran-coated charcoal treatment of FBS to evaluate the treatment efficiency. Analyze also the levels of 5-alpha-dihydrotestosterone by radioimmunoassay (RIA). Note that this check should be performed by Pathology Laboratories.

Cells are maintained in this medium for 3 days and then left untreated or treated for 3 min with 10 nM R1881. Untreated cells are challenged with vehicle (ethanol) alone. Thereafter, Petri dishes are transferred to an ice bucket and cell monolayer is rinsed with cold PBS. For each experimental condition, cells from four dishes are gently scraped, collected in a 15 ml centrifuge tube (Falcon), then centrifuged at 1,200 rpm for 4 min at 4°C. Supernatant is removed and cold lysis buffer (0.5 ml) is added. The lysate is transferred to 1.5 ml tubes, which are shaken in a cold room at 4°C for 30 min. Tubes are subsequently centrifuged for 30 min at 14,000 rpm at 4°C and the supernatant assayed for protein using the Bio-Rad assay. Five milliliters of 1:5 diluted Bio-Rad is added to 5 and 10 μl of samples (containing about 1 mg of proteins) and 95 or 90 μl water, respectively. The remaining lysate is divided into 2 aliquots. One is incubated with anti-Src antibody (10 μl), whereas the other is incubated with the mouse IgG antibody (10 μl of 0.1 mg/ml; control). All the samples are gently shaken overnight in the cold room. To each sample, 45 μl of A/G protein suspension (Santa Cruz) is added and shaking is continued for 60–90 min in the cold. After centrifugation at 14,000 rpm for 5 min, beads are collected and washed with 0.5 ml of lysis buffer three times in the cold room. Beads are finally resuspended in Laemmli sample buffer (50 μl) and boiled for 5 min. The supernatant is collected, centrifuged at 14,000 rpm for 3 min, then used for SDS-PAGE.

3.2. SDS-PAGE

This protocol refers to the preparation of a minigel with Mini-PROTEAN® II Electrophoresis Cell, Bio-Rad. It can be easily adapted to other systems.

1. Prepare a 1.5 mm thick, 12% separating gel monomer solution by mixing 4.0 ml of 30% acrylamide-bis solution with 2.5 ml Tris-HCl pH 8.8, 3.3 ml distilled, deionized water, 100 μl 10% SDS solution, 100 μl APS solution (10%). Finally, add 10 μl TEMED. Pour the gel into the cell, leaving space for a stacking gel and immediately overlay the monomer solution with water. The separating gel will polymerize in 45–60 min.
2. Pour off the water from the gel.
3. Prepare the 4% stacking gel monomer solution by mixing 665 μl of 30% acrylamide-bis solution with 1.25 ml Tris-HCl pH 6.8, 3 ml distilled, deionized water, 50 μl 10% SDS solution, 50 μl of 10% APS solution. Finally, add 5 μl TEMED. Pour the stacking gel on the top of the polymerized separating gel and insert the comb. The stacking gel should polymerize in about 45 min.
4. Prepare 1 l of running buffer.
5. Carefully remove the comb from the stacking gel and wash the wells with the running buffer using a 5 ml syringe fitted with a 22 gauge needle.

6. Load 20–30 μl of each sample in a well and include one well for prestained molecular weight markers.
7. Fill each well carefully with running buffer.
8. Add the running buffer to the upper and the lower chambers of the gel unit.
9. Complete the assembly of the gel unit and connect to a power supply. Run the gel at 120 V for 60–90 min until the dye front has run off the gel.

3.3. Western Blotting for AR

The samples that have been separated by SDS-PAGE are transferred to nitrocellulose membrane electrophoretically.

These instructions refer to the use of a Mini Trans-Blot® Electrophoretic Transfer Cell, Bio-Rad.

1. Prepare the transfer solution (1 l final volume).
2. Disconnect the gel unit from the power supply and disassemble it.
3. Cut the nitrocellulose membrane (wearing gloves to prevent contamination) and the 3MM filter paper to the dimension of the separating gel. Equilibrate the gel and soak two fiber pads, four sheets of 3MM filter paper, and one membrane in transfer buffer for about 15 min.
4. Prepare the gel sandwich by putting the cassette of the Mini Trans-Blot® Electrophoretic Transfer Cell, with the gray side down, on a clean surface and placing, in order, one pre-wetted fiber pad, two sheets of filter paper, the nitrocellulose membrane, the equilibrated gel, two sheets of filter paper, and one pre-wetted fiber pad. It is crucial to remove any air bubbles between the equilibrated gel and nitrocellulose membrane or between the equilibrated gel and filter paper. Close the cassette firmly, being careful not to move the sandwich. Lock the cassette closed with the white latch and place the cassette in the module with its gray side toward the red side of the electrode module.
5. Place the electrode module in the tank and completely fill the tank with transfer buffer. Put the lid on the tank and connect to a power supply. Transfer overnight at 20 V at room temperature.
6. Upon completion of the run, disassemble the blotting sandwich and remove the membrane for development.
7. Incubate the nitrocellulose in freshly made BB1 for 1 h at room temperature on a rocking platform.
8. Prepare BB2.
9. Discard BB1 and wash the membrane three times for 5 min each with TBS-T 0.1%.

10. Incubate the nitrocellulose with N20 anti-AR antibody diluted 1:1000 in freshly made BB2 by gently shaking for 90 min at room temperature.
11. Remove the BB2 containing the primary antibody and wash the membrane three times for 5 min each with TBS-T 0.1%.
12. Incubate the nitrocellulose filter in freshly diluted secondary anti-rabbit antibody (1:6000 in BB2) for 45 min at room temperature on a rocking platform.
13. Discard the secondary antibody solution and wash the membrane three times for 5 min each with TBS-T 0.1% enough to cover the sheet.
14. During the last wash, warm the ECL reagents to room temperature.
15. Remove final wash, mix together the ECL reagents, and immediately add them to the blot. Rotate the blot by hand for 90 s to ensure even coverage.
16. Put the membrane between two leaves of an acetate sheet protector.
17. Put the acetate-containing membrane in an X-ray film cassette with film for a few minutes (about 5 min – this step is done in a dark room) (*see* **Note 3**).

3.4. Stripping and Re-probing Blot for Src

When the Western blot for AR is complete, the membrane is stripped and re-probed with Src antibody to check that an equal amount of the samples was immunoprecipitated.

1. Pre-heat a water bath to 60°C.
2. Prepare 50 ml of stripping buffer by mixing 6.25 ml Tris-HCl pH 6.8, 1 g SDS, and 350 μl β-mercaptoethanol in distilled water.
3. Incubate the blot with stripping buffer in the water bath for 20 min with occasional agitation.
4. Rinse the membrane under running tap water for 1–2 h (*see* **Note 4**)
5. Wash extensively for 5 min with TBS-T 0.1% and incubate the nitrocellulose in BB1 for 1 h at room temperature on a rocking platform.
6. Discard the buffer and wash the membrane three times for 5 min each with TBS-T 0.1%.
7. Incubate the nitrocellulose with anti-Src antibody diluted 1:100 in BB2 in agitation for 2 h at room temperature (*see* **Note 5**).
8. Remove the primary antibody and wash the membrane three times for 5 min each with TBS-T 0.1%.

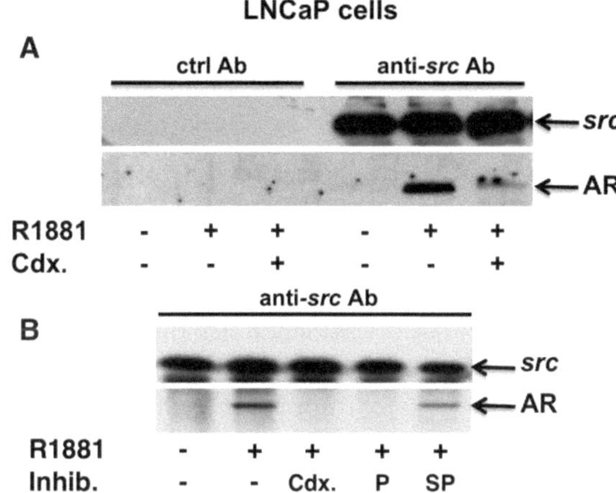

Fig. 21.1. AR/Src association in androgen-treated human prostate cancer cells. Resting LNCaP cells were treated for 3 min with 10 nM synthetic androgen, R1881, in the absence or presence of the indicated compounds: 10 μM of the androgen antagonist, Casodex, Cdx; 1 nM of the synthetic peptide derived from the proline-rich 377–386 sequence of the human AR (P), and 1 nM of a peptide with an identical amino acid composition to that of P but with a scrambled sequence (SP). Cell lysates were processed by Src immunoprecipitation, followed by detection of Src and AR by Western blot analysis.

9. Incubate the nitrocellulose in freshly prepared secondary anti-mouse antibody diluted 1:3000 in BB2 for 45 min on a rocking platform.
10. Discard the secondary antibody solution and wash the membrane three times for 5 min each with TBS-T 0.1%.
11. Treat with ECL and develop as above.

An example of the results of the described procedures is given in **Fig. 21.1**.

4. Notes

1. The pH value is crucial for an optimal antibody activity. Adjust pH to 7.5–8.0 with a solution of NaOH if necessary.
2. For the analysis of steroid-triggered rapid action, including the androgen-induced AR/Src association, the use of hormone-responsive cells is mandatory. Therefore, cells at low passage and 40–50% confluence are used. They should be cultured in phenol red-free medium containing charcoal-stripped fetal calf serum (cell medium B) for no more than 3 days before the hormonal treatment. Cells can no longer be cultured in cell medium B because of cyclin D1 accumulation and the consequent loss of hormone responsiveness.

3. To obtain a correct alignment of the subsequent film with the nitrocellulose filter, we use a strip of luminescent tape (Amersham, GE Healthcare) as a marker. This helps identify the signals with the lanes.
4. Traces of β-mercaptoethanol might damage the antibody.
5. For convenience, we prepare only 2 or 3 ml of diluted (1:100 in BB2) Src-antibody solution and incubate the nitrocellulose filter by fixing it on a parafilm sheet, which in turn is attached to the workbench with tape. We then cover the nitrocellulose filter with the primary antibody solution. It is important at this stage that the nitrocellulose filter is entirely covered and that the solution does not dry. To this end, the filter can be covered by a small box upside down during the incubation time and re-incubated with recovered Src antibody solution if necessary.

Acknowledgments

We gratefully thank Pia Giovannelli for her advice in editing this chapter. This work was supported by grants from Associazione Italiana per la Ricerca sul Cancro (National and Regional Grants).

References

1. Peterziel H, Mink S, Schonert A, Becker M, Klocker H, Cato AC. (1999). Rapid signalling by androgen receptor in prostate cancer cells. *Oncogene* **18**, 6322–9.
2. Migliaccio A, Castoria G, Di Domenico M, de Falco A, Bilancio A, Lombardi M, Barone MV, Ametrano D, Zannini MS, Abbondanza C, Auricchio F. (2000). Steroid-induced androgen receptor-oestradiol receptor beta-Src complex triggers prostate cancer cell proliferation. *EMBO J.* **19**, 5406–17.
3. Migliaccio A, Di Domenico M, Castoria G, Nanayakkara M, Lombardi M, de Falco A, Bilancio A, Varricchio L, Ciociola A, Auricchio F. (2005). Steroid receptor regulation of epidermal growth factor signaling through Src in breast and prostate cancer cells: steroid antagonist action. *Cancer Res.* **65**, 10585–93.
4. Migliaccio A, Varricchio L, De Falco A, Castoria G, Arra C, Yamaguchi H, Ciociola A, Lombardi M, Di Stasio R, Barbieri A, Baldi A, Barone MV, Appella E, Auricchio F. (2007). Inhibition of the SH3 domain-mediated binding of Src to the androgen receptor and its effect on tumor growth. *Oncogene* **26**, 6619–29.
5. Castoria G, Lombardi M, Barone MV, Bilancio A, Di Domenico M, Bottero D, Vitale F, Migliaccio A, Auricchio F. (2003). Androgen-stimulated DNA synthesis and cytoskeletal changes in fibroblasts by a non-transcriptional receptor action. *J. Cell Biol.* **161**, 547–56.

Chapter 22

Analysis of Androgen-Induced Increase in Lipid Accumulation in Prostate Cancer Cells

Jørgen Sikkeland, Torstein Lindstad, and Fahri Saatcioglu

Abstract

Increased metabolic activity is a hallmark of proliferating cancer cells. One common deregulated metabolic pathway in prostate cancer is de novo lipogenesis which is highly increased in prostate cancer and is linked to poor prognosis and metastasis. Male sex hormones play an essential role in prostate cancer growth and have been shown to increase the expression and activity of several lipogenic factors, such as fatty acid synthase (FASN) and sterol regulatory element-binding proteins (SREBPs), leading to accumulation of neutral lipids in prostate cancer cells. These factors are being evaluated as potential prognostic markers and therapeutic targets in prostate cancer. Here we describe methods to directly detect and quantify accumulation of neutral lipids and assess concomitant changes in lipogenic gene expression in LNCaP prostate cancer cells.

Key words: Prostate cancer, de novo lipogenesis, neutral lipids, Oil red O, confocal microscopy, spectrophotometric analysis.

1. Introduction

Normal tissues, except liver and adipose tissue, synthesize low levels of long-chain fatty acids. However, in rapidly proliferating cancer cells, fatty acids can be synthesized de novo in order to provide lipids for increased membrane formation, energy production via β-oxidation and lipid modification of proteins (1). Increased lipogenesis is an important hallmark of cancer, including prostate cancer (2). Lipogenesis is observed in the early, androgen responsive stages of prostate cancer, i.e., the prostate intraepithelial neoplasia (PIN) lesions, where staining intensity correlates with grade (3, 4). This androgen-regulated lipid accumulation persists or

F. Saatcioglu (ed.), *Androgen Action*, Methods in Molecular Biology 776,
DOI 10.1007/978-1-61779-243-4_22, © Springer Science+Business Media, LLC 2011

re-emerges with the development of androgen-independent cancer (5); this indicates that lipogenesis is an important aspect of prostate cancer progression and may thus be a target for therapy (5, 6).

Verhoeven and co-workers were the first to report that androgens regulate lipid metabolism in the LNCaP prostate cancer cell line by direct detection of neutral lipid increase (triglycerides and cholesteryl esters) (7, 8). Later, molecular evidence was provided showing that androgens stimulate lipogenesis through activation of the sterol regulatory element-binding protein (SREBP) pathway (9, 10). For example, FASN expression is upregulated by androgens through potentiation of SREBP signaling (11).

Here we provide a detailed protocol to directly detect and quantify accumulation of neutral lipids in prostate cancer cells upon androgen treatment using three different lipid specific dyes: Oil Red O, HCS LipidTOX Red, and AdipoRed. In addition, protocols are provided for concomitant assessment of changes in lipogenic gene expression.

2. Materials

2.1. Cell Culture and Androgen Treatment

1. LNCaP cells are obtained from the American Type Culture Collection (ATCC) and maintained in medium supplemented with 10% FBS. Cells are propagated in 100 mm plastic dishes at 37°C in a humidified 5% CO_2 + 95% air incubator.
2. Regular medium: RPMI 1640 with L-glutamine supplemented with 5 mg/ml penicillin/streptomycin.
3. Fetal bovine serum (FBS).
4. 1X phosphate-buffered saline (PBS), pH 7.4: Mix 150 mM sodium chloride (NaCl), 6.7 mM di-sodium hydrogen phosphate dihydrate, 1.85 mM sodium dihydrogen phosphate monohydrate in MQ water, adjust to pH 7.4, and sterilize by autoclaving.
5. 200 mg/l trypsin/EDTA.
6. Synthetic androgen R1881 (Dupont-NEN), 1000X stock (10^{-5} M) dissolved in 100% ethanol.
7. Activated charcoal (Sigma) – treated FBS (CT-FBS).

2.2. RNA Isolation

1. Trizol (Invitrogen).
2. Chloroform.

3. Isopropyl alcohol.
4. Diethylpyrocarbonate (DEPC)-treated water.
5. 70% ethanol in DEPC-treated water.
6. NanoDrop (Thermo Scientific) or equivalent spectrophotometer.

2.3. cDNA Synthesis

1. Superscript II Reverse Transcriptase (RT) and 5X RT buffer (Invitrogen).
2. 0.1 M 1,4-dithiothreitol (DTT).
3. Poly(dT) oligonucleotides dissolved in sterilized distilled H_2O (sdH_2O) to a final concentration of 0.5 mg/ml.
4. dNTP Mix (10 mM each).
5. RNasin (Promega).

2.4. Quantitative PCR Amplification

1. Lightcycler 480 SYBR green I Master mix (Roche Diagnostics).
2. Lightcycler 480 Instrument (Roche Diagnostics).
3. Lightcycler 480 96-well plates (Roche Diagnostics).
4. Oligonucleotide primers are dissolved in sdH_2O to a stock concentration of 10 μM. Primer sequences are *RPLP0*: Forward 5′-CAA TGT GGG CTC CAA GCA GAT G-3′, Reverse 5′-GGC ACA GTG ACT TCA CAT GGG G -3′. *FASN*: Forward 5′-GCA CCT CTC AGG CAT CGA C-3′, Reverse 5′-CTG TGG TCC CAC TTG ATG AG-3′. PSA: Forward 5′- CCC TGA GCA CCC CTA TCA AC-3′, Reverse 5′-TGA GTG TCT GGT GCG TTG TG-3′.

2.5. Fixing Cells

1. 4% (w/v) paraformaldehyde (PFA): Make fresh. Prepare 1 ml per well (6-well plate). Dissolve PFA in PBS at 70°C with a few drops of 1 M NaOH. When all is dissolved, cool down to room temperature and neutralize solution with 1 M HCl to pH 7.4 (*see* **Note 1**).
2. 1X PBS, pH 7.4.

2.6. Mounting of Cover Slip

1. Mowiol mounting medium:
   - Mix 2.4 g poly(vinyl alcohol) (Mowiol) with 6 g glycerol in a 50 ml tube.
   - Add 6 ml sterile MQ water and rotate at room temperature for 2 h.
   - Add 12 ml 200 mM Tris–HCl, pH 8.5, and incubate at 50°C until the Mowiol is dissolved.
   - Clarify by centrifugation at 1900 RCF for 20 min at room temperature.

- Add 1,4-diazabicyclo[2.2.2]octane (DABCO) from a 20% (w/v) stock (dissolved in sterile MQ water) to a final concentration of 0.2% (w/v).
- Aliquot and store at –20°C.

2. Tissue paper.
3. Glass slides.

2.7. Oil Red O Staining

1. Oil Red O solution stock: Dissolve 0.5 g Oil Red O (Sigma) in 200 ml isopropyl alcohol at 56°C for 1 h. Cool down before further use. Store at room temperature.
2. Oil Red O working solution: Make fresh the day of use. Mix 4 parts MQ water with 6 parts of stock solution. Let solution rest for at least 15 min. Precipitate will form. Filter solution through a filter paper, e.g., 3MM Chr Whatman paper. Keep at room temperature until use.
3. 1X PBS, pH 7.4.
4. 60% (v/v) isopropyl alcohol in MQ water.

2.8. Oil Red O Extraction

1. Isopropyl alcohol.
2. Spectrophotometer.

2.9. HCS LipidTOX Red Staining

1. LipidTOX Red staining solution: Dilute HCS LipidTOX Red (Invitrogen) 1:200 in PBS. Prepare 40 μl per cover slip; allow 10% extra for pipetting errors (*see* **Note 2**).
2. 0.2% Triton X-100 in PBS (v/v).
3. Incubation chamber: Prepare a dark chamber big enough to store all cover slips. On base of the chamber place a parafilm layer and secure it with tape. In the inner sides of the chamber place moist tissue paper (*see* **Note 3**).
4. 1X PBS, pH 7.4.
5. 4′,6-diamidino-2-phenylindole (DAPI) staining solution: Prepare a stock dissolving DAPI in sterile MQ water at room temperature at 1 mg/ml, aliquot in small volumes (e.g., 200 μl) to prevent unnecessary freeze/thawing and light exposure and store at –80°C. DAPI working solution is a 1:1000 dilution of the stock in PBS. Prepare 1 ml per well (6-well plate) + 10% to allow for pipetting errors.
6. Any light-proof cover, e.g., aluminum foil.
7. Tissue paper.

2.10. AdipoRed Assay

1. 1X PBS, pH 7.4.
2. AdipoRed reagent (Lonza) diluted 3:100 in PBS. Prepare 2 ml per well (6-well plate); allow 10% for pipetting errors. Make fresh just before use.

3. Methods

3.1. Cell Culture

1. Grow sufficient amount of LNCaP cells. Wash with PBS, trypsinize, and determine the cell number using a hemocytometer.
2. Plate out 1 × 10^5 cells into each well of a 6-well plate. If needed, add cover slips to the bottom of each well before plating the cells (*see* **Note 4**).
3. After 1 day, change medium to regular medium supplemented with 5% CT-FBS.
4. Two days later, add R1881 to a final concentration of 10^{-8} M R1881 and culture the cells for 96 h before collection.

3.2. RNA Isolation

1. Wash cells with PBS, collect into 1.5 ml microcentrifuge tubes using a cell scraper, and spin down at 350 RCF for 10 min.
2. After removal of PBS, add 0.5 ml Trizol to each tube. Repeated mixing up and down by a pipette easily dissolves the cell pellet.
3. Add 0.1 ml chloroform and mix by vortexing for 15 s.
4. After a short incubation time of 2–3 min, centrifuge tubes for 15 min at 12000 RCF at 4°C.
5. Transfer the phenol-free transparent top layer, which contains the RNA, to a new tube and mix with 0.25 ml 100% isopropyl alcohol.
6. After 10 min of incubation at room temperature, centrifuge the samples for 10 min at 12000 RCF at 4°C.
7. Wash the cell pellets, which should be visible, with 70% ethanol and air-dry for approximately 5 min.
8. Dissolve the pellets in DEPC-treated water and determine the total RNA concentration by spectrophotometric analysis at 260 nm.

3.3. cDNA Synthesis

1. Mix 2 μg of total RNA, 0.5 μl poly(dT) primers, and 0.5 μl dNTP mix in PCR tubes to a final volume of 5 μl in DEPC-treated water.
2. Incubate samples at 65°C for 5 min using a thermal cycler and cool down on ice.
3. Prepare a master mix containing 0.25 μl RT (50 units), 2 μl 5X RT buffer, 0.25 μl RNasin, and 1 μl 0.1 M DTT, and 1.5 μl DEPC-treated water per sample and add to the tubes from above followed by 50 min incubation at 42°C (*see* **Note 5**).

4. Inactivate the enzymatic activity by 15 min incubation at 70°C. Store at –20°C until use.

3.4. Quantitative PCR

1. Make a cDNA pool using 2 μl non-diluted cDNA from each sample and make serial dilutions of 1:5, 1:20, 1:100, and 1:500 with sdH_2O.
2. Dilute the remaining cDNA 1:10 with sdH_2O.
3. Make a mixture of 5 μl SYBR green I Master mix, 0.5 μl forward primer, 0.5 μl reverse primer, 2 μl sdH_2O, and 2 μl cDNA (or 2 μl sdH_2O as a negative control) and add to a 96-well plate. Load the plate in the PCR machine and run with the following program:

Denaturation	95°C for 5 min
Amplification (45 cycles)	95°C for 10 s
	62°C for 20 s
	72°C for 30 s, single acquisition
Melting curve analysis	95°C for 5 s
	65°C for 1 min
	Continuous acquisition until 97°C
Cooling	40°C for 30 s

4. To confirm exclusion of non-specific PCR by-products, perform a melting curve analysis for each reaction. The change in fluorescence of SYBR Green dye in every cycle gives the crossing point (CP) values, defined as the points at which the fluorescence increased appreciably above background. Use the CP values to determine the relative expression of the gene of interest. Use the standard curve method to determine the absolute cDNA levels. Calculate for each sample, the ratio between the target gene and the housekeeping gene (in this experiment RPLP0) to compensate for variations in the quantity or quality of the starting mRNA as well as for differences in RT efficiency. **Figure 22.1** shows the relative mRNA expression of the androgen-regulated gene *PSA* and *FASN* in LNCaP cells in the absence or presence of R1881 (10^{-8} M).

3.5. Fixing Cells

1. Wash cells very carefully with PBS as they are easy to come off the plate.
2. Add 1 ml of 4% PFA per well and incubate for 20 min at room temperature on a level surface with no shaking or tilting.
3. Discard the PFA in a hazardous waste chamber and wash cells for 5 min in PBS (*see* **Note 6**).

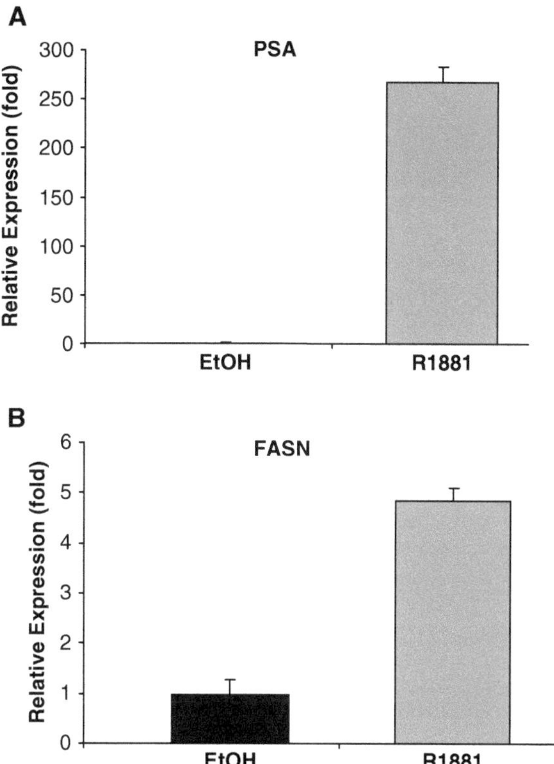

Fig. 22.1. Androgen treatment increases mRNA expression of PSA (**a**) and FASN (**b**) in LNCaP cells. LNCaP cells were cultured in medium containing 5% CT-FBS for 48 h and treated with 10^{-8} M R1881 or vehicle for 96 h and RNA was isolated. Relative mRNA levels were determined by qPCR and normalized to the expression of the ribosomal protein RPLPO.

3.6. Mounting of Cover Slip

1. Add a 40 μl drop of mounting solution in the middle of a glass slide (*see* **Note** 7).
2. Remove the cover slip from the 6-well plate using a tweezer and dry the edges by blotting with tissue paper. Then place the cover slip on the mounting solution (side with cells facing down) taking care not to generate bubbles.
3. Incubate 2 h at room temperature in the dark.
4. Wipe off any precipitate on the cover slip with a moist tissue paper.

3.7. Oil Red O Staining

1. Fix cells and remove PBS (see **Section 3.5**).
2. Wash cells with 60% isopropyl alcohol.
3. Add 1.5 ml of Oil Red O working solution per well and incubate for 15–30 min. Observe under microscope until cells are properly stained. Make sure to stop the staining

when precipitate starts to form. Wash cells with 60% isopropyl alcohol (*see* **Note 8**).

4. Wash cells with 1 ml PBS and leave in PBS until further use.
5. Mount cover slips on glass slides (see **Section 3.6**) or use the stained cells for quantification by Oil Red O extraction (see **Section 3.8**). The mounted Oil Red O-stained cells can be photographed in a standard light microscope or higher magnification images can be obtained using any microscope equipped with a RGB camera. **Figure 22.2a** shows a representative image of Oil Red O-stained LNCaP cells that were either left untreated or treated with R1881(10^{-8} M) for 96 h.

3.8. Oil Red O Extraction and Quantification

1. Remove PBS from stained cells.
2. Add 1 ml of isopropyl alcohol and incubate on a tilting board for 5 min.
3. Transfer the isopropyl alcohol solution containing the extracted Oil Red O solution to a 1.5 ml tube and spin down at 2300 RCF for 1 min at room temperature. Transfer supernatant to new tube.
4. Quantify Oil Red O levels by spectrophotometric analysis at 518 nm. Use isopropyl alcohol as blank control. **Figure 22.2b** shows the relative absorbance of Oil Red O in LNCaP cells that were either left untreated or treated with R1881(10^{-8} M) for 96 h.

3.9. HCS LipidTOX Red Staining

1. Fix cells (see **Section 3.5**) and remove the PBS.
2. Keep the cells in 0.2% Triton X-100 on a tilting board for 15 min for permeabilization (*see* **Note 9**).
3. Wash cells two times for 5 min with PBS.
4. Place 40 μl of LipidTOX Red staining solution on the parafilm in the incubation chamber.
5. Remove the cover slip from the well using a tweezer and dry by blotting the edges with tissue paper.
6. Place the cover slip on the drop of LipidTOX Red staining solution (cells facing down). Close the box and incubate at room temperature for 1 h. Minimize the light exposure to the cells from this point on.
7. Bring the cover slip out of the chamber and back to the 6-well plate (cells facing up).
8. Wash cells for 5 min in PBS.
9. Add 1 ml of DAPI staining solution per well and stain for 5 min in the dark.

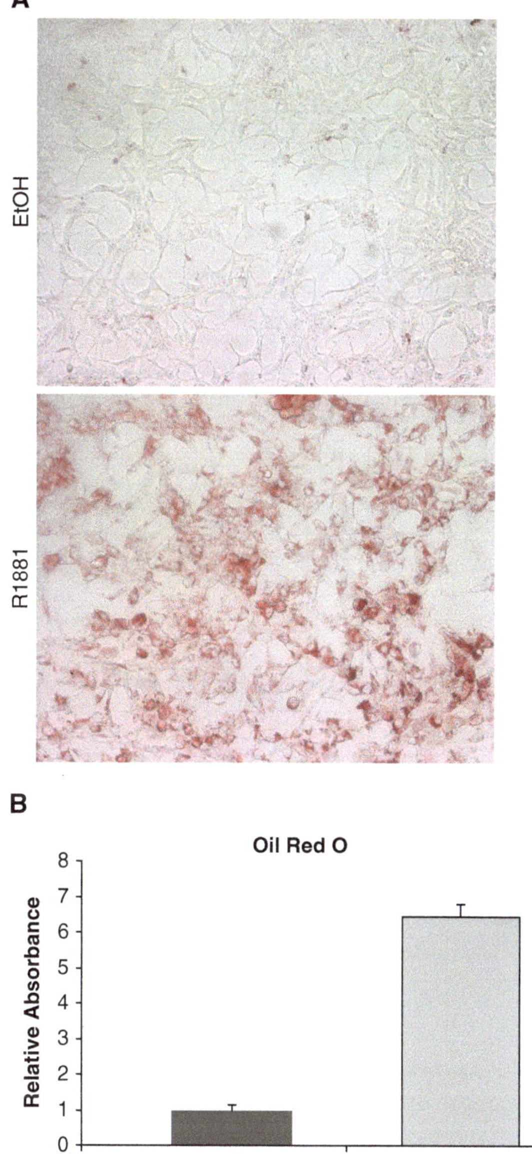

Fig. 22.2. Increased lipid accumulation in LNCaP cells in response to androgen treatment detected by Oil Red O. LNCaP cells were cultured in medium containing 5% CT-FBS for 48 h and treated with 10^{-8} M R1881 or vehicle for 96 h, fixed with PFA, and stained with Oil Red O. Cells were subsequently mounted for imaging (**a**) or Oil Red O staining was quantified using isopropyl alcohol extraction followed by measurement of absorbance at 518 nm (**b**).

10. Wash cells three times for 5 min in PBS with a cover.
11. Mount the cover slips on glass slides (see **Section 3.6**).
12. Use a fluorescence microscope to observe the stained cells. HCS LipidTOX Red has absorption and emission

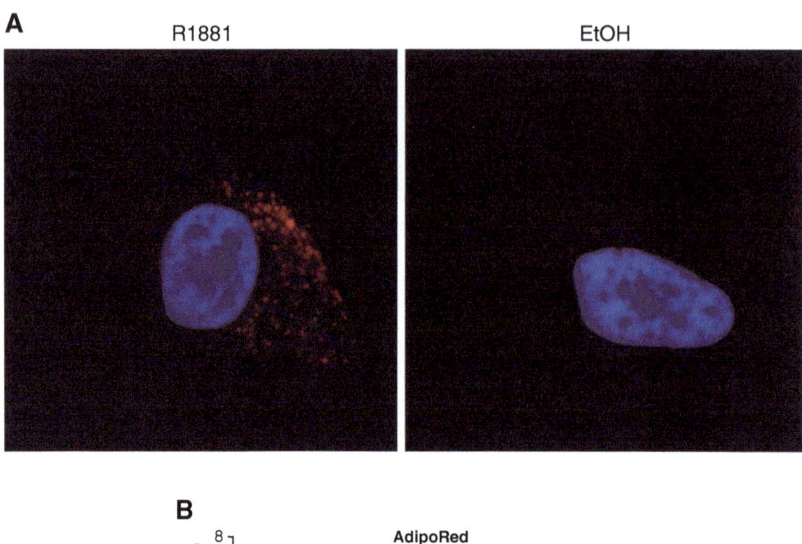

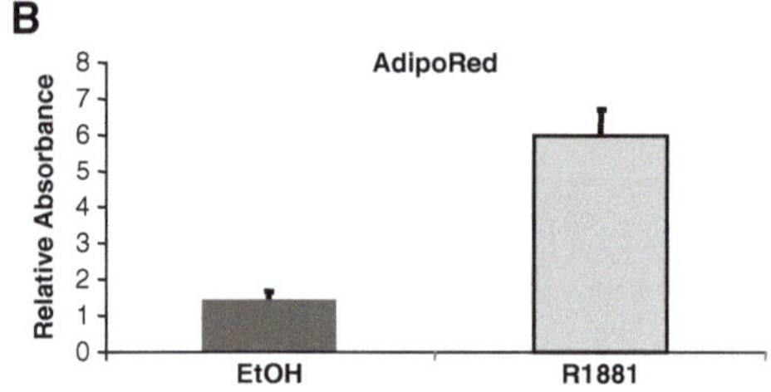

Fig. 22.3. Increased lipid accumulation in LNCaP cells in response to androgen treatment detected by HCS LipidTOX Red and AdipoRed. LNCaP cells were cultured in medium containing 5% CT-FBS for 48 h and treated with 10^{-8} M R1881 or vehicle for 96 h. Cells were then either fixed with PFA, stained with HCS LipidTOX Red and DAPI, mounted, and visualized using a fluorescence confocal microscopy (**a**) or subjected to AdipoRed assay with fluorimetric detection at 485/535 nm (**b**).

wavelengths that are similar to Alexa Fluor 594, of 590 and 617 nm, respectively. DAPI has absorption and emission maximum at 358 and 491 nm, respectively. **Figure 22.3a** shows an overlay of DAPI and HCS LipidTOX Red-stained LNCaP cells that were either left untreated or treated with R1881(10^{-8} M) for 96 h.

3.10. AdipoRed Assay

1. Wash cells carefully with PBS.
2. Add 2 ml of diluted AdipoRed reagent mix per well and incubate for 20–30 min at room temperature (*see* **Note 10**).
3. Quantify AdipoRed staining of neutral lipids using a fluorimeter with excitation at 485 nm and emission at 572 or 535 nm. **Figure 22.3b** shows the AdipoRed staining in LNCaP cells that were either left untreated or treated with R1881(10^{-8} M) for 96 h (*see* **Note 11**).

4. Notes

1. PFA is a possible carcinogen. One should work with it in a ventilation hood and discard it safely.
2. We have been successfully using this reagent supplied by Invitrogen. Alternative colors of the same reagent and similar reagents from other sources are available and should work just as well.
3. How the chamber looks and what it is made of is not so important. The main point is that it is sealed from light and that it is relatively air-tight so that the air inside will stay humid during an hour of incubation at room temperature. We typically use a 30 × 30 × 3 cm plastic chamber wrapped with aluminum foil. This can easily contain 24 cover slips on a sheet of parafilm at one time. The humid air inside the chamber is maintained by placing moist tissue paper in the inside walls.
4. When using 12- or 24-well plates, LNCaP cells may clump at the center of the well and may influence the outcome of the experiment. We therefore recommend using 6-well plates.
5. A control sample containing no RT should be included to test for genomic DNA contamination.
6. After the cells are fixed with PFA and properly washed, they can be stored for a few days in 2.5 ml PBS per well at 4°C before further steps. Seal the plate with parafilm.
7. Before applying the mounting solution it is useful to cut the end of the pipette tip. This makes it easier to avoid bubbles since it is a very viscous solution.
8. Be sure to remove all Oil Red O precipitate and excess working solution when washing with 60% isopropyl alcohol and PBS. The precipitate not only impairs the clarity in pictures, but poorly washed cells can of course give erroneous results.
9. If DAPI stain is not needed along with HCS LipidTOX Red staining, then it is not necessary to permeabilize the cells.
10. Less than half of the reagent used in the manufactures' protocol is sufficient. The assay is designed for triglyceride detection mainly in adipocytes and generally requires a longer incubation time in LNCaP cells.
11. Oil Red O may be a better choice for staining and quantifying lipid droplets in the LNCaP cells as it is less expensive

than the various quantification kits that are available in the market. On the other hand, the advantage of using a kit, such as AdipoRed (which is a solution of the hydrophilic stain Nile Red), is that the assay can be performed directly on live cells, making it easier and faster to perform which are criteria to consider for high-throughput experiments.

Acknowledgments

This work was supported by grants from the Norwegian Cancer Society and the Norwegian Research Council.

References

1. Flavin, R., Peluso, S., Nguyen, P. L., and Loda, M. (2010) Fatty acid synthase as a potential therapeutic target in cancer, *Future Oncol* **6**, 551–562.
2. Menendez, J. A., and Lupu, R. (2007) Fatty acid synthase and the lipogenic phenotype in cancer pathogenesis, *Nat Rev Cancer* **7**, 763–777.
3. Swinnen, J. V., Roskams, T., Joniau, S., Van Poppel, H., Oyen, R., Baert, L., Heyns, W., and Verhoeven, G. (2002) Overexpression of fatty acid synthase is an early and common event in the development of prostate cancer, *Int J Cancer* **98**, 19–22.
4. Rossi, S., Graner, E., Febbo, P., Weinstein, L., Bhattacharya, N., Onody, T., Bubley, G., Balk, S., and Loda, M. (2003) Fatty acid synthase expression defines distinct molecular signatures in prostate cancer, *Mol Cancer Res* **1**, 707–715.
5. Swinnen, J. V., Heemers, H., van de Sande, T., de Schrijver, E., Brusselmans, K., Heyns, W., and Verhoeven, G. (2004) Androgens, lipogenesis and prostate cancer, *J Steroid Biochem Mol Biol* **92**, 273–279.
6. Zadra, G., Priolo, C., Patnaik, A., and Loda, M. (2010) New strategies in prostate cancer: targeting lipogenic pathways and the energy sensor AMPK, *Clin Cancer Res* **16**, 3322–3328.
7. Swinnen, J. V., Esquenet, M., Goossens, K., Heyns, W., and Verhoeven, G. (1997) Androgens stimulate fatty acid synthase in the human prostate cancer cell line LNCaP, *Cancer Res* **57**, 1086–1090.
8. Swinnen, J. V., Van Veldhoven, P. P., Esquenet, M., Heyns, W., and Verhoeven, G. (1996) Androgens markedly stimulate the accumulation of neutral lipids in the human prostatic adenocarcinoma cell line LNCaP, *Endocrinology* **137**, 4468–4474.
9. Swinnen, J. V., Ulrix, W., Heyns, W., and Verhoeven, G. (1997) Coordinate regulation of lipogenic gene expression by androgens: evidence for a cascade mechanism involving sterol regulatory element binding proteins, *Proc Natl Acad Sci USA* **94**, 12975–12980.
10. Heemers, H., Maes, B., Foufelle, F., Heyns, W., Verhoeven, G., and Swinnen, J. V. (2001) Androgens stimulate lipogenic gene expression in prostate cancer cells by activation of the sterol regulatory element-binding protein cleavage activating protein/sterol regulatory element-binding protein pathway, *Mol Endocrinol* **15**, 1817–1828.
11. Heemers, H. V., Verhoeven, G., and Swinnen, J. V. (2006) Androgen activation of the sterol regulatory element-binding protein pathway: Current insights, *Mol Endocrinol* **20**, 2265–2277.

Index

A

F. Saatcioglu (ed.), *Androgen Action*, Methods in Molecular Biology 776,
DOI 10.1007/978-1-61779-243-4, © Springer Science+Business Media, LLC 2011

B

C

D

E

F

G

H

I

W

X

MIX
Papier aus verantwortungsvollen Quellen
Paper from responsible sources
FSC® C105338

If you have any concerns about our products,
you can contact us on
ProductSafety@springernature.com

In case Publisher is established outside the EU,
the EU authorized representative is:
Springer Nature Customer Service Center GmbH
Europaplatz 3, 69115 Heidelberg, Germany

Printed by Libri Plureos GmbH
in Hamburg, Germany